CMOS Circuits for Passive Wireless Microsystems

Fei Yuan

CMOS Circuits for Passive Wireless Microsystems

Springer

Fei Yuan
Department of Electrical and Computer Engineering
Ryerson University
350 Victoria Street
Toronto, Ontario, Canada M5B 2K3
Email: fyuan@ee.ryerson.ca

ISBN 978-1-4419-7679-6 e-ISBN 978-1-4419-7680-2
DOI 10.1007/978-1-4419-7680-2
Springer New York Dordrecht Heidelberg London

Printed on acid-free paper

Springer is part of Springer Science+Business Media (www.springer.com)

This book is dedicated to
Jing

Preface

Passive wireless microsystems harvest their operational power from radio-frequency waves or other energy sources such as vibration and solar. The absence of bulky batteries not only minimizes the physical dimension and implementation cost of these microsystems, it also removes the need for maintenance. As a result, passive wireless microsystems can be embedded in products or implanted in living bodies permanently to provide the unique identification of the products and living bodies in which they reside, to carry out precision measurement of the parameters of the products or living bodies, or to perform micron-scale control actions that otherwise cannot be performed.

Attributive to their small, wireless accessibility and programmability, and maintenance-free operation, passive wireless microsystems have found a broad range of emerging applications in biomedical implants such as cochlear implants and retinal prosthetic implants, swallowable capsule endoscopy, multi-site pressure sensors for wireless arterial flow characterization, embedded micro-strain sensors for product performance and safety monitoring, wireless temperature sensors for human body and environmental monitoring, and radio-frequency identification tags for logistic automation, to name a few. Passive wireless microsystems are also a viable choice for low-cost high-security product authentication to replace existing product authentication methods such as holograms, water-marks, invisible barcodes, security threads, chemical, and DNA markers that are often too costly to be used for general goods.

This book provides a comprehensive treatment of CMOS circuits for passive wireless microsystems. It focuses on the design of the key blocks of passive wireless microsystems. These blocks include radio-frequency power harvesters, demodulators, low-power precision voltage references, system clock generation and calibration, and ultra-low power analog-to-digital converters. The materials presented in the book are compiled from recently published work in this fast-evolving field. The book is organized as the follows :

Chapter 1 provides an overview of passive wireless microsystems and highlights the key considerations and design challenges of these microsystems.

Chapter 2 begins with a brief examination of the parameters that characterize the performance of radio-frequency power harvesters. Our focus is then turned to power-matching and gain-boosting using LC networks and step-up transformers to increase the voltage at the input of voltage multipliers so as to boost their power efficiency. An emphasis is given to the power efficiency of the power-matching and gain-boosting network itself as the overall power efficiency of a power harvester is determined by the power efficiency of its power-matching and gain-boosting network, that of its voltage multiplier, and the efficiency of its antenna. The design of voltage multipliers for passive wireless microsystems is then investigated in detail.

Chapter 3 examines the pros and cons of commonly used data encoding schemes for wireless communications and explores their suitability for passive wireless microsystems. Non-return-to-zero encoding popular in high-speed data communications over wire channels is studied first. It is followed by an investigation of return-to-zero encoding. Manchester encoding and its characteristics are examined. Miller encoding and Miller-modulated sub-carrier encoding are also studied. An emphasis is given to the distinct characteristics of Miller-modulated sub-carrier encoding and its usefulness in encoding data to be backscattered to base stations. FM0 encoding and pulse interval encoding that are widely used in radio-frequency identification systems are then explored. The chapter is concluded with a comparison of the performance of the encoding schemes studied in the chapter.

Chapter 4 deals with modulation and demodulation. The chapter starts with a close look at the three basic modulation schemes, namely amplitude-shift-keying (ASK), frequency-shift-keying (FSK), and phase-shift-keying (PSK). The pros and cons of these modulation schemes are studied and compared in detail. ASK modulators and demodulators for passive wireless microsystems are investigated. A significant portion of this section is devoted to CMOS circuits for ASK demodulators. FSK modulators and demodulators are then examined in detail. An in-depth study of the advantages and design constraints of FSK demodulators for biomedical implants is provided. The advantages of PSK modulation over FSK modulation in biomedical implants and the design challenges encountered are investigated. Both coherent and non-coherent demodulation of BPSK-modulated signals are presented. The performance of recently published ASK, FSK, and PSK demodulators for passive wireless microsystems is compared.

Chapter 5 is concerned with temperature-independent precision voltage references for passive wireless microsystems. The chapter starts with a brief examination of the figure-of-merits that characterize the performance of voltage references. It is followed by a detailed investigation of the temperature-dependent characteristics of semiconductors. First-order voltage references are studied in a great detail. An in-depth investigation of high-order voltage references follows. The performance of first-order voltage reference and that of high-order voltage references are compared. Ultra low-power voltage references where devices operate in weak inversion are also studied.

Chapter 6 deals with the generation and calibration of the system clock of passive wireless microsystems. The generation of the system clock of passive wireless microsystems directly from the carrier of the received RF signal is investigated first. It is followed by a close look at the generation of the system clock from the envelope of the received RF signal. Direct generation of the system clock from the received RF signal using injection-locked frequency division to take the advantage of its low power consumption and high frequency accuracy is investigated. The system clock of passive wireless microsystems can also be generated using a local oscillator directly. Since the frequency of the local oscillator is subject to the effect of process, voltage, and temperature (PVT) variations, calibrating the frequency of the local oscillator prior to any data communications becomes indispensable. Calibration of the frequency of the system clock using injection-locking with the carrier as the injection signal is investigated first. Frequency calibration of the local oscillator using digital trimming techniques is followed. Our focus is then turned to the presentation of frequency calibration using either phase-locked loops or frequency-locked loops. The chapter further explores the calibration of the frequency of the local oscillator using injection-locking with the envelope as the injection-locking signal. Integrating feedback is employed to increase the lock range of frequency calibration using injection-locking with the envelope as the injection-locking signal.

Chapter 7 focuses on the architecture and design of low-power analog-to-digital converters (ADCs). The fundamentals of ADCs are studied first. It is followed by a close examination of the figure-of-merits used to quantify the performance of ADCs. Integrating ADCs are investigated. Both single-slope and dual-slope integrating ADCs, and their advantages and disadvantages are examined and compared. The design of oscillation-based ADCs for temperature measurement is explored. A close attention is paid to both relaxation oscillator and ring oscillator based temperature ADCs. Time-to-digital converter based ADCs for temperature measurement is also investigated. As compared with oscillator-based temperature ADCs, these ADCs have the advantage of low power consumption. The chapter then moves on to investigate frequency-to-digital based ADCs for temperature measurement. Charge redistribution successive approximation ADCs are investigated in a great detail. Our focus is given to the design of charge-scaling digital-to-analog converters (DACs) used in charge redistribution successive approximation ADCs. Three configurations of charging-scaling capacitor arrays, namely single-stage binary-weighted capacitor arrays, two-stage binary-weighted capacitor arrays, and C-2C capacitor arrays are studied in detail and their pros and cons are examined.

Since the objective of the book is to provide readers with the state-of-the-art of CMOS circuits for passive wireless microsystems, the details of the operation of semiconductor devices and basic microelectronic circuits are omitted due to a space constraint. Readers are assumed to have the basic knowledge of electrical networks, semiconductor devices, microelectronic circuits, signals and systems, and digital communications. A rich collection of recently published work on passive wireless microsystems is provided at the end of the book for readers to seek further information on the subjects.

Although an immense amount of effort was made in preparation of the manuscript, flaws and errors will still exist due to erring human nature and the limited knowledge of the author on the subjects. Suggestions and corrections from readers will be gratefully appreciated by the author.

FEI YUAN

JULY 2010

Acknowledgments

The author is grateful to the Natural Science and Engineering Research Council of Canada, Ryerson University, CMC Microsystems, and other research partners for their support to our research. The support from the Department of Electrical and Computer Engineering of Ryerson University where I introduced and taught graduate courses EE8501 (CMOS analog integrated circuits), EE8502 (VLSI systems), EE8503 (VLSI circuits for data communications), and EE8604 (Special topics on electrical engineerings - CMOS circuits for wireless communications) is gratefully acknowledged. The students of these classes constantly motivated me to explore new things in the field of microelectronics that is mature and yet fast-evolving. Special thanks go to my graduate students, both past and present, for fruitful discussion in our research meetings where many new topologies and configurations emerged. Mr. Jason Naughton, our System Administrator, deserves a special thank-you for keeping CAD tools up-to-date and running all the time.

The editorial staff of Springer, especially Mr. Charles Glaser, Senior Acquisitions Editor, have been warmly supportive from the invitation of the book proposal to the publish of the book. This is the 4th book that I have published with Springer. It has been a wonderful experience of working with Springer.

Finally and most importantly, this book could not have been possible without the unconditional support of my family. I am indebted to my wife Jing for her love, patient, and understanding. I also want to thank our daughter and son, Michelle and Jonathan, for the joy and happiness they have brought to our life, and for their forbearance of my bad temper due to the stress of writing and proof-reading.

Contents

Chapter 1

PASSIVE WIRELESS MICROSYSTEMS

Passive wireless microsystems such as wireless transponders, radio-frequency identification (RFID) tags, wireless microsensors, and biomedical implants, harvest their operational power either from radio-frequency waves emitted by their base station or from other energy sources such as vibration and solar. The absence of bulky batteries not only minimizes the physical dimension and implementation cost of these microsystems, it also removes the need for maintenance. As a result, passive wireless microsystems can be embedded in products or implanted in living bodies permanently to provide the unique identification of the products or living bodies in which they reside, to provide the precision measurement of the parameters of the products or living bodies, and to carry out control actions in a micron scale that otherwise cannot be performed. The key intrinsic attributes of passive wireless microsystems include small size, battery-less and maintenance-free operation, programmability, and wireless accessibility.

Passive wireless microsystems have found a broad range of emerging applications include implantable bio-microelectromechanical-systems (MEMS) pressure sensors [1, 2], retinal prosthetic devices [3–6], swallowable capsule endoscopy [7–9], multi-site pressure sensors for wireless arterial flow characterization [10], embedded micro-strain sensors for product performance and safety monitoring, wireless temperature sensors for human bodies and environmental monitoring [11–14], radio-frequency identification tags for object tracking in logistics automation [15, 16] and replacing bar codes in retailing, warehouse inventory automation, e-tickets, e-passports, and low-cost high-security product authentication keys to replace existing product authentication means such as holograms, water-marks, invisible barcodes, security threads, chemical, and DNA markers that are often too costly to be used for general goods [17].

F. Yuan, *CMOS Circuits for Passive Wireless Microsystems*,
DOI 10.1007/978-1-4419-7680-2_1, © Springer Science+Business Media, LLC 2011

1.1 The Spectrum

Communications between a passive wireless microsystem and its base station typically take place in ISM (Industrial, Scientific, and Medical) bands. ISM bands are open frequency bands that allow for operation without a license. The most widely used ISM bands are 13.553-13.57 MHz, 902-928 MHz, 2.4-2.4825 GHz, and 5.725-5.850 GHz [18]. The maximum power of the radio-frequency waves emitted by an antenna in ISM bands is regulated by Federal Communications Commission (FCC) in the United States. Effective isotropic radiated power (EIRP), which is defined as the power that would have to be supplied to an ideal antenna that radiates uniformly in all directions in order to get the same electrical field strength that the device under test produces at the same distance, is widely used to quantify the amount of the radiation power from an antenna. EIRP is computed from [19]

$$\mathrm{EIRP} = 10\log\left(\frac{4\pi E^2 r^2}{0.377}\right), \tag{1.1}$$

where E is the electrical field strength and r is the distance from the antenna. FCC part 15 rules govern the transmission power permitted in ISM bands. For example, the maximum transmitter output power fed into an antenna is 30 dBm (1 Watt) and the maximum EIRP is 36 dBm (4 Watts) in 902-928 MHz and 2.4-2.4825 GHz ISM bands.

Wireless communications between a passive wireless microsystem and its base station can take place either in the near field or the far field of the antennas of the base station. Near-field coupling is viable for frequencies up to a few tens of MHz with the characteristics of a low data rate, a large antenna dimension, and a short link distance. The low upper bound of the frequency is mainly due to the low self-resonant frequency of the coupling coils between base stations and passive wireless microsystems. Far-field coupling occurs at ultra-high frequencies (UHF) and microwave frequencies and offers the advantages of a high data rate, a small antenna dimension, and a long link distance. Biomedical implants are located in the near field mainly due to the high loss of electromagnetic waves in living bodies at high frequencies. Other near-field passive wireless microsystems include smart cards, access cards, e-passports, etc. Transponders and RFID tags are typically located in the far field of the antenna of their base station to take the advantages of the low loss of electromagnetic waves in air.

1.2 The Challenges

Although the emerging applications of passive wireless microsystems are quite broad, the performance of these microsystems is commonly affected by a number of fundamental issues. These fundamental issues include a low power conversion efficiency in power harvest from radio-frequency waves, fluctuating supply voltages, the low level of data security arising from the difficulty to implement encryption algorithms on passive wireless microsystems due to their limit power resource, signal collisions when multiple microsystems co-exist in a close proximity, the drift of the frequency of the system clock of passive wireless microsystems that controls the operation of their baseband blocks and the backscattering communications from the microsystems to their base station, and the need for precision voltage resources and ultra-low power analog-to-digital converters.

1.2.1 Efficiency of Radio-Frequency Power Harvest

The efficiency of power harvest from radio-frequency waves determines the amount of power available for microsystems to operate. This, in turn, sets the degree of the complexity subsequently the functionality of passive wireless microsystems. The amount of the power that can be harvested by a passive wireless microsystem from a radio-frequency wave is dominated by the efficiency of the antenna, the accuracy of impedance matching between the antenna and the impedance transformation network, the efficiency of the impedance transformation network, and the efficiency of the voltage multiplier of the microsystem. The widely used diode-bridge rectification in near-field power telemetry performs poorly when passive wireless microsystems are located in the far field of the antenna of its base station simply because the amplitude of the received RF signal is small [4, 6]. To obtain a sufficiently large supply voltage and at the same time to maximize the power efficiency of voltage multipliers, many novel voltage multipliers evolved from the classic Cockcroft-Walton voltage multiplier [20] and Dickson voltage multiplier [21] emerged [22–28]. The power efficiency of voltage multipliers can be best improved by boosting the voltage at the input of the voltage multipliers. This is achieved by inserting a passive resonant network such as a LC resonator [25] or a step-up transformer resonator [29] between the antenna and the voltage multiplier of the microsystem. The ohmic loss of the impedance transformation network, however, limits the overall power efficiency of the power harvesters.

1.2.2 Fluctuating Supply Voltage

Aside from RF power harvest, a unique characteristic of passive wireless microsystems is their poor supply voltage stability, arising from the time-varying nature of the power harvest from radio-frequency waves. The fluctuation of the supply voltage warrants the deployment of a voltage regulator, aiming at stabilizing the supply voltage. This, however, is at the expense of an additional amount of power consumption. Both diode-based voltage regulators [30] and feedback op-amp voltage regulators [31, 32] are available. The former consumes less power while the latter yields a better voltage stability. In addition to voltage regulation, the circuits of passive wireless microsystems must also be designed in such a way that their performance is less sensitive to fluctuating supply voltages. Such a constraint often excludes the circuits with which we are familiar. LC oscillators and regulated cascodes are examples of these circuits.

1.2.3 Sensitivity to Changing Environment

In addition to a time-varying supply voltage, the performance of a passive wireless microsystem is also severely affected by the condition of the environment in which the microsystem resides. For example, the frequency of the local oscillator of a microsystem, which provides the system clock for its baseband units and controls the backscattering communications with its base station, is sensitive to temperature variation [33, 34]. This is on top of the effect of process variation and supply voltage fluctuation. Although the effect of temperature variation can be minimized using complex compensation circuitry, the practicality of this approach largely diminishes once the limited power resource of passive wireless microsystems is considered [35, 36]. An alternative and yet more effective way to ensure the accuracy of the frequency of the system clock of a passive wireless microsystem is to use an external calibrating signal sent by its base station to the microsystem as a reference to calibrate the frequency of the local oscillator of the microsystem prior to the commence of any communications between the microsystem and its base station. The key advantage of this approach is that the reference is from the base station and is therefore independent of the condition of the environment in which the passive wireless microsystem resides. Also, since the calibration is conducted remotely, the frequency of the passive wireless microsystem can be calibrated even though the microsystem has already been embedded in products or implanted in living bodies.

1.2.4 Precision Voltage References

Temperature-insensitive precision voltage references are as important as a stable system clock to the operation of passive wireless microsystems. This is because temperature-insensitive precision voltage references play a critical role in biasing analog circuitry, providing constant currents for relaxation oscillators and other blocks, and providing a precision voltage reference for analog-to-digital converters. Although bandgap voltage references have been widely studied and many bandgap voltage references are available, the low and fluctuating supply voltage, and the low-power consumption requirement of passive wireless microsystems impose stringent constraints on the design of voltage references. As a result, many of the known bandgap voltage references can not be used for passive wireless microsystems either due to their high power consumption or due to their need for a large supply voltage. A detailed study of the principles and design of high precision low-power CMOS voltage references is therefore an essential part of any text on passive wireless microsystems.

1.2.5 Ultra-Low Power Analog-to-Digital Converters

Analog-to-digital conversion is essential to passive wireless microsensors. Although analog-to-digital converters (ADCs) are perhaps the most widely studied mixed analog-digital subsystems and there are many ways to perform analog-to-digital conversion such as flash ADCs, pipelined ADCs, oversampled sigma-delta ADCs, to name a few, the low and fluctuating supply voltage and the low power consumption constraints of passive wireless microsystems disqualifies many architectures of ADCs. Only a few architectures of ADCs, such as integrating ADCs, charge-redistribution successive approximation ADCs, oscillation-based ADCs, time-to-digital ADCs, and frequency-to-digital ADCs made to the the short list.

1.2.6 Encryption and Authentication

The applications of passive wireless microsystems, especially those located in the far-field of the antenna of base stations, are hindered by their vulnerability to interception by third parties, mainly due to the limited computing power of these microsystems. The large-scale deployment of these microsystems mandates the equipment of encryption for authentication. Existing authentication algorithms for computer and cellular networks are generally too complex and power-consuming to be adopted for passive wireless microsystems. High-security authentication protocols with a reconfiguration capability are desirable. Mutual authentication protocols shared between a passive wireless microsystem and its base station and the optimal partition of authentication protocols between the microsystem and its base station can achieve the required level of security without excessively increasing the complexity subsequently power consumption of the passive wireless microsystem.

1.2.7 Signal Collision

Signal collision occurs when multiple wireless microsystems coexist in a close proximity, arising from the fact that passive wireless microsystems typically operate in the same frequency channel. Two spread spectrum approaches, namely frequency hopping and direct sequence spread spectrum, are perhaps the most widely used techniques in dealing with multi-access wireless communications. The former avoid signal collision by allocating an unused channel in the selected ISM band while the latter spreads the signal to be transmitted over a large frequency range of the selected ISM band using a pseudo random sequence and recovers the transmitted signal at the receiving end using de-sequencing. Frequency hopping is less attractive for passive wireless microsystems due to its high power consumption and its need for synchronization in both time and frequency domains. The intrinsic characteristics of direct sequence spread spectrum including reduced crosstalk interferences, better data integrity, low susceptibility to multi-path fading, an increased operating distance, hard to detect, intercept, and jam, make it particularly attractive for passive wireless microsystems. Although this approach has been widely used in cellular phones in the form of code-devision-multi-access (CDMA), challenges exist in applying it for passive wireless microsystems mainly due to the limited power resource, small memory, and finite computing power of these microsystems.

1.2.8 Dimension of Antennas

The dimension of the antenna of microsystems also imposes a great challenge in embedding these microsystems in products and living bodies. The antenna of a passive wireless microsystem should be small in size such that the microsystem can be embedded in products or implanted in living bodies with a little difficulty. The efficiency of the antenna, on the other hand, should be high so that the voltage at the output of the antenna is sufficiently large. Recent efforts on minimizing the physical dimensions of the antennas of microsystems have been made in circular loop antennas with multiple stubs [37], text-based meander line dipole antennas [38, 37], folded-slot antennas [39–41] with an omni-directional radiation pattern in UHF ISM bands.

Since the focus on this book is on CMOS circuits for passive wireless microsystems, the design of the antennas of passive wireless microsystems is clearly beyond the scope of the book. The in-depth treatment of frequency hopping and direct sequence spread spectrum is readily available in standard texts on digital communications and will therefore not be covered.

Chapter 2

RADIO-FREQUENCY POWER HARVEST

Passive wireless microsystems harvest their operational power from the radio-frequency waves emitted from their base stations. Based on the characteristics of the wireless links with base stations, passive wireless microsystems are loosely classified as inductively-coupled also known as near-field-coupled and electromagnetically-coupled also known as far-field-coupled. The boundary that separates the near-field and far-field is defined as $\frac{\lambda}{2\pi}$ where λ is the wavelength of the signals. Near-field-coupling is viable for frequencies up to a few ten MHz, mainly due to the low resonant frequency of the planar coupling coils. Near-field-coupling has been widely used in applications such as biomedical implants where a high degree of the absorption of electromagnetic waves by living bodies exists at high frequencies. The key characteristics of a near-field-coupled passive wireless microsystem include a large voltage at the coupling coils of the microsystem and weak interferences from neighboring devices due to the close distance between the base station and the passive wireless microsystem. Far-field-coupling, on the other hand, is used at ultra-high frequencies (UHF) and microwave frequencies, such as ISM 900 MHz and 2.4 GHz bands. A high data rate, a small antenna dimension, and a long link distance are the key characteristics of far-field-coupled passive wireless microsystems. The fact that the maximum EIRP of base stations in North America can not exceed 4 W in UHF bands limits the maximum distance between a far-field-coupled microsystem and its base station to a few meters [42]. The efficiency of power harvest from RF waves determines the maximum distance over which a reliable wireless link between a base station and a passive wireless microsystem can be established. The efficiency of radio-frequency power harvest is determined by a number of factors including the efficiency of the antenna of the microsystem, the accuracy of power matching between

F. Yuan, *CMOS Circuits for Passive Wireless Microsystems*,
DOI 10.1007/978-1-4419-7680-2_2,

the antenna and the voltage multiplier, and the power efficiency of the voltage multiplier that converts the received RF signal to a dc voltage from which the microsystem is powered.

This chapter deals with power harvest from radio-frequency waves. The chapter is organized as the follows : Section 2.1 investigates the figure-of-merits that characterize the performance of RF power harvesters. Section 2.2 focuses on the design of voltage multipliers for passive wireless microsystems in the far field of the antenna of the base stations. In Section 2.3, power-matching and gain-boosting using a LC network is investigated. Section 2.4 presents power-matching and gain-boosting using a step-up transformer. Frequency tuning mechanisms for power-matching and gain-boosting using a LC network and that using a step-up transformer are also addressed. The measurement results of the proposed power-matching network, together with the measurement results of a LC power-matching network are compared. The chapter is concluded in Section 2.5.

2.1 Characterization of Radio-Frequency Power Harvest

The efficiency of power harvest from a radio-frequency wave determines the maximum distance over which a reliable link between a base station and a passive wireless microsystem can be established. It also sets the complexity subsequently the functionality of the microsystem. The efficiency of a radio-frequency power harvesting system is determined by the efficiency of the antenna of the microsystem, the accuracy of impedance matching between the antenna and the voltage multiplier of the microsystem, and the power efficiency of the voltage multiplier that converts a received RF signal to a dc voltage from which the microsystem is powered.

2.1.1 Power Matching

The radiation resistance of the antenna of passive wireless microsystems has a typical value of 50Ω at the desired frequency. The input impedance of voltage multipliers typically has a reactance component, owing to the capacitance of rectifying diodes or MOSFETs. An impedance transformation network is therefore required to transform the input impedance of the voltage multiplier to 50Ω.

Consider Fig.2.1 where an impedance transformation network is inserted between the antenna represented by voltage source V_a and radiation resistance R_a and the voltage multiplier represented by resistor R_L. Note that to simplify analysis, the input impedance of the voltage multiplier is assumed to be purely resistive. Note that a pure resistive input impedance of voltage multipliers can be obtained by employing a shunt inductor that resonates out the reactive part of the input impedance of the voltage multiplier [42]. The function of the impedance transformation network is two-fold : (i) It provides a matching impedance to the antenna to maximize the power transmission from the antenna to the voltage multiplier and to minimize the reflection of the signals. (ii) It provides a large voltage gain such that the voltage at the input of the voltage multiplier or the output of the impedance transformation network is maximized. If we assume that the impedance transformation network is lossless, the power delivered to the impedance transformation network will be the same as that delivered to the load. A large voltage at the input port of the voltage multiplier will reduce the power loss at the voltage multiplier. As a result, the overall power efficiency of the power harvesting path is improved. The power delivered to the load R_L is given by [43]

$$P_L = \frac{V_L^2}{R_L} = \frac{A_v^2 V_a^2}{R_L}, \tag{2.1}$$

where A_v is the voltage gain provided by the impedance transformation network.

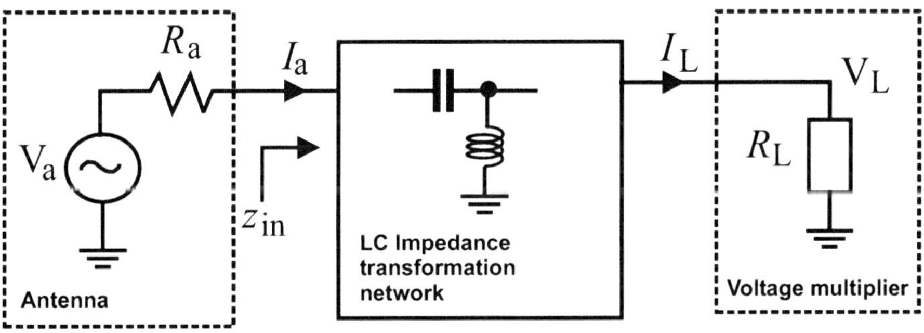

Figure 2.1. Power-matching and gain-boosting using an impedance transformation network.

The maximum power will be delivered from the antenna to the impedance transformation network when

$$R_a = z_{in}^* \tag{2.2}$$

is satisfied, where z_{in} is the input impedance of the impedance transformation network and the superscript * denotes complex conjugation. Since in this case

$$I_a = \frac{V_a}{R_a + z_{in}} = \frac{V_a}{2R_a}, \tag{2.3}$$

we obtain the maximum power delivered from the antenna to the impedance transformation network

$$P_{L,max} = |I_a^2|\Re e\left[z_{in}\right] = \frac{V_a^2}{4R_a}. \tag{2.4}$$

Eq.(2.4) is also the maximum power delivered to the load provided that the impedance transformation network is lossless. Equating (2.1) and (2.4) yields the relation between the voltage gain of the impedance transformation network and the load resistance at which the power delivered to the load is maximized

$$A_v = \frac{1}{2}\sqrt{\frac{R_L}{R_a}}. \tag{2.5}$$

Eq.(2.5) reveals that the voltage gain of the lossless impedance transformation network is proportional to the square-root of the load resistance.

2.1.2 Power Efficiency

The power efficiency of a voltage multiplier is defined as the ratio of the output power of the voltage multiplier, denoted by P_{out}, to the power at the input of the voltage multiplier, denoted by P_{in}

$$\eta_V = \frac{P_{out}}{P_{in}}. \tag{2.6}$$

The power efficiency of voltage multipliers is less than 100% due to the power consumption of rectifying devices of voltage multipliers.

The power efficiency of an impedance transformation network is defined as the ratio of the power delivered to the multiplier, denoted by P_L, to the power available at the input of the impedance transformation network, denoted by P_{in}

$$\eta_I = \left.\frac{P_L}{P_{in}}\right|_{z_{in}=R_a}. \tag{2.7}$$

Note that power-matching condition

$$z_{in} = R_a \tag{2.8}$$

must be met at the input of the impedance transformation network for the maximum power transfer from the antenna to the impedance transformation network. The power efficiency of impedance transformation networks is less than 100% due to the resistive loss of these networks.

The global power efficiency of a RF power harvester is defined as the ratio of the incident power of the RF signal to the dc power at the output of the voltage multiplier

$$\eta = \frac{\text{DC output power}}{\text{Incident RF power}}. \tag{2.9}$$

The incident RF power, denoted by R_{RF}, is quantified by Friis relation [44]

$$P_{RF} = P_B G_B G_M \left(\frac{\lambda}{4\pi r}\right)^2, \tag{2.10}$$

where P_B is the amount of the power that the base station provides to its antenna, G_B is the gain of the antenna of the base station, G_M is the gain of the antenna of the passive wireless microsystem, λ is the wave length, and r is the distance between the passive wireless microsystem and its base station. Often, the effective isotropically radiated power, denoted by P_{EIRP}, is used. It is obtained from

$$P_{EIRP} = P_B G_B. \tag{2.11}$$

2.2 Voltage Multipliers

Wireless communications between a near-field passive wireless microsystem, such as a biomedical implant or a smart card, and its base station is established using an inductive link, much like a transformer with the base station connected to the primary winding of the transformer and the passive wireless microsystem connected to the secondary winding of the transformer. One of the key characteristics of this inductive link is the large voltage at the secondary winding. As a result, RF-to-DC conversion can be carried out using a diode bridge even with the voltage loss across the diodes accounted for. The dc voltage at the output of the diode bridge is sufficiently large to power the passive wireless microsystem. To minimize the voltage loss across the diodes so as to improve RF-to-DC conversion efficiency, Schottky diodes, which typically have a low forward conduction voltage, are widely used [4, 6]. Schottky diodes, however, are not available in standard CMOS processes. Instead, MOSFET-based diodes formed by connecting the gate and drain together can be used for rectification such that the voltage rectifier can be implemented using standard CMOS technologies. Not that there is a voltage loss of at least one device threshold voltage when MOSFET-diodes are used.

Wireless communications between a far-field passive wireless microsystem, such as a RFID tag or a wireless microsensor, and its base station is established using a radio-frequency wave. Unlike near-field inductive links, the voltage at the antenna of the passive wireless microsystem is small, typically a few hundred mV. Diode bridge-based rectification approaches become very inefficient as the voltage loss across the diodes is significant as compared with the amplitude of the incoming RF signal. Voltage multipliers that are evolved from the well-know voltage doubler are required to perform RF-to-DC conversion and at the same time to yield a dc voltage that is many times the amplitude of the incoming RF signal.

This section investigates design techniques for voltage multipliers of passive wireless microsystems. The design constraints of voltage multipliers and the techniques that improve the power efficiency of voltage multipliers are studied. The principle and operation of diode bridges, both half-wave and full-wave diode bridges, are readily available in standard texts on microelectronics and will therefore not be presented here.

2.2.1 Voltage Doubler

Shown in Fig.2.2 is the schematic of a widely used voltage doubler. It consists of a voltage peak detector formed by D2 and C2 and a voltage clamper formed by D1 and C1. If we assume that the diodes are ideal, i.e. the forward conduction voltage is zero, and let the amplitude of the input ac voltage be V_m, it can be shown that voltage of the output of the voltage doubler is $2V_m$. To demonstrate this, let us assume initially $V_{C1} = 0$, $V_{C2} = 0$, and $C_1 = C_2$.

During the first negative half-cycle of the input voltage, D1 will be forward biased. C1 in this case will be charged to a voltage equal to the peak amplitude V_m of the input. During the following positive half cycle of the input voltage, D1 will be reverse biased and therefore will not conduct current. The voltage across C1 will remain unchanged and will add on to the input voltage, in other word, $V_1 = v_{in} + V_m$ during this half cycle. Since D2 is forward biased, C_2 will be changed all the way to $2V_m$.

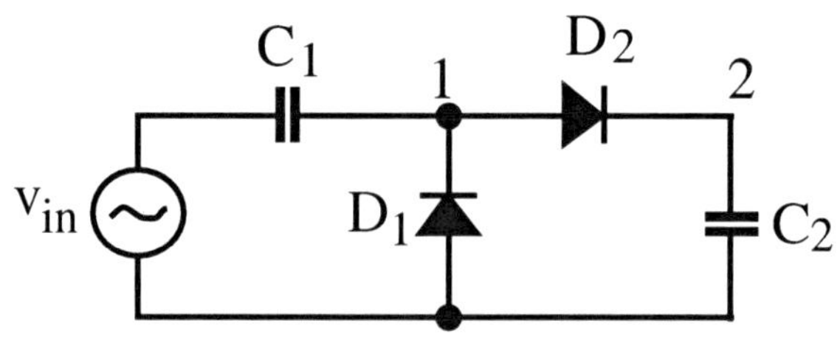

Figure 2.2. Voltage doubler.

2.2.2 Cockcroft-Walton Voltage Multiplier

To obtain a dc voltage that is more than $2V_m$, a multi-stage configuration of voltage doublers is needed. Perhaps the most cited early implementation of voltage multipliers is Cockcroft-Walton voltage multiplier shown in Fig.2.3 [20]. The efficient multiplication of Cockcroft-Walton voltage multiplier will only occur if the capacitance of the coupling capacitors C is much larger as compared with the stray capacitance C_s at the coupling nodes [21]. This is because the clocking signals ϕ and $\overline{\phi}$ only drive the first two coupling capacitors. All other coupling capacitors are connected in series with the stray capacitors. The effectiveness of Cockcroft-Walton voltage multiplier largely diminishes in monolithic integration where stray capacitance C_s and C become comparable.

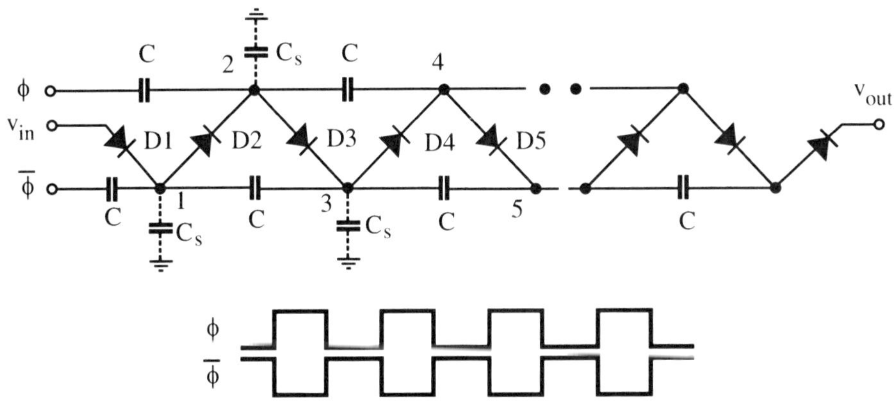

Figure 2.3. Cockcroft-Walton voltage multiplier [20].

2.2.3 Dickson Voltage Multipliers

Dickson modified Cockcroft-Walton voltage multiplier by injecting clocking signals ϕ and $\overline{\phi}$ to all the coupling nodes, as shown in Fig.2.4, such that both the coupling and stray capacitors are driven by the clocking signals directly [21]. The drawback of Cockcroft-Walton voltage multiplier is therefore eliminated. Because these capacitors are connected in parallel, the shunt capacitor connected to the output node of the Dickson voltage multiplier must withstand the full output voltage. It was shown in [34] that the output voltage of a N-stage diode-based Dickson voltage multiplier is given by

$$V_{DC} = N(V_m - V_T), \tag{2.12}$$

where V_T is the forward conduction voltage of the diodes and N is the number of stages. To boost the output voltage, the threshold voltage of the rectifying diodes must be minimized. Schottky diodes are widely used in Dickson voltage multipliers due to their low forward conduction voltage, large saturation current, low junction capacitance, and small series resistance [34].

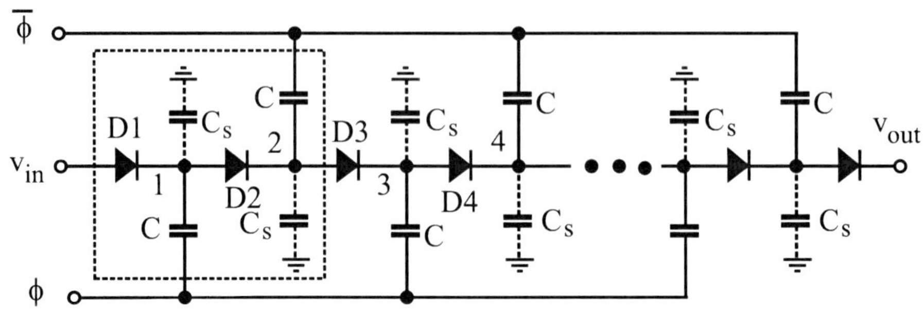

Figure 2.4. Dickson voltage multiplier with diode switches [21].

Dickson voltage multipliers can also be implemented using MOSFET diodes, as shown in Fig.2.5. pMOS Dickson voltage multipliers can also be constructed in a similar way. The advantage of these configurations is their full compatibility with standard CMOS technologies with a main drawback of the voltage loss across the MOSFET devices of at least one threshold voltage. This is accompanied with a low power efficiency, especially when the amplitude of the input voltage is low.

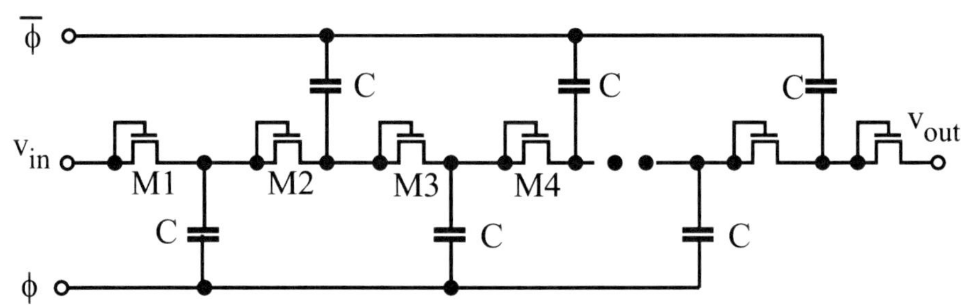

Figure 2.5. Dickson voltage multiplier with nMOS diodes.

The preceding Dickson voltage multiplier with MOSFET diodes suffers from the drawback of the voltage loss of at least one threshold voltage across the MOSFETs. The observation that the voltage drop across of the drain and source of a MOSFET is low if the device is operated in the triode region suggests that the voltage loss of MOSFET diodes can be minimized by connecting a MOSFET working in the triode in parallel with each of the MOSFET diodes, as shown in Fig.2.6 [22]. We term this voltage multiplier Dickson voltage multiplier with static charge transfer switches. The operation of this voltage multiplier is depicted as follows : When $\phi = 0$, V_1 is initially zero and transistor M1A is on. C_1 is charged by the input voltage. Note that without transistor M1B, the maximum voltage of V_1 will only reach $V_m - V_T$, where V_m is the amplitude of V_{in} and V_T is the threshold voltage of MOSFETs. Since M2A is off, the gate voltage of M1B is at V_{DD} and M1B is in the triode. As a result, $V_{1,max} = V_m - V_{ds1} \approx V_m$. The drawback of the voltage loss of Dickson voltage multipliers is therefore removed. It was demonstrated in [22] that with $v_{in} = 1.5$V and $I_{out} = 10\mu$A, the output voltage of a 4-stage Dickson voltage multiplier with static charge transfer switches is approximately 4 V. The output voltage of a corresponding conventional Dickson voltage multiplier is only 2 V.

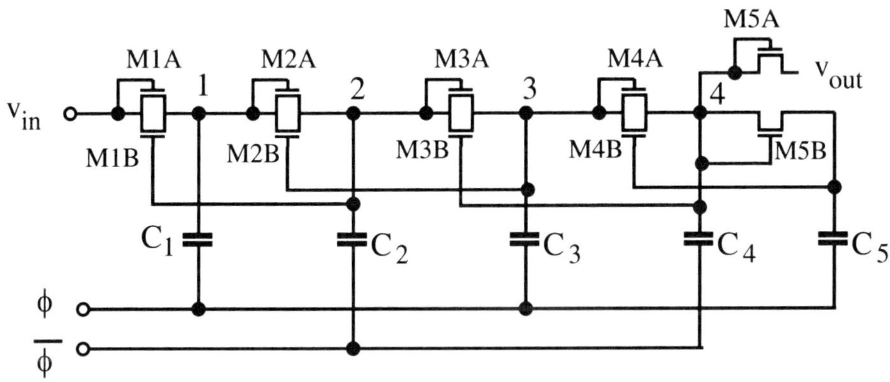

Figure 2.6. Dickson voltage multiplier with static charge transfer switches [22].

San *et al.* proposed bootstrapped gate transfer switches to replace MOSFET diodes of Dickson voltage multiplier, as shown in Fig.2.7 [23]. For each transistor, there are five additional transistors $M_{1a,...,5a}$ and one capacitor C_a are added. When $\phi = 1$, $M_{1a,3a,5a}$ are ON while $M_{2a,4a}$ are OFF. As a result, $V_{B,C} = 0$, $V_A = V_m$, and C_a is charged to V_m. In the following phase where $\phi = 0$, $M_{1a,3a,5a}$ are OFF and $M_{2a,4a}$ are ON. The voltage of the capacitor C_a is applied between the gate and drain of M_2. For MOSFETs in the triode, $V_{DS} \leq V_{GS} - V_T$ must be satisfied. Referring to Fig.2.7(b), re-write the preceding condition for the MOSFET in the triode

$$V_D - V_S \geq V_D - V_b - V_S - V_T, \tag{2.13}$$

where V_b is the voltage applied between the gate and the drain of the MOSFET. It follows that $V_b - V_T > 0$. It becomes evident that if $V_b > V_T$, M_2 will be in the triode region and V_{DS2} will be small. In was shown in [23] that the output voltage of a 4-stage Dickson voltage multiplier with bootstrapped gate transfer switches, $C_{1-4} = 15$ pF, $C_{out} = 30$ pF, $f = 5$ MHz, and $V_m = 2$V, is 9 V approximately. The output voltage is only 4 V approximately with static charge transfer switches. The power conversion efficiency is increased from 40% approximately with static charge transfer switches to above 90% with bootstrapped gate transfer switches.

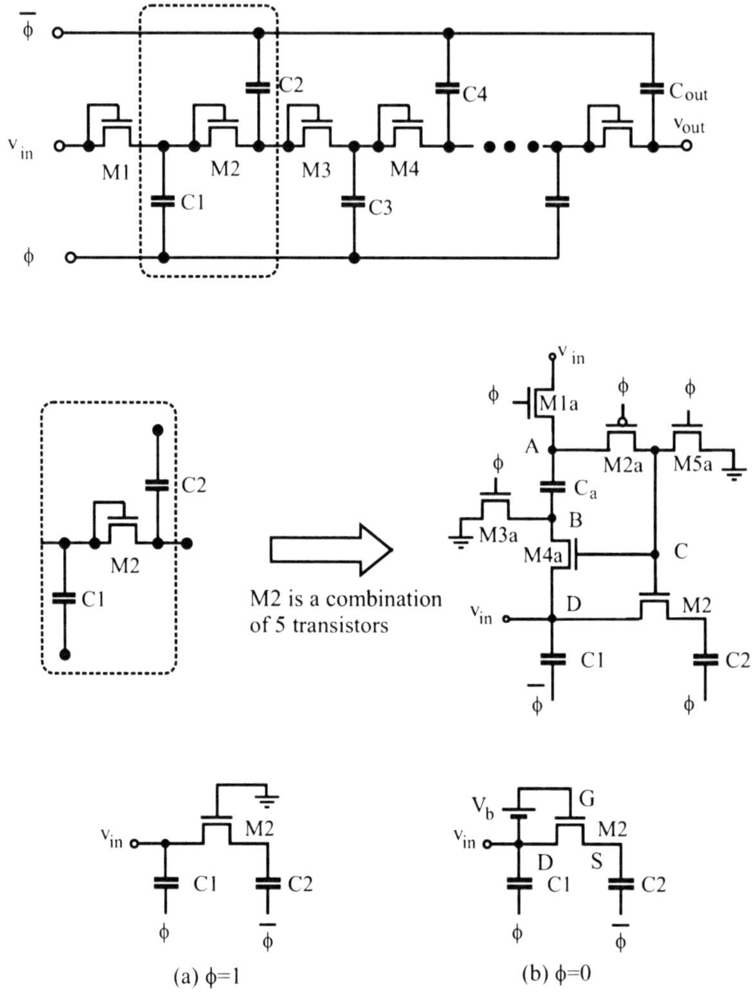

Figure 2.7. Dickson voltage multiplier with bootstrapped gate transfer switches [23].

2.2.4 Modified Dickson Voltage Multipliers

Dickson voltage multiplier requires non-overlapping clock signals ϕ and $\overline{\phi}$. For power telemetry, only one RF signal is available. Dickson voltage multiplier therefore can not be used directly. The fact that the clocking signal

ϕ and $\overline{\phi}$ are applied to every coupling node of Dickson voltage multiplier suggests that if $\overline{\phi}$ and v_{in} terminals are grounded and the RF input is connected to ϕ terminal, as shown in Fig.2.8, each section, as highlighted in the figure, becomes a voltage doubler. This configuration is termed modified Dickson voltage multiplier, in distinction from the original Dickson voltage multiplier studied earlier. Because the input signal is coupled to every other node of the diode chain, the effect of the stray capacitance is suppressed effectively. Clearly if a fully differential input is available, v_{in+} and v_{in-} can be coupled to $C_{1,3,5,...}$ and $C_{2,4,6,...}$ to further improve the performance.

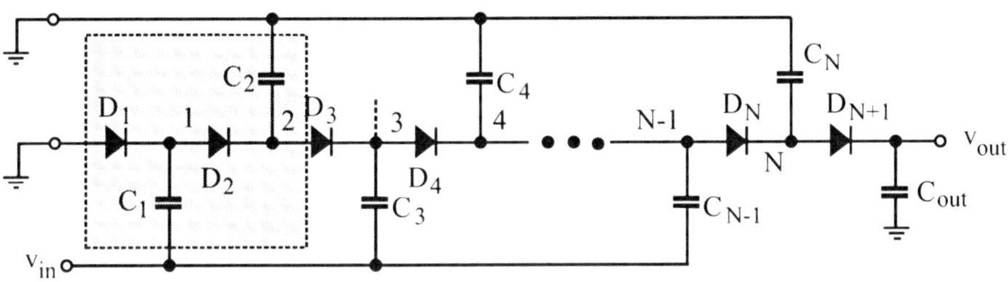

Figure 2.8. Modified Dickson voltage multiplier with diode switches.

The schematic of modified Dickson voltage multipliers with nMOS-diodes is shown in Fig.2.9 and that with pMOS-diode is shown in Fig.2.10. Because the voltage drop across the drain and source of the MOSFETs is at least one threshold voltage, the efficiency of this voltage multiplier is lower as compared with that of its Schottky-diode counterpart. To overcome this drawback, native nMOS transistors whose threshold voltage is approximately zero have been used [24, 25]. The main drawback is that native MOS structure is not generally supported. Also, the large channel resistance of native MOSFETs deteriorates the performance.

2.2.5 Mandal-Sarpeshkar Voltage Multiplier

To overcome the drawback of modified Dickson voltage multiplier with MOSFET-diodes, Mandal and Sarpeshkar proposed a low-power high power efficiency voltage multiplier with its configuration shown in Fig.2.11 [26]. It is ready to verify that once a load is connected between nodes 1 and 2, the current flowing through the load is always in the same direction. A key advantage of this voltage multiplier is the low voltage drop across switching MOSFETs. As a result, a large voltage exists at the output of the rectifying cell and is given by $V_{out,max} = V_m - (V_{DS,n} + V_{SD,p})$. The modular configuration of Mandal-Sarpeshkar voltage multiplier offers the flexibility of adjusting the size of each stage to obtain optimal performance.

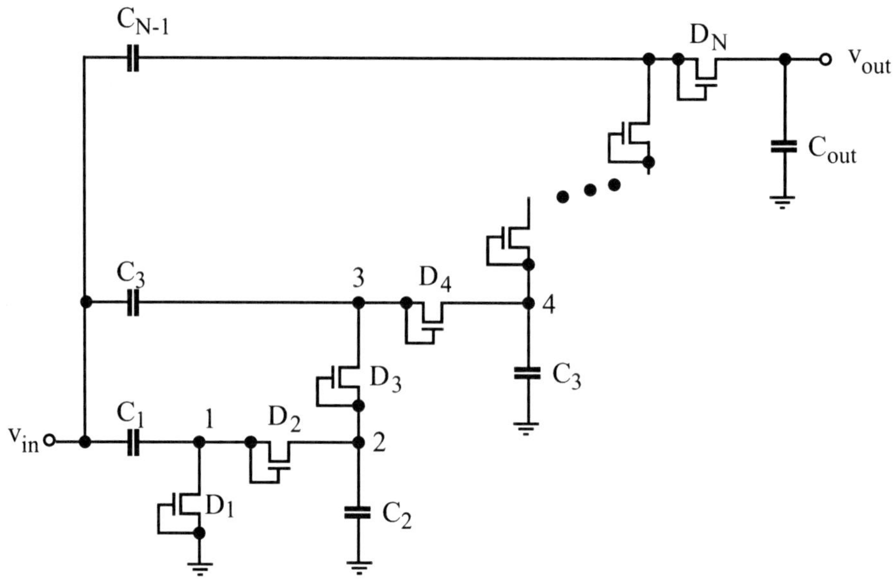

Figure 2.9. Modified Dickson voltage multiplier with nMOS transistors.

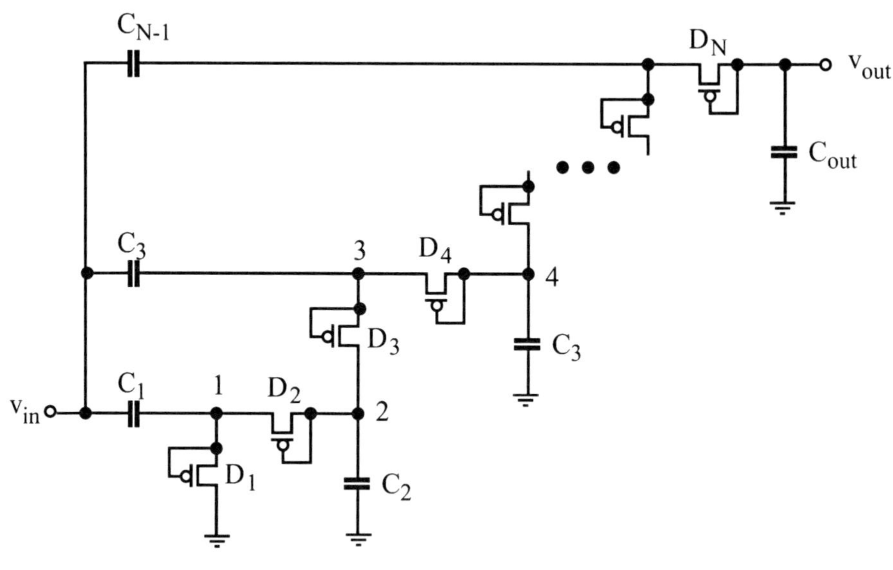

Figure 2.10. Modified Dickson voltage multiplier with pMOS transistors.

2.2.6 Voltage Multiplier with V_T-Cancellation

Umeda *et al.* proposed an elegant mechanism shown in Fig.2.12 to minimize the voltage drop across MOSFET switches so as to increase the power efficiency of voltage multipliers [27]. For M_1, because

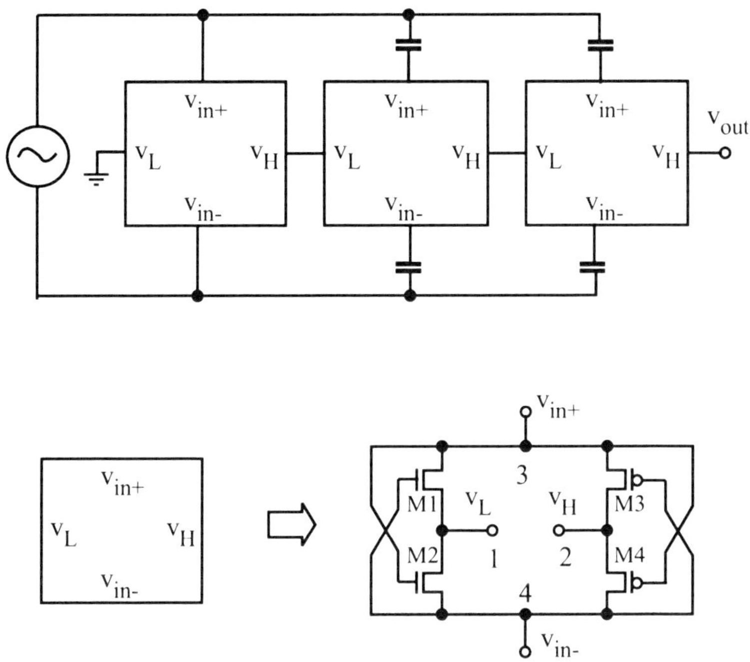

Figure 2.11. Mandal-Sarpeshkar voltage multiplier. If $v_{in+} = V_m$ and $v_{in-} = 0$, $M_{2,3}$ are in the triode (provided that V_m is large enough) and $M_{1,4}$ are off. The voltage drop across $M_{2,3}$ is low [26].

$$V_{G1} = V_m + V_b, \tag{2.14}$$

where V_{G1} is the gate voltage of M1, we have

$$V_{out} = 2(V_{GS,max} - V_T) = 2(V_m + V_b - V_T). \tag{2.15}$$

If we set $V_b = V_T$, then $V_{out} = 2V_m$ follows. The power efficiency loss caused by the threshold voltage of MOSFETs is eliminated completely. The required compensation voltage V_b can be obtained in various ways. The approach given in [27] used an external voltage source and a switched capacitor array to generate a set of V_b for all the transistors.

Nakamoto *et al.* proposed an internal threshold voltage generation mechanism to eliminate the voltage drop across MOSFETs without the need for an external voltage source, as shown in Fig.2.13 [28]. The voltage dividers formed by R_1 and M_{1a}, and R_2 and M_{2a} provide the required gate voltages for M_1 and M_2, respectively. These voltages are held by C_{1a} and C_{2a}, respectively. The values of R_1 and R_2 should be made large to minimize the static

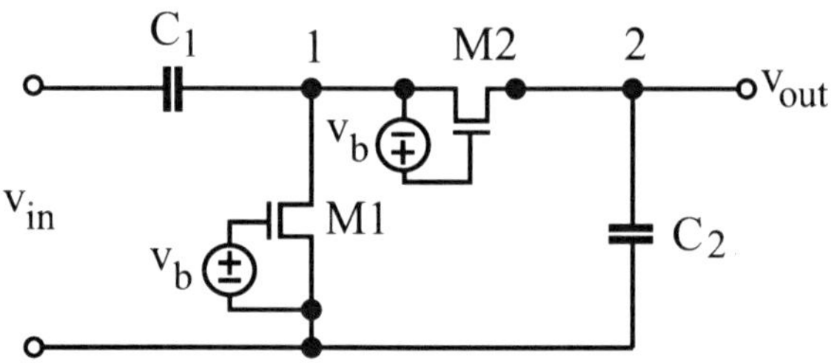

Figure 2.12. Umeda voltage multiplier with external threshold voltage cancellation [27].

power consumption of the compensation transistors $M_{1a,2a}$. Implemented in a 0.35μm CMOS technology with an input at 953 MHz, the power efficiency of Nokamoto voltage multiplier at 4-meter distance from a 4 W base station is 36.6% while that of Umeda voltage multiplier is only 16.6%.

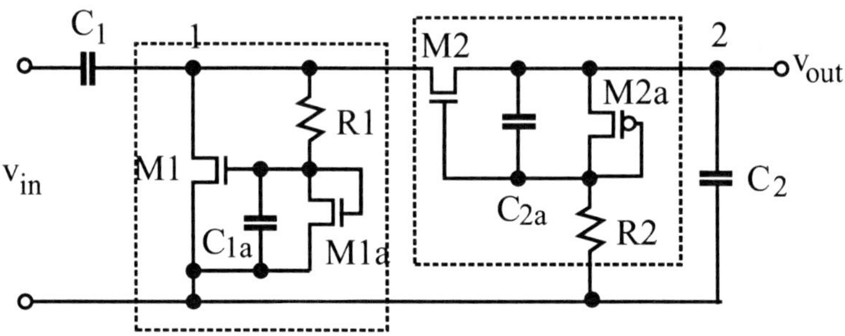

Figure 2.13. Nakamoto voltage multiplier with internal threshold voltage cancellation [28].

2.2.7 Bergeret Voltage Multiplier

Bergeret *et al.* pointed out that an important reason of the low power efficiency of Dickson voltage multipliers including modified Dickson is the propagation of high-frequency signals throughout the circuits [45]. The large area associated with multi-stage voltage multipliers gives rise to a higher substrate loss, subsequently a low power efficiency. Bergeret *et al.* modified the configuration of conventional voltage multiplier by only using a single-stage rectifier to generate a dc voltage. This voltage is then used to power a low-frequency VCO whose outputs, together with the output of the single-stage rectifier, are used to drive a high-efficiency voltage multiplier proposed in [46, 47]. It was demonstrated that this voltage multiplier improved the power efficiency by 14% over the conventional modified Dickson voltage multiplier and the output voltage is 1.5 times that of the modified Dickson voltage multiplier.

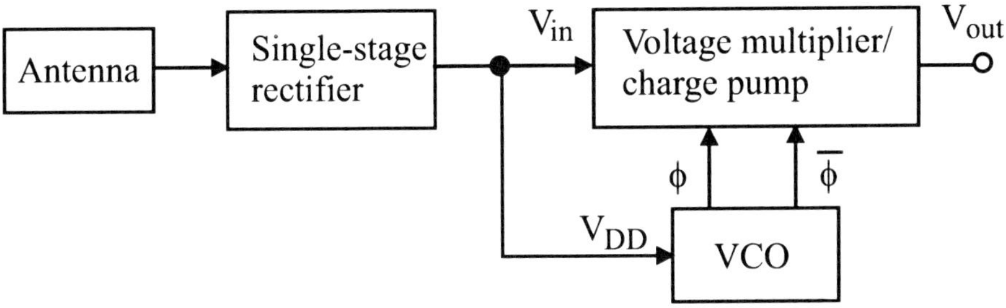

Figure 2.14. Voltage multiplier proposed by Bergeret *et al.* [45].

2.3 Power-Matching and Gain-Boosting Using LC Tanks

Voltage multipliers implemented in standard CMOS technologies suffer from a low efficiency. It was shown in [34] that to boost the power efficiency of the voltage multiplier, the amplitude of the voltage from the antenna of passive wireless microsystems must be maximized. The radiation resistance of the antenna of a passive wireless microsystem is determined by the dimension and type of the antenna. The finite antenna dimension of the passive wireless microsystem limits the voltage across the antenna to be small. As a result, an impedance transformation network that converts the input impedance of the downstream voltage multiplier to the matching impedance of the antenna for the maximum power transmission is inserted between the antenna and the voltage multiplier. In [48], a shunt inductor power-matching network between the antenna and voltage multiplier was employed to resonate out the capacitive part of the input impedance of the voltage multiplier. No attempt, however, was made to match the real part of the input impedance of the multiplier to the radiation resistance of the antenna, leaving the task of power-matching entirely to the voltage multiplier. De Vita and Iannaccone proposed a LC power-matching network that consists of one floating inductor, a shunt capacitor, and a grounded inductor [42]. The grounded inductor is used to resonate out the input capacitance of the downstream voltage multiplier while the LC network provides the matching impedance and voltage gain. The LC power-matching network used by Shameli *et al.* consists of a grounded inductor and a floating capacitor [25].

Power-matching and gain-boosting can be achieved simultaneously by inserting a passive impedance transformation network consisting of a spiral inductor and a metal-insulator-metal (MIM) capacitor between the antenna and the multiplier , as shown in Fig.2.1 [25]. The impedance transformation network provides a matching impedance to the antenna in order to maximize the power transmission from the antenna to the impedance transformation network at the carrier frequency. At the same time, it resonates at the carrier frequency such that the voltage at the output of the impedance transformation network or the input of the following voltage multiplier is maximized. Since spiral inductors suffer from both a resistive loss mainly due to the ohmic loss of the spiral and a capacitive loss due to the shunt capacitance between the spiral and the substrate, power matching, power loss, and voltage gain of the impedance transformation network must be considered simultaneously in design.

To maximize the amount of the power transferred from the antenna to the impedance transformation network, the impedance transformation network in Fig.2.1 must be designed in such a way that

$$z_{in} = R_a. \tag{2.16}$$

Fig.2.15 shows a simplified schematic of a power-matching and gain-boosting network using a shunt spiral inductor and a series MIM capacitor. To simplify analysis, the MIM capacitor is assumed to be ideal and is represented by an ideal capacitor C. The spiral inductor is modeled using the RLC network with R_s and R_p the series and shunt parasitic resistances, respectively, and C_p the parasitic shunt capacitance. The voltage multiplier is modeled using resistor R_L in parallel with capacitor C_L.

To facilitate analysis, the branch consisting R_s and L_p is replaced with its equivalent parallel $R_s'{\sim}L_p'$ network shown in Fig.2.15 with R_s' and L_p' given by [49]

$$\begin{aligned} L_p' &= L_p\left[1+\left(\frac{R_s}{\omega L_p}\right)^2\right], \\ R_s' &= R_s\left[1+\left(\frac{\omega L_p}{R_s}\right)^2\right]. \end{aligned} \tag{2.17}$$

For practical spiral inductors, $\omega L_p \gg R_s$ holds, i.e. the reactance of the inductor is much larger than the resistance of the inductor. As a result, $L_p' \approx L_p$ follows.

It is conveniently to show that the matching condition for the maximum power transfer at node A can be shifted to node B. Moving the impedance matching point from node A to node B will greatly simplify analysis, as to be seen shortly. The impedance matching condition in this case becomes

$$Z_{in} = Z_s^*. \tag{2.18}$$

Let

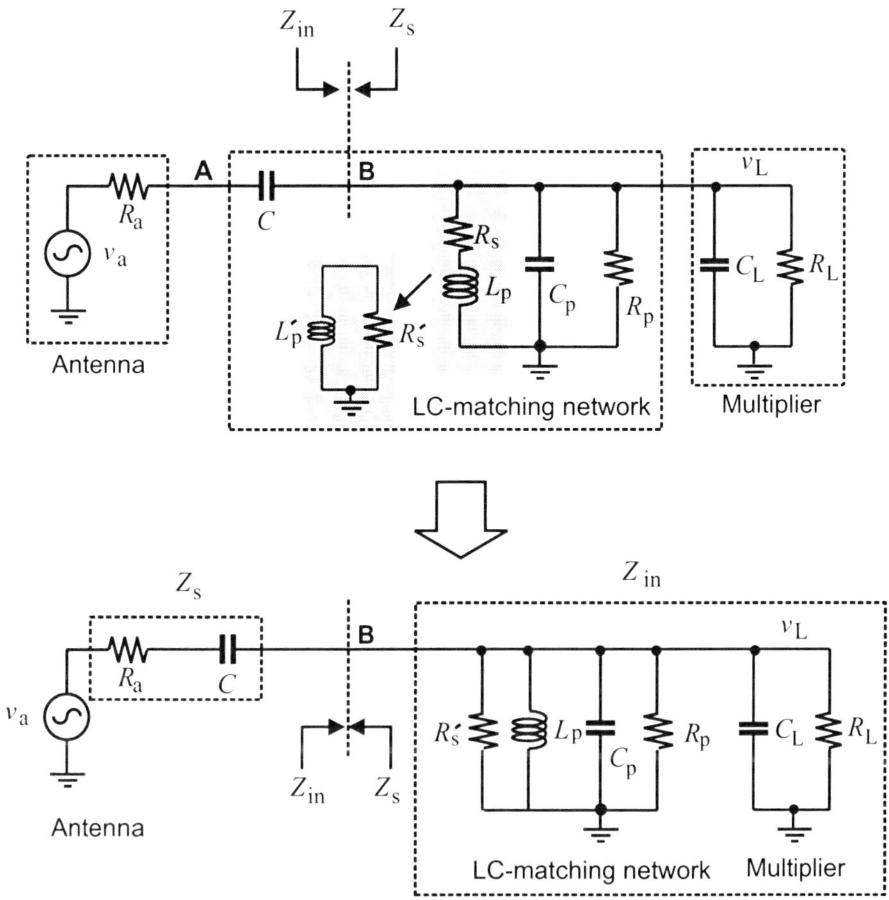

Figure 2.15. Power-matching and gain boosting network using a shunt spiral inductor and a series MIM capacitor.

$$Z_{in} = R_{in} + jX_{in}, \tag{2.19}$$

and

$$Z_s = R_s + jX_s, \tag{2.20}$$

where R_{in} and X_{in} are the resistance and reactance of the impedance looking into the impedance transformation network, R_s and X_s are the resistance and reactance of the impedance looking into the source network. Submitting these results into (2.18) yields

$$\begin{aligned} R_{in} &= R_s, \\ X_{in} &= -X_s. \end{aligned} \tag{2.21}$$

Since

$$Y_{in} = \frac{1}{Z_{in}} = \frac{R_{in} - jX_{in}}{R_{in}^2 + X_{in}^2}, \tag{2.22}$$

and

$$Y_s = \frac{1}{Z_s} = \frac{R_s - jX_s}{R_s^2 + X_s^2}, \tag{2.23}$$

if $R_{in} = R_s$ and $X_{in} = X_s$, we have

$$Y_L = Y_s^*. \tag{2.24}$$

Eq.(2.24) confirms that impedance matching is the same as admittance matching in maximizing power transfer.

The input impedance Z_{in} is given by

$$\begin{aligned} Z_{in} &\approx R_{eq}||j\omega L_p \\ &= \frac{j\omega R_{eq}^2 L_p + \omega^2 R_{eq} L_p^2}{R_{eq}^2 + (\omega L_p)^2}, \end{aligned} \tag{2.25}$$

where

$$R_{eq} = R_s'||R_p||R_L \tag{2.26}$$

is the total shunt resistance. Note that we have neglected the shunt capacitances in (2.25) to simplify analysis because the gain provided by the impedance transformation network is mainly due to R_{eq}, as to be seen shortly. Matching the impedance at node B yields

$$R_a = \frac{\omega^2 R_{eq} L_p^2}{R_{eq}^2 + (\omega L_p)^2}, \tag{2.27}$$

and

$$\frac{1}{\omega C} = \frac{\omega R_{eq}^2 L_p}{R_{eq}^2 + (\omega L_p)^2}. \tag{2.28}$$

It follows from (2.27) that

$$\frac{R_{eq}}{R_a} = Q_p^2 + 1, \tag{2.29}$$

where

$$Q_p = \frac{R_{eq}}{\omega L_p} \tag{2.30}$$

is the quality factor of the shunt $R{\sim}L$ network consisting of the spiral inductor and the voltage multiplier with the shunt capacitances neglected. Similarly, one can show from (2.28) that the frequency at which the impedance matching condition is satisfied is given by

$$\omega = \omega_o \sqrt{1 + \frac{1}{Q_p^2}}, \tag{2.31}$$

where

$$\omega_o = \frac{1}{\sqrt{L_p C}} \tag{2.32}$$

is the resonant frequency of the ideal LC impedance transformation network.

Let us now matching the admittance at node B. Because

$$Y_{in} = \frac{1}{R_{eq}} + j\left[\omega C_p' - \frac{1}{\omega L_p}\right], \tag{2.33}$$

and

$$Y_s = \frac{1}{R_a + \frac{1}{j\omega C}} = \frac{\omega^2 C^2 R_a + j\omega C}{1 + (\omega R_a C)^2}, \tag{2.34}$$

where

$$C_p' = C_L + C_p. \tag{2.35}$$

we have

$$\frac{1}{R_{eq}} = \frac{\omega^2 R_a C^2}{1 + (\omega R_a C)^2}, \tag{2.36}$$

and

$$\omega C'_p - \frac{1}{\omega L_p} = -\frac{\omega C}{1 + (\omega R_a C)^2}, \tag{2.37}$$

Solving (2.36) yields

$$\frac{R_{eq}}{R_a} = Q_a^2 + 1, \tag{2.38}$$

where

$$Q_a = \frac{1}{\omega R_a C} \tag{2.39}$$

is the quality factor of $R_a \sim C$ network. Solving (2.37) and noting that $Q_a \gg 1$ yield the frequency at which the admittance matching is satisfied

$$\omega \approx \frac{\omega_p}{\sqrt{1 + \frac{C}{C'_p}}}, \tag{2.40}$$

where

$$\omega_p = \frac{1}{\sqrt{L_p C'_p}} \tag{2.41}$$

is the resonant frequency of the shunt network.

The impedance transformation network is lossy due to the power dissipation of R'_s and R_p. To maximize the amount of the power transferred from the antenna to the multiplier, the power loss of the impedance transformation network must be minimized. Since $\omega L_p \gg R_s$ holds for spiral inductors, we have from (2.17)

$$R'_s \approx \frac{(\omega L_p)^2}{R_s}. \tag{2.42}$$

Further from (2.40) we have

$$(\omega L_p)^2 = \frac{1}{\omega^2 \left(C + C'_p\right)^2}. \tag{2.43}$$

Substitute (2.43) into (2.42)

$$R'_s = \frac{1}{R_s \omega^2 C^2 \left(1 + \frac{C'_p}{C}\right)^2}. \tag{2.44}$$

Further from (2.38) with $Q_a \gg 1$, we have

$$R_{eq} R_a = \frac{1}{(\omega C)^2}. \tag{2.45}$$

Making use of (2.45), (2.44) becomes

$$R'_s = \frac{R_{eq} R_a}{R_s \left(1 + \frac{C'_p}{C}\right)^2}. \tag{2.46}$$

Substituting (2.26) into (2.46) yields

$$R'_s = \left[\frac{R_a}{R_s}\left(\frac{C}{C + C'_p}\right) - 1\right](R_L || R_p). \tag{2.47}$$

It is seen from (2.47) that C'_p lowers R'_s. This is echoed with an increase in the ohmic loss of the inductor.

The power efficiency of the impedance transformation network is obtained from

$$\eta_I = \frac{P_L}{P_{in}}, \tag{2.48}$$

where P_L is the amount of power delivered to the voltage multiplier and P_{in} is the amount of power available at the input of the impedance transformation network. By assuming that there is no loss in capacitor C, P_{in} is the amount of power at node B, i.e. the input port of the shunt network consisting of the spiral inductor and the voltage multiplier.

$$P_{in} = \frac{V_L^2}{R_{eq}}, \tag{2.49}$$

and

$$P_L = \frac{V_L^2}{R_L}. \tag{2.50}$$

It follows that

$$\eta_I = \frac{P_L}{P_{in}} = \frac{R_{eq}}{R_L}. \tag{2.51}$$

Since

$$\frac{1}{R_{eq}} = \frac{1}{R_s'} + \frac{1}{R_p} + \frac{1}{R_L}, \tag{2.52}$$

making use of (2.47), we can write (2.52) as

$$\frac{1}{R_{eq}} = \left(\frac{1}{R_p} + \frac{1}{R_L}\right) \frac{\dfrac{R_a}{R_s}\left(\dfrac{C}{C + C_p'}\right)^2}{\dfrac{R_a}{R_s}\left(\dfrac{C}{C + C_p'}\right)^2 - 1}. \tag{2.53}$$

Substituting (2.53) into (2.48) yields

$$\eta_I = \frac{1 - \dfrac{R_s}{R_a}\left(1 + \dfrac{C_p'}{C}\right)}{1 + \dfrac{R_L}{R_p}}. \tag{2.54}$$

Since $\dfrac{R_L}{R_p} \ll 1$ typically holds, making use of

$$\frac{1}{1 + x} \approx 1 - x, \tag{2.55}$$

when $|x| \ll 1$, we can write (2.54) as

$$\eta_I \approx 1 - \frac{R_s}{R_a}\left(1 + \frac{C_p'}{C}\right) - \frac{R_L}{R_p}. \tag{2.56}$$

It is evident from (2.56) that the power efficiency of the impedance transformation network is less than 100% due to the non-zero parasitic series resistance R_s, the finite parasitic shunt resistance R_p, and the shunt capacitance C'_p. To increase η_I, R_s and C'_p should be minimized while R_p should be maximized.

As pointed out earlier, the overall power efficiency of the power harvester can be improved if a large voltage gain is provided by the impedance transformation network. The power delivered to the impedance transformation network is computed from

$$P_L = \frac{V_L^2}{R_{eq}} = \frac{A_v^2 V_a^2}{R_{eq}}, \tag{2.57}$$

where

$$A_v = \frac{V_L}{V_a} \tag{2.58}$$

is the voltage gain of the impedance transformation network. The maximum power delivered to the impedance transformation network is given by

$$P_{L,max} = \frac{V_a^2}{4R_a}. \tag{2.59}$$

Equating (2.57) and (2.59) yields the optimal voltage gain of the impedance transformation network at which the maximum power transfer takes place

$$A_v = \frac{1}{2}\sqrt{\frac{R_{eq}}{R_a}}. \tag{2.60}$$

Substituting (2.26) and (2.47) into (2.60), we arrive at

$$A_v = \frac{1}{2}\sqrt{\frac{R_p||R_L}{R_a}\left[1 - \frac{R_s}{R_a}\left(1 + \frac{C'_p}{C}\right)\right]}. \tag{2.61}$$

We comment on the preceding development :

- If $R_L > R_p$, $R_p||R_L$ will be dominated by R_p. Increasing R_L beyond R_p will no longer improve $R_p||R_L$ subsequently the voltage gain.
- To improve the voltage gain of the impedance transformation network, R_s must be made much smaller than R_a. Minimizing the resistive loss of the spiral inductor is critical.

- Parasitic shunt capacitances of the spiral inductor C'_p increases the effect of conductive loss by a factor of $\left(1+\frac{C'_p}{C}\right)$. To minimize its effect, $C'_p \ll C$ is essential.

- The power efficiency of the impedance transformation network can be further analyzed by neglecting the shunt capacitances C_p and C_L for simplicity. The maximum power delivered to the impedance transformation network is given by

$$P_{in} = \frac{V_a^2}{4R_a} \tag{2.62}$$

and the current flowing from the antenna to the impedance transformation network at the maximum power transfer is given by

$$I_{in} = \frac{V_a}{2R_a}. \tag{2.63}$$

The power delivered to the voltage multiplier is obtained from

$$P_L = R_L I_L^2. \tag{2.64}$$

Since

$$\begin{aligned} I_L &= \left| \frac{R'_p || (j\omega L_p)}{R'_p || (j\omega L_p) + R_L} \right| I_{in} \\ &= \frac{R'_p}{\sqrt{Q_L^2 + \left(1 + \frac{R_L}{R'_p}\right)^2}}, \end{aligned} \tag{2.65}$$

where $R'_p = R_p || R'_s$ and

$$Q_L = \frac{R_L}{\omega L_p}. \tag{2.66}$$

Note that since R_L is typically smaller than R_p and R'_s, $R_L \approx R_{eq}$ holds. As result, $Q_L \approx Q_p$. The power efficiency of the impedance transformation network is obtained from

$$\eta_I = \frac{P_L}{P_{in}} = \frac{R_L}{R_a}\left[\frac{1}{Q_L^2 + \left(1+\frac{R_L}{R_p}\right)^2}\right]. \tag{2.67}$$

When R_L is small, i.e. $R_L \ll \omega L_p$ and $R_L \ll R_p$, we have

$$\eta_I \approx \frac{R_L}{R_a}. \tag{2.68}$$

The power efficiency of the LC impedance transformation network in this case is directly proportional to R_L. It is interesting to note from (2.56) that

$$\eta_I \approx 1 - \frac{R_s}{R_a}\left(1+\frac{C_p'}{C}\right) - \frac{R_L}{R_p} = \frac{R_L}{R_a}\left[\frac{R_a}{R_L} - \frac{R_s}{R_L}\left(1+\frac{C_p'}{C}\right) - \frac{R_a}{R_p}\right]. \tag{2.69}$$

Since $R_a \ll R_L, R_p$ and $R_s \ll R_L$ typically hold, (2.69) is simplified to

$$\eta_I \approx \frac{R_L}{R_a}. \tag{2.70}$$

When R_L is large, i.e. $R_L \gg \omega L_p$, we have

$$\eta_I \approx \frac{(\omega L_p)^2}{R_a R_L}. \tag{2.71}$$

The power efficiency in this case is inversely proportional to R_L. Fig.2.16 plots η_I at 2.4 GHz with an ideal inductor of inductance $L_p = 10.6$ nH and $R_a = 50\Omega$. As can be seen that η_I rises with R_L approximately linearly when R_L is small and decreases with R_L when R_L is large. An optimal R_L thus exists.

To quantify the dependence of the output voltage and power efficiency of the LC impedance transformation network on the resistive load, the power harvester with a LC impedance transformation network is analyzed. The voltage

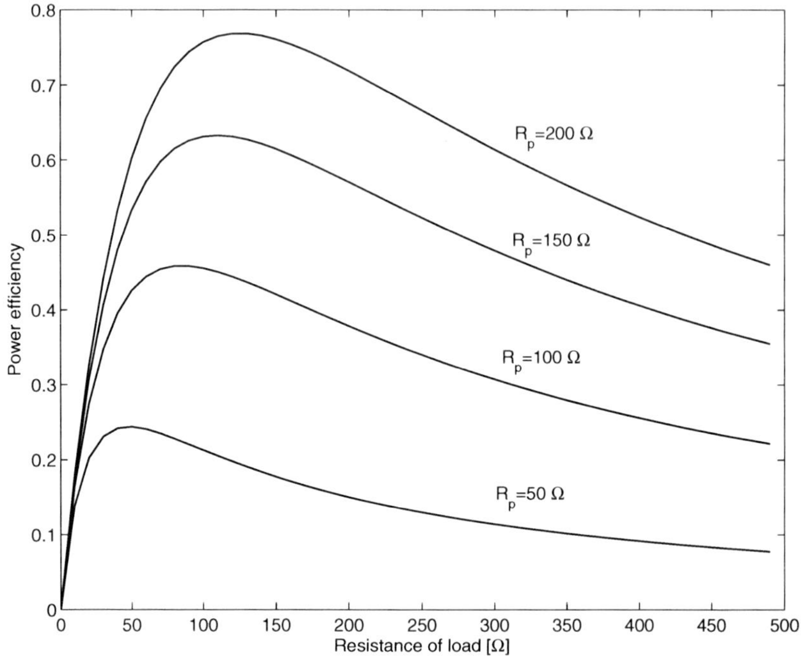

Figure 2.16. Power efficiency of LC impedance transformation network.

multiplier is modeled as an ideal resistor for simplicity. The inductor is an octagonal spiral inductor of 5.5 turns with its outer radium 222 μm and spiral width 15 μm. Fig.2.17 shows the dependence of the output voltage and power efficiency of the LC impedance transformation network on the resistance of the load. It is seen that the output voltage increases with the load resistance in the nonlinear fashion that follows a square-root profile. The power efficiency arises with the load resistance when the load resistance is low and levels off when the load resistance becomes large. Also, the profile of the power efficiency agrees with that given in Fig.2.16. A trade-off between the power efficiency of the impedance transformation network and its output voltage, which will affect the power efficiency of the downstream voltage multiplier, is needed.

2.4 Power-Matching and Gain-Boosting Using Transformers

A step-up transformer is characterized by a small voltage and a large current in the primary winding and a large voltage and a small current in the secondary winding . The relation between the current and voltage of the primary winding

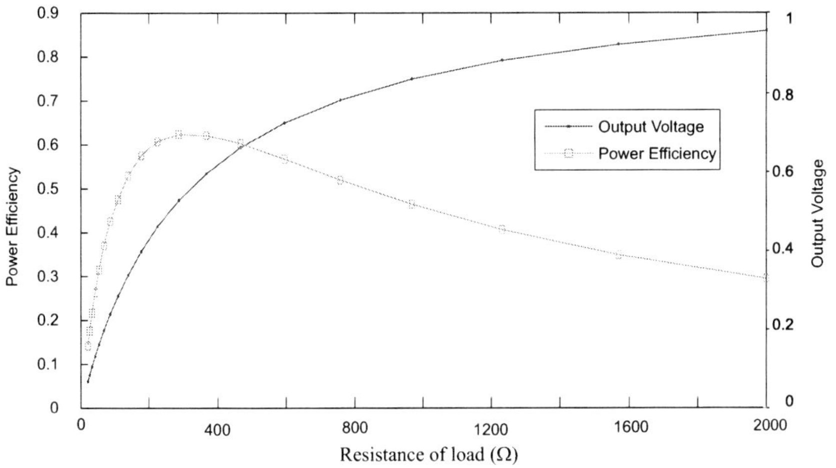

Figure 2.17. Dependence of the power efficiency of LC impedance transformation network on the resistance of the load. The spiral inductor is a 5.5-turn octagonal spiral implemented in TSMC-0.18μm CMOS technology with its outer radius 222 μm and spiral width 15 μm. The input voltage is a 2.4-GHz 1-V sinusoid.

and those of the secondary winding of a lossless step-up transformer are given by

$$\frac{I_1}{I_2} = \frac{n_2}{n_1} \tag{2.72}$$

and

$$\frac{V_1}{V_2} = \frac{n_1}{n_2}, \tag{2.73}$$

where I_1, I_2 and V_1, V_2 are the current and voltage of the primary winding and secondary winding, respectively, n_1 and n_2 are the turns of the primary winding and that of the secondary winding, respectively. With $n_2 > n_1$, we have $I_2 < I_1$ and $V_2 > V_1$. By employing a step-up transformer, the same power can be delivered from the primary winding to the secondary winding with a higher voltage at the secondary winding. It is evident that by letting n large, the voltage of the secondary winding of the transformer will be larger than that of its primary winding while the current of the secondary winding will be smaller than that of its primary winding. The loss of the primary winding will therefore be dominated by its ohmic loss due to its large current while the loss of the secondary winding will be dominated by its spiral-substrate loss due to its large number of turns. These observations are critical as they reveal that the series loss of a step-up transformer with a large turn ratio is dominated by that of the primary winding while its shunt loss is dominated by that of the secondary winding. Soltani and Yuan pointed out that since the primary winding has a fewer turns, its series loss can be effectively reduced by increasing the width of the spiral of the primary winding. The width of the primary winding can be set to such a value that both windings will have approximately the same silicon area. The shunt loss of the secondary winding, on the other hand, can be lowered effectively by reducing the width of the spiral of the secondary winding. Note that since the current of the secondary winding is small, reducing the width of the spiral of the secondary winding will not overly increase its resistive loss [29].

Fig.2.18 shows the equivalent circuit of a power harvester with a step-up transformer. The transformer is represented using the narrow-band model given in [50, 51]. Capacitors C_1 and C_2 are used to resonate out the self-inductance of the primary winding and that of the secondary winding, respectively.

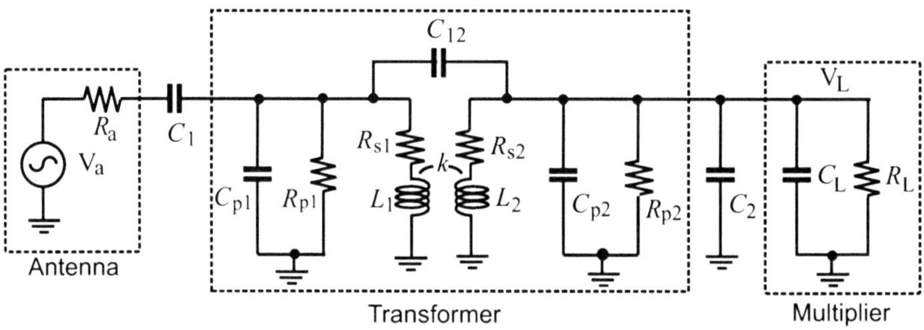

Figure 2.18. Equivalent circuit of a power harvester with a step-up transformer power-matching and gain-boosting network. C_{p1} and R_{p1} are the parasitic shunt capacitance and shunt resistance of the primary winding, respectively, R_{s1} is the parasitic series resistance of the primary winding, L_1 is the inductance of the primary winding, C_{p2} and R_{p2} are the parasitic shunt capacitance and shunt resistance of the secondary winding, respectively, R_{s2} is the parasitic series resistance of the secondary winding, L_2 is the inductance of the secondary winding, C_1 and C_2 are the series capacitance at the primary winding and the shunt capacitance at the secondary winding, respectively. They are used form LC resonant networks with L_1 and L_2, respectively.

Because the voltage of the primary winding is small, the resonance of the primary winding will only create a small voltage gain from the antenna to the primary winding. The large current of the primary winding demands that the cross-sectional area of the spiral of the primary winding be large. This can be achieved by using multiple metal layers that are connected using vias for the primary winding. Further, since the primary winding has a fewer turns, its spiral can be implemented using lower metal layers without encountering a significant substrate loss. Of course, this arrangement is technology-dependent.

The preceding arrangement is for TSMC-0.18μm CMOS technology used to fabricate the power harvester with a step-up transformer. The objective of any configuration of the primary winding is to minimize the resistive loss of the winding. The resonance at the secondary winding, on the other hand, is significant due to its large self-inductance needed to produce a large voltage gain.

To analyze the voltage gain obtained from the resonance of the secondary winding, we follow the approach used for analysis of the power harvester with a LC impedance transformation network given in Section 2.3, specifically L_2 is separated from the rest of the transformer, as shown in Fig.2.19 . A Thévenin equivalent circuit is used to represent the overall effect of the matching network excluding L_2 and C_2 . To further simplify analysis, we use

$$C_2' = C_2 + C_{p2} + C_L \tag{2.74}$$

and

$$R_2' = R_{p2}||R_L \tag{2.75}$$

to account for all shunt capacitances and resistances at the secondary winding, respectively. Note that the simplified circuit has the same topology as that of Fig.2.15. The approach used for the analysis of the circuit in Fig.2.15 can thus be followed to analyze the circuit in Fig.2.19.

Thévenin voltage V_T is obtained by open-circuiting the secondary winding and deriving the voltage across the series resistance R_{s2} and mutual inductance M due to V_a with L_1 and C_1 resonated out, i.e.

$$j\omega L_1 + \frac{1}{j\omega C} = 0, \tag{2.76}$$

$$V_T = \frac{j\omega M}{R_a + R_{s1}} V_a, \tag{2.77}$$

where

$$M = K\sqrt{L_1 L_2} \tag{2.78}$$

is the mutual inductance and K the coupling coefficient of the transformer. Thévenin impedance is obtained by short-circuiting V_a and applying a test voltage source over R_{s2} and M with L_1 and C_1 resonated out

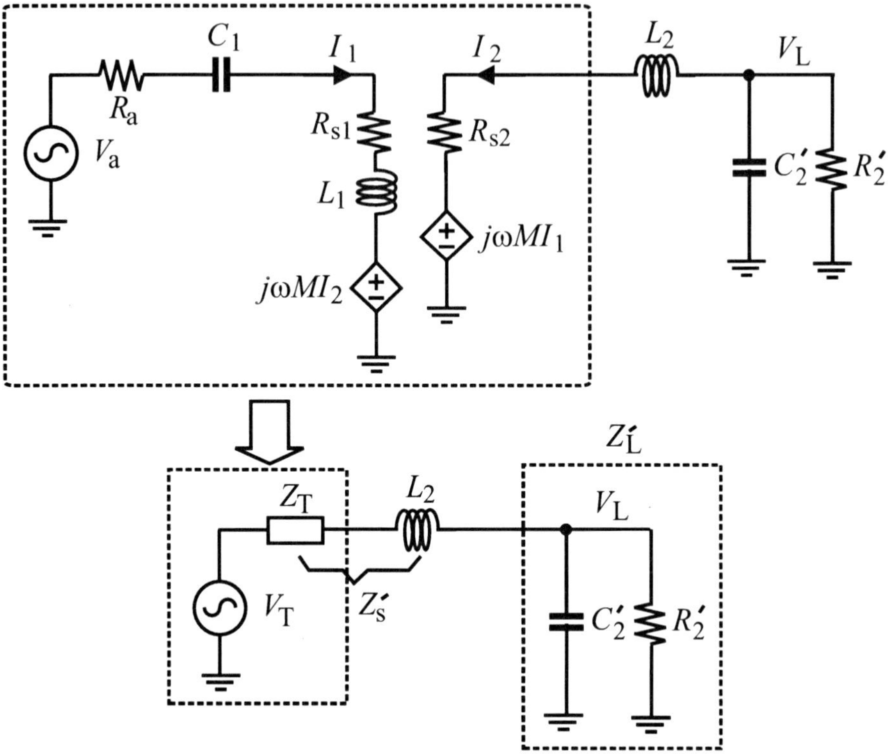

Figure 2.19. Thévenin equivalent circuit of step-up transformer power-matching and gain-boosting network.

$$Z_T = R_{s2} + \frac{(\omega M)^2}{R_a + R_{s1}}. \tag{2.79}$$

It is seen from (2.79) that Thévenin impedance is purely resistive. Also, it has two components: the series resistance of the secondary winding and the resistance of the primary winding referred to the secondary winding.

To find out the voltage gain from V_T to V_L, we notice that for the maximum power transfer,

$$Z_L' = (Z_s')^* \tag{2.80}$$

is required. Since

$$Z_L' = R_2' || \frac{1}{j\omega C_2'} \tag{2.81}$$

and

$$Z'_s = Z_T + j\omega L_2, \tag{2.82}$$

the magnitude of the voltage gain is obtained from

$$\left|\frac{V_L}{V_T}\right| = \left|\frac{Z'_L}{Z'_L + Z'_s}\right| = \frac{1}{2}\sqrt{1 + \left(\frac{\omega L_2}{Z_T}\right)^2}. \tag{2.83}$$

Substituting (2.79) into (2.83) yields

$$\left|\frac{V_L}{V_T}\right| = \frac{1}{2}\sqrt{1 + \left[\frac{R_{s2}}{\omega L_2} + \frac{\omega^2 M^2}{\omega L_2(R_a + R_{s1})}\right]^{-2}}. \tag{2.84}$$

The quality factor of the secondary winding satisfies

$$Q_2 = \frac{\omega L_2}{R_{s2}} \gg 1 \tag{2.85}$$

and

$$Q_1 = \frac{\omega L_1}{R_a + R_{s1}} \tag{2.86}$$

is the equivalent quality factor of the primary winding [52]. Making use of (2.85) and (2.86), we can write (2.84) as

$$\left|\frac{V_L}{V_T}\right| = \frac{1}{2}\sqrt{1 + \left(\frac{Q_2}{k^2 Q_1 Q_2 + 1}\right)^2}. \tag{2.87}$$

It is seen from (2.87) that a large voltage gain of the step-up transformer matching network can be obtained by (i) boosting the mutual inductance M, (ii) lowering the series resistance of the primary winding, (iii) lowering the series resistance of the secondary winding, and (iv) increasing the self inductance of the secondary winding. There are two ways to increase the mutual inductance M : increase the coupling coefficient or increase the turn ratio. The former can only be achieved by using a stacked configuration while the latter requires a large number of the turns of the secondary winding. The series resistance of the primary winding can be lowered effectively by increasing the spiral width of the primary winding and by using multiple metal layers connected together using vias. The self-inductance of the secondary winding and the coupling coefficient can be increased simultaneously by reducing the width of the spiral of the secondary winding so that more turns can be accommodated for a given silicon area. This, however, is at the cost of the increased series resistance of the secondary winding. Fortunately, the current of the secondary winding of the step-up transformer is small as compared with that of the primary winding, its resistive winding loss is not of a critical concern.

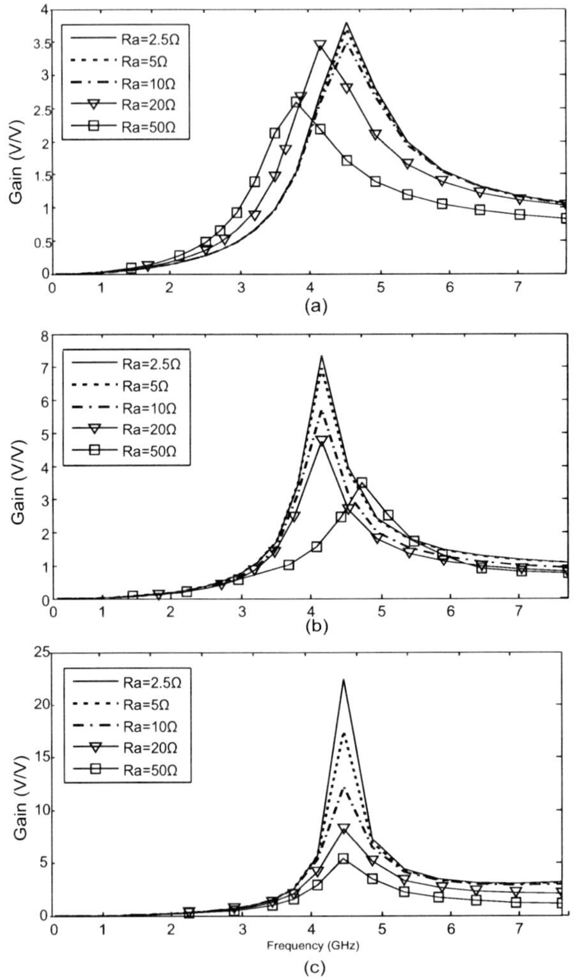

Figure 2.20. Simulated voltage gain of power-matching and gain-boosting networks for different values of radiation resistance. (a) LC matching network (spiral width =5μm). (b) LC matching network with reduced spiral width (spiral width =1.5μm). (c) Step-up transformer matching network (spiral width in primary winding : 5μm, spiral width in secondary winding : 1.5μm. The load resistance is fixed at $R_L = 40k\Omega$. The power harvesters are implemented in TSMC-0.18μm CMOS technology (Copyright (c) IEEE).

To find out the overall voltage gain, i.e. the voltage gain from the antenna to the output of the impedance transformation network, we make use of Fig.2.19. Since this circuit itself is a LC impedance transformation network, we can use (2.60) to write the voltage gain from V_T to V_L

$$A_v = \frac{V_L}{V_T} = \frac{1}{2}\sqrt{\frac{R_2'}{Z_T}}. \tag{2.88}$$

Note that Z_T is purely resistive. Substituting R_2' and Z_T with their values, the overall voltage gain of the transformer matching network is obtained

$$\begin{aligned} A_v &= \frac{1}{2}\left(\frac{\omega M}{R_a + R_{s1}}\right)\sqrt{\frac{R_2'}{Z_T}} \\ &= \frac{1}{2}\sqrt{\frac{R_L || R_{p2}}{\frac{R_{s2}}{A_{ind}^2} + R_a + R_{s1}}}, \end{aligned} \tag{2.89}$$

where

$$A_{ind} = \frac{\omega M I_1}{V_a} = \frac{\omega M}{R_a + R_{s1}} \tag{2.90}$$

is the voltage gain from V_a to V_T. Let us now comment on the preceding development :

The effect of the series resistance of the secondary winding is scaled down by A_{ind}^2. This finding is significant as it allows us to use more turns with reduced spiral width in the secondary winding to boost the voltage gain without a large shunt loss. To increase A_{ind}, the mutual inductance of the transformer M must be increased and R_{s1} must be decreased. As mentioned earlier that there are two ways to increase M: increase the coupling coefficient or increase the turn ratio. The former can be achieved by using a stacked configuration while the latter requires more turns of the secondary winding.

The series resistance of the primary winding R_{s1} is low due to the small number of the turns of the primary winding and the use of multi-layer spirals connected using vias, the effect of the series loss of the primary winding is small.

The step-up transformer matching technique is particularly attractive for antennas with a low R_a. Since in (2.89) all terms except R_a are small, reducing R_a will significantly increase A_v. Reducing the source impedance below $R_{s1} + \dfrac{R_{s2}}{A_{ind}^2}$ will not contribute further to the voltage gain.

When R_a is large, from (2.90) A_{ind} will be small. In this case, a large voltage gain can be achieved by reducing the spiral width of the secondary. This is because reducing the spiral width will reduce the winding area subsequently the shunt capacitance of the secondary winding [5]. The series loss of the secondary winding, however, is also increased. This will in turn lower the overall voltage gain. When R_a is significantly larger than $R_{s1} + \frac{R_{s2}}{A_{ind}^2}$, it will dominate the denominator of (2.89). An increase in R_{s2} in this case will have a less impact on the overall voltage gain. Decreasing the spiral width of the secondary winding is beneficial to the overall gain when the radiation resistance is large. It should be noted that an additional gain obtained is due to the resonance of the secondary winding with C'_{p2} as quantified by (2.87).

Although the step-up transformer impedance transformation network should be designed in such a way that it resonates preciously at the carrier frequency, parameter spreading from process variation will cause the resonant frequency of the impedance transformation network to deviate from the carrier frequency. It is therefore highly desirable that the resonant frequency of impedance transformation network can be tuned. The resonant frequency of the LC impedance transformation network can be tuned by placing a variable capacitor C_v in parallel with the voltage multiplier, as shown in Fig.2.21. Rewriting (2.40)

$$\omega_o \approx \frac{1}{\sqrt{L_p(C + C'_p)}}, \tag{2.91}$$

we observe that both C and C'_p can be adjusted to tune the resonant frequency of the LC impedance transformation network. Varying C will affect the voltage gain of the impedance transformation network. This is clearly undesirable. On the other hand, it is observed from (2.61) that the dependence of the voltage gain on the shunt capacitance C'_p in Fig.2.15 is governed by a much weaker function. This observation reveals that the variable capacitor required to tune the resonant frequency of the LC impedance transformation network should be placed in parallel with C'_p, as shown in Fig.2.21.

Similarly, the resonant frequency of the transformer impedance transformation network can be tuned by replacing C_1 and C_2 in Fig.2.18 with variable capacitors C_{v1} and C_{v2} respectively, as shown in Fig.2.22.

To tune the resonant frequency of the step-up transformer impedance transformation network of Fig.2.22, both capacitors C_{v1} and C_{v2} have to be adjusted. Lowering the resonant frequency using capacitors will result in a roll-off in the voltage gain. A key advantage of using a step-up transformer instead of an inductor for power-matching and gain-boosting is the reduced effect of the resistive loss of the secondary winding, as described by (2.89). By reducing the resonant frequency using capacitors, the mutual inductance M will remain unchanged while the circuit will operate at a lower frequency. This will result

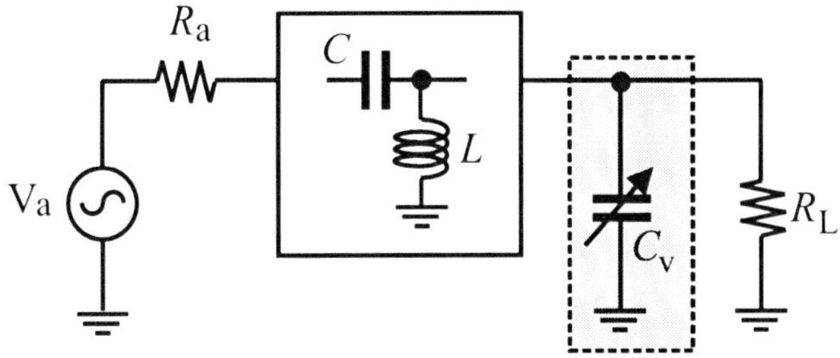

Figure 2.21. Frequency tuning in power-matching and gain-boosting network using a LC network.

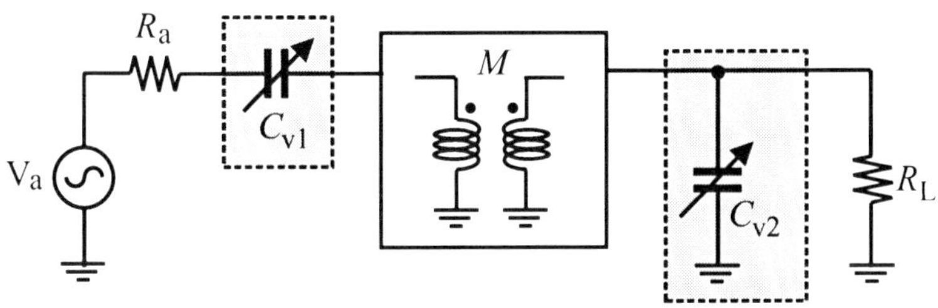

Figure 2.22. Frequency tuning in power-matching and gain-boosting using a step-up transformer network.

in a reduction in A_{ind}, as expected from (2.90). Fig.2.23(a) shows the gain roll-off caused by tuning down the resonant frequency of the step-up transformer. Fig.2.23(b) shows the dependence of the resonant frequency and voltage gain on C_{v2}. As can be seen that the voltage gain drops from 5.0 at 3.7 GHz to 4.35 at 2 GHz.

To validate the preceding findings, a step-up transformer impedance matching and gain boosting network is designed in TSMC-0.18μm 1.8V 6-metal CMOS technology with thick metal options. The primary winding has 1.5 turns and is implemented using metal layers 3 to 5 with identical thickness of 1 μm. These metal layers are connected using vias to minimize the winding resistance. The secondary winding has 5.5 turns and is implemented using the top metal layer with thickness 2.3 μm. The width of the spiral in the primary winding is 8 μm and that in the secondary winding is 1.5 μm. For the purpose of comparison, a LC matching network with the same loading condition is also implemented. The inductor of the LC matching network has a 4.5-turn spiral implemented using the top metal layer of width 5 μm. On-wafer probing is conducted using a Cascade Microtech RF-1 probe station with four MH5 positioners, RF and DC probes. Since the design is to be tested using an off-chip RF source, the choice of the source impedance is set to 50Ω. To measure the output voltage, the power-matching circuits are connected to dual-half-wave

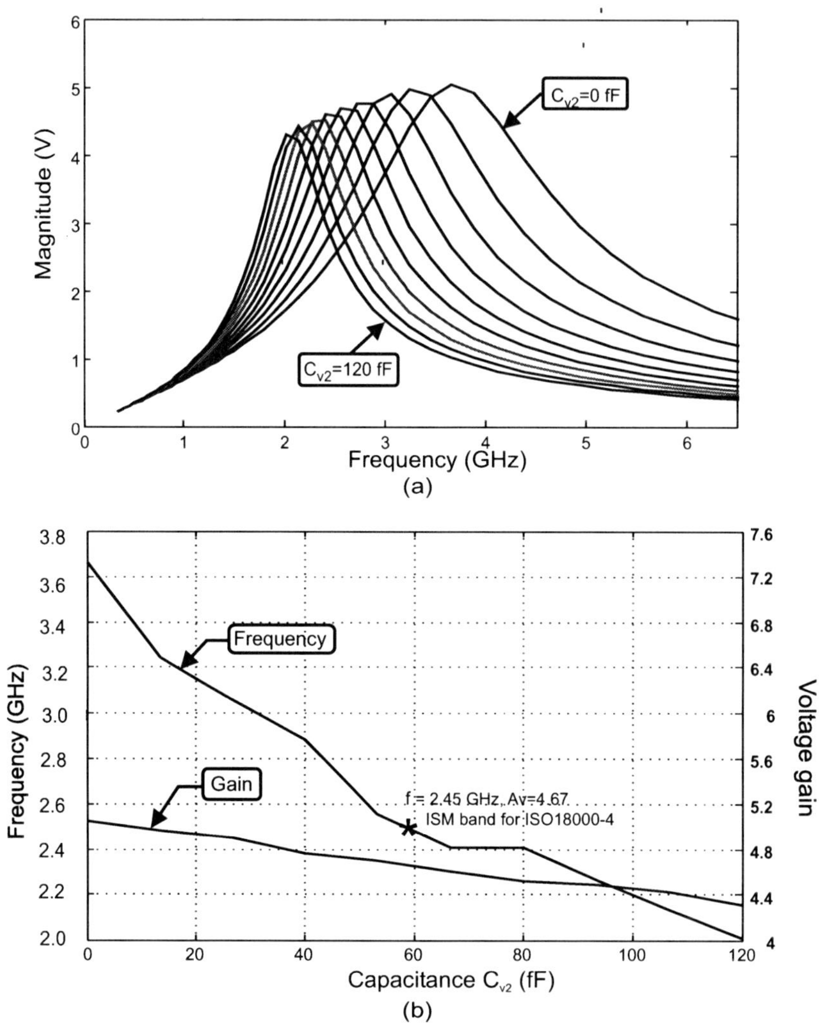

Figure 2.23. (a) Simulated dependence of the voltage gain of transformer impedance transformation network on C_{v2} and C_{v1}. (b) Simulated dependence of resonant frequency and voltage gain of transformer impedance transformation network on C_{v2}. note that C_{v1} is also varied with C_{v2} to maintain that the resonant condition is on the primary side (Copyright (c) IEEE).

rectifiers, each consisting of two diode-connected standard PMOS devices, one rectifying the positive and the other rectifying the negative half of the signal coming from the output of the matching network.

The sensitivity measurement of the power harvester with the step-up transformer impedance transformation network and that with the LC impedance transformation network is performed by measuring the dc output voltage for different levels of input signal power and the results are shown in Fig.2.24. It is observed that the larger the load resistance, the larger the output voltage. The sensitivity of the power harvester with the step-up transformer impedance transformation network is more than twice that with the LC matching network.

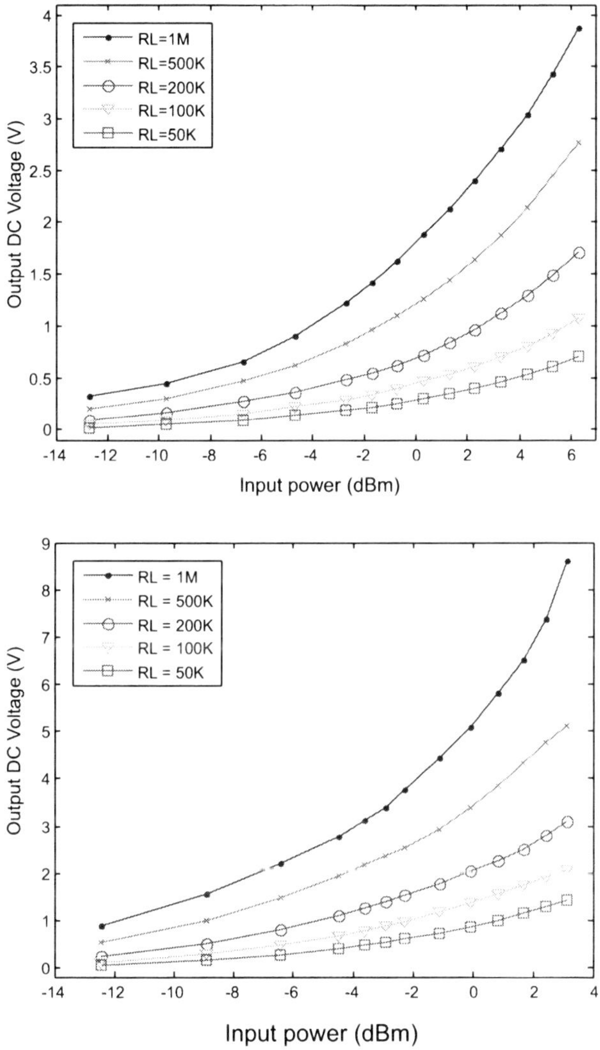

Figure 2.24. (a) Measured output voltage of the power harvester with a LC power-matching and gain-boosting network. (b) Measured output voltage of the power harvester with a step-up transformer power-matching and gain-boosting network (Copyright (c) IEEE).

Fig.2.25 plots the output DC voltage of both power harvesters. The output voltage of the power harvester with the LC matching network peaks at 3.5

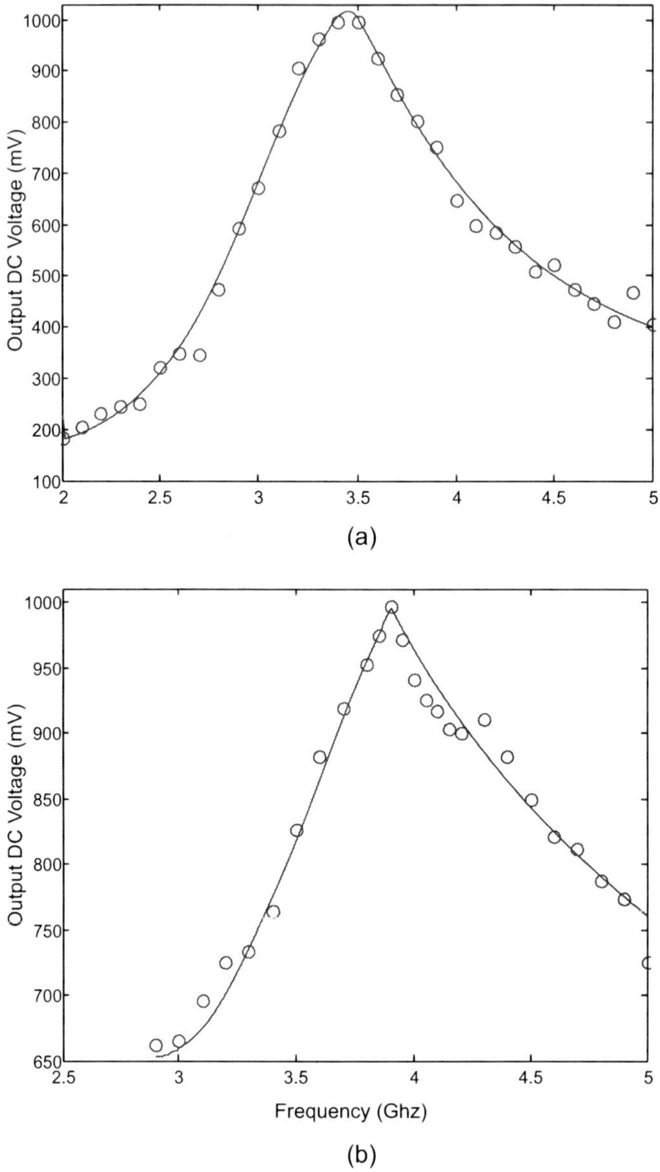

Figure 2.25. (a) Measured DC output voltage of the power harvester with a LC power-matching and gain-boosting network (Input power -4 dBm). (b) Measured DC output voltage of the power harvester with a step-up transformer power-matching and gain-boosting network (Input power -11.9 dBm) (Copyright (c) IEEE).

GHz approximately whereas that with the step-up transformer matching network peaks at 3.8 GHz approximately. Although the the impedance matching networks are designed to peak at 2.45 GHz, parameter spreading caused by process variations gives rise to a large deviation of the resonant frequency. These observations re-affirm the importance of having the capability to tune the resonant frequency of power-matching and gain-boosting networks.

2.5 Chapter summary

The chapter started with the characterization of radio-frequency power harvesting systems. Power matching conditions have been derived and the maximum amount of the power delivered to the load has been obtained. Figure-of-merits such as power efficiency have been introduced. We have shown that the overall efficiency of a RF power harvester is determined by the efficiency of the antenna of the microsystems, the accuracy of power matching between the antenna and the voltage multiplier of the microsystem for the maximum power transmission, and the power efficiency of the voltage multiplier that converts a received RF signal to a dc voltage from which the microsystem is powered.

We have shown that power-matching and gain-boosting can be achieved simultaneously by inserting a passive impedance transformation network consisting of a spiral inductor and a metal-insulator-metal capacitor between the antenna and the multiplier. The impedance transformation network provides a matching impedance to the antenna to maximize the power transmission from the antenna to the impedance transformation network at the carrier frequency. At the same time, it resonates at the carrier frequency such that the voltage at the output of the impedance transformation network or the input of the voltage multiplier is maximized. The power efficiency of the LC power-matching and gain-boosting network is affected by the resistive loss mainly due to the ohmic loss of the spiral and capacitive loss due to the shunt capacitance between the spiral and the substrate of the spiral inductor. To increase the voltage gain of the LC impedance transformation network, the inductance must be increased. This, however, is echoed with the increased series and shunt losses of the inductor, and will result in the reduction of the overall power efficiency of the power harvester.

To further increase the voltage gain, a step-up transformer impedance-matching and gain-boosting to improve the efficiency of power harvest of passive wireless microsystems has been presented. A step-up transformer is characterized by a small voltage and a large current in the primary winding and a large voltage and a small current in the secondary winding. By employing a step-up transformer, the same power can be delivered from the primary winding to the secondary winding with a higher voltage at the secondary winding. Since the voltage of the secondary winding of the transformer will be larger than that of its primary winding while the current of the secondary winding will be smaller than that of its primary winding. The loss of the primary winding is dominated by its ohmic loss due to its large current while the loss of the secondary winding is dominated by its spiral-substrate loss due to its large number of turns. Because the primary winding has a fewer turns, its series loss can be effectively reduced by increasing the width of the spiral of the primary winding. The width of the primary winding can be set to such a value that both windings will occupy approximately the same silicon area. The shunt loss of the secondary winding, on the other hand, can be lowered effectively by reducing the width of the spiral of the secondary winding. Note that since the current of the secondary winding is small, reducing the width of the spiral of the secondary winding will not overly increase its resistive loss.

The design techniques of voltage multipliers for power harvest of wireless microsystems have been investigated. For a passive wireless microsystem in the near field of the antenna of its base station, because inductive coupling is used, RF-to-DC conversion can be carried out using a diode bridge and the dc voltage at the output of the diode bridge is sufficiently large to power the passive wireless microsystem. Schottky diodes are widely used to further improve power efficiency. For a passive wireless microsystem in the far field of the antenna of its base station, the voltage at the antenna of the passive wireless microsystem is small and diode bridge-based rectification approaches become very inefficient. Voltage multipliers that perform both RF-to-DC conversion and voltage boosting are needed. Cockcroft-Walton voltage multiplier is a multi-stage configuration of voltage doublers. The effectiveness of Cockcroft-Walton voltage multiplier diminishes in monolithic integration where stray capacitances are large. Dickson voltage multiplier remove the drawback of Cockcroft-Walton voltage multiplier by injecting clocking signals into all the coupling nodes. Dickson voltage multiplier with MOSFET diodes, however, suffers from the drawback of low power efficiency due to the voltage loss of one threshold voltage across each MOSFET. Dickson voltage multiplier with static charge transfer switches minimizes this voltage loss and achieves a better power efficiency. Dickson voltage multiplier with bootstrapped gate transfer switches further improves power efficiency. Dickson voltage multipliers require a non-overlapping clock, which is not available in passive wireless microsystems. Modified Dickson voltage multipliers evolved from Dickson voltage multipliers can perform both RF-to-DC conversion and voltage boosting, and are therefore widely used in passive wireless microsystems. Modified Dickson voltage multipliers with MOSFET-diodes exhibit a low power efficiency due to the voltage loss across MOSFET diodes. The voltage multiplier by Mandal and Sarpeshkar exhibits a high power efficiency. The voltage multiplier with threshold voltage cancellation proposed by Umeda *et al.* utilizes an external voltage source to minimize the voltage drop across MOSFET switches so as to increases the power efficiency of the voltage multiplier. The voltage multiplier proposed by Nakamoto *et al.* employs an internal threshold voltage generation mechanism to eliminate the voltage drop across MOSFETs without the need for an external voltage source.

Chapter 3

DATA ENCODING

Binary data transmitted between passive wireless microsystems and their base stations are encoded as line codes, i.e. a stream of symbols, prior to their transmission. Although a variety of data encoding schemes are available for data communications over wireless channels, the limited power resource of passive wireless microsystems and the fact that the operational power of these systems is harvested from the radio-frequency waves emitted by their base stations impose a stringent constraint on the choice of data encoding schemes. The different characteristics of down-links (from base stations to microsystems) and up-links (from microsystems to base stations) arising from the limited power resource of passive wireless microsystems and the virtually unlimited power resource of their base stations further require that different data encoding schemes be used for down links and up links to achieve optimal performance. As an example, EPC class-1 generation-2 UHF RFID protocols use pulse interval encoding (PIE) to encode data transmitted from interrogators to tags with the objective to maximize the power flow from the interrogators to the tags during data transmission. For up-links, Miller-modulated sub-carrier encoding or PM0 encoding are used, aiming at maximizing the number of the transitions in the data so as to improve the performance of backscattering.

Data encoding schemes for passive wireless microsystems are selected with the following considerations :

- Power flow - Since passive wireless microsystems harvest their operational power from the radio-frequency waves emitted by their base stations, maximizing the power flow from the base stations during the initial charge-up and maintaining a continuous power flow from the base stations to the passive wireless microsystems during data transmission from the base stations to the passive wireless microsystems are critical to ensure that the mi-

F. Yuan, *CMOS Circuits for Passive Wireless Microsystems*,
DOI 10.1007/978-1-4419-7680-2_3,

crosystems will have sufficient and steady power to operate over the entire communication interval.

- Backscattering - Backscattering transmits data from a passive wireless microsystem, such as a RFID tag, to its base station, such as a RFID interrogator, by modulating the termination resistance of the antenna of the tag between R_a, the characteristic resistance of the antenna, and infinity. The former gives a zero reflection while the latter yields a full reflection. By detecting the presence of the reflected wave, the interrogator will be able to determine whether the received data is a binary 1 or a binary 0. In practice, the value of the transmitted data is determined by the number of the times that the termination resistance is modulated. The data encoding schemes used in backscattering must therefore contain different number of transitions for binary 1 and binary 0.

- Bandwidth - Although the data rate of passive wireless microsystems is typically low, it is highly desirable to minimize the bandwidth requirement of both up-links and down-links such that the data rate can be increased. Fig.3.1 shows the spectrum of a bit stream with bit time T_b and period T_b. It is seen that the shorter the bit time, the more widely the spectrum of the bit stream will spread subsequently the larger the bandwidth required to transmit the data.

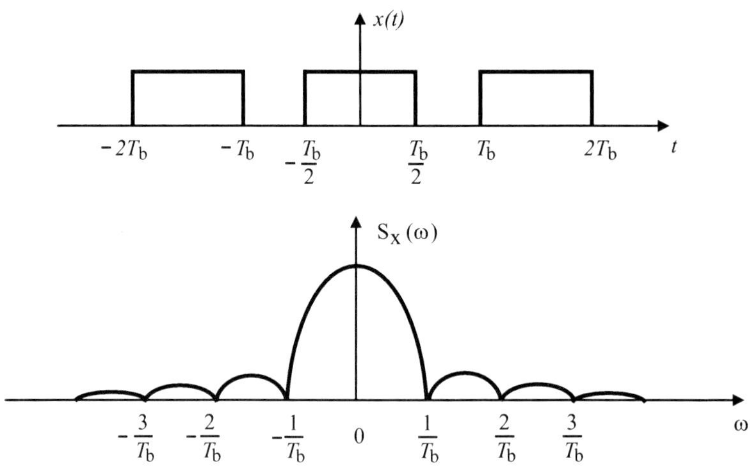

Figure 3.1. Spectrum of a bit stream with bit time T_b and period T_b.

In this chapter, we briefly investigate the pros and cons of commonly used data encoding schemes and their suitability for passive wireless microsystems. A complete mathematical treatment of data encoding schemes is readily available in standard texts on digital communications and is therefore not the focus of this book.

The chapter is organized as the follows : Section 3.1 studies non-return-to-zero (NRZ) encoding. NRZ is popular in high-speed data communications over wire channels. Return-to-Zero encoding is investigated in Section 3.2. Manchester encoding and its characteristics are the focus of Section 3.3. In Section 3.4 and Section 3.5, Miller encoding and Miller-modulated sub-carrier encoding are studied, respectively. An emphasis is given to the distinct characteristics of Miller-modulated sub-carrier encoding. Section 3.6 deals with FM0 encoding. Pulse interval encoding is dealt with in Section 3.7. The performance of the encoding schemes studied in this chapter is also compared. The chapter is summarized in Section 3.8.

3.1 Non-Return-to-Zero Encoding

In non-return-to-zero (NRZ) encoding, a binary 1 is represented by a full-power voltage across the entire bit window and a binary 0 is represented by a zero voltage occupying the full bit window, as shown in Fig.3.2. NRZ encoding is the preferred choice for data communications over wire channels such as Gbps serial links primarily due to its need for smaller bandwidth [49]. NRZ encoding is not particularly suitable for encoding data transmitted from base stations to passive wireless microsystems. Let us use a RFID tag to illustrate this. Since the data rate of passive wireless microsystems is typically low, the bandwidth requirement can be largely relaxed. As a result, bandwidth-consuming encoding schemes such as pulse interval encoding (PIE) and Miller encoding can be used for passive wireless microsystems. The operational power of a tag is harvested from the radio-frequency waves emitted from its interrogator. Since the maximum amount of power transmitted from an interrogator to a RFID tag is upper bounded by government regulations, the amount of the power that the tag can harvest from the radio-frequency wave emitted by its interrogator in the charge-up phase prior to its operation is rather limited and is often not sufficient to support the entire operation of the tag. It is therefore highly desirable to have a continuous flow of RF power from the interrogator to the tag during data transmission in down links. This requires that the encoded data should have a long full-power interval in which RF power can flow from the interrogator to the tag. If NRZ coding is used and the data transmitted from the interrogator to the tag contain a long string of 0s, the tag will receive no power during the transmission of these 0s. As a result, the operation voltage of the tag will drop below the minimum threshold voltage required to sustain the operation of the tag. NRZ encoding is therefore not particularly suitable for encoding data transmitted from base stations to passive wireless microsystems.

NRZ encoding is also not particularly suitable for encoding data in uplinks. This is because passive wireless microsystems use backscattering to transmit data to their base stations. Backscattering transmits data from a tag to its interrogator by modulating the termination resistance of the antenna of the tag between the characteristic resistance of the antenna of the interrogator and infinity. The former gives a zero reflection while the latter yields a full reflection. The value of the transmitted data is determined by the number of the times that the termination resistance is modulated. For example, in EPC class-1 generation-2 UHF RFID protocols, Miller-modulated sub-carrier encoding is used to encode data transmitted from tags to interrogators. Each binary 1 and binary 0 have multiple transitions in each data bit window but the number of transitions for binary 1 and that for binary 0 differ. The interrogator differentiates binary 1 and binary 0 on the basis of the number of transitions it received in each bit time. If NRZ is used to encode the data transmitted from the tags to the interrogators, no transition will exist in each bit. As a result, the interrogator will not be able to distinguish binary 1 and binary 0.

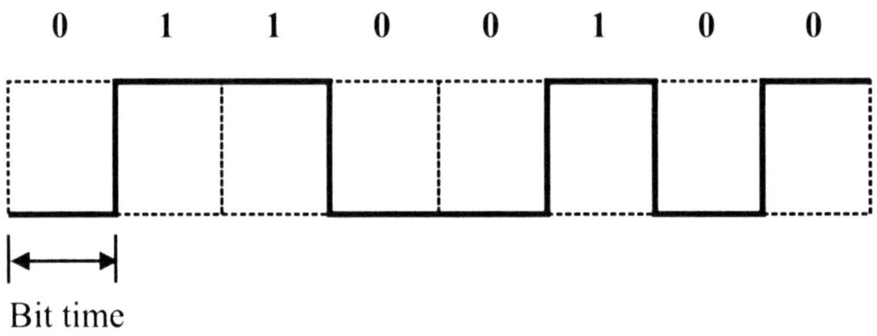

Figure 3.2. Baseband waveform of a NRZ-encoded data stream. The minimum pulse width is one bit time T_b. The number of transitions is 5, and the normalized average value of the encoded bit stream is 0.5.

3.2 Return-to-Zero Encoding

In return-to-zero (RZ) encoding, a binary 1 is represented by a full-power voltage in the first half of the bit time and a zero voltage in the second half of the bit time. A binary 0 is represented by a zero voltage during the entire bit time, as shown in Fig.3.3. As compared with the NRZ-encoded data stream shown in Fig.3.2, the RZ-encoded data stream not only has more transitions, the pulse width is also smaller. As a result, it requires more bandwidth to transmit the same amount of data as compared with NRZ. A distinct characteristic of RZ encoding is that timing information is inherently embedded in data, specifically in binary 1 where a transition exists. As a result, timing information can be transmitted with data by embedding it in the data.

Similar to NRZ encoding, if the data transmitted from an interrogator to a tag contain a long string of 0s, the tag will receive no power during the transmission of these 0s. The operation voltage of the tag will drop below the minimum threshold voltage, resulting in the shutdown of the tag. Also, the amount of power delivered from the interrogator to the tag is strongly data-dependent. RZ encoding is therefore not particularly attractive for encoding data to be transmitted from base stations to passive wireless microsystems. For the same arguments as those for NRZ encoding, RZ encoding is not particularly suitable for up-links as well.

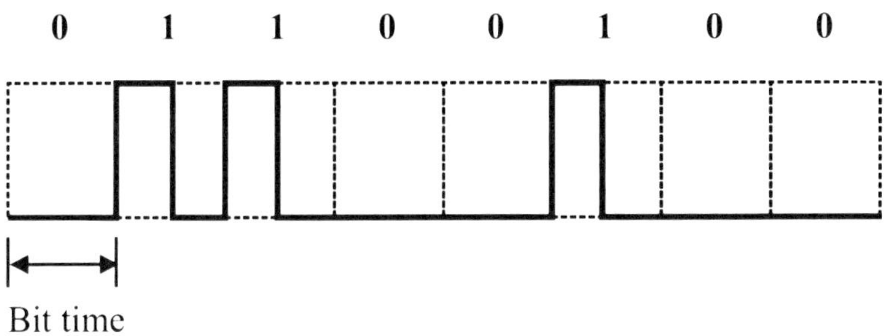

Figure 3.3. Baseband waveform of a RZ-encoded data stream. The minimum pulse width is $T_b/2$. The number of transitions is 6, and the normalized average value of the encoded bit stream is 1.5/8.

3.3 Manchester Encoding

Manchester encoding is also known as bi-phase-level encoding. Fig.3.4 shows the wave form of a Manchester-encoded data stream. A binary 1 in Manchester encoding is represented by a high-to-low transition occurring in the middle of the bit window while a binary 0 is presented by a low-to-high transition also occurring in the middle of the bit window. A unique characteristic of Manchester encoding is that there is a guaranteed transition in each data bit. This differs distinctly from NRZ and RZ encoding depicted earlier. As a result, a Manchester-encoded data stream not only has a zero DC component but is also truly self-clocking. The zero-dc characteristics of Manchester encoding allows a Manchester-encoded data stream to be coupled over inductively or capacitively while the self-clock characteristics of Manchester encoding allow the timing information to be embedded in the Manchester data stream and recovered at the receiving end conveniently. A drawback of Manchester coding is the existence of a large number of transitions, which demands a large bandwidth to transmit data. Since half of each bit time is filled with a full-power carrier, Manchester encoding can be used for encoding data transmitted from base stations to passive wireless microsystems, ensuring that power will flow from the base stations to the microsystems during each down-link data transmission.

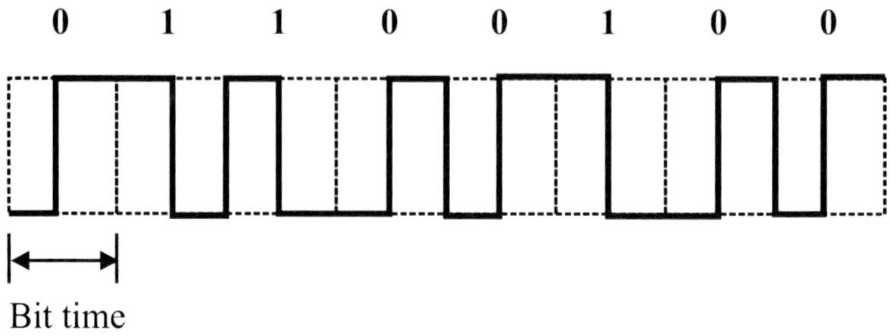

Figure 3.4. Baseband waveform of a Manchester-encoded data stream. The minimum pulse width is $T_b/2$. The number of transitions is 11, and the normalized average value of the encoded bit stream is 0.5.

3.4 Miller Encoding

In Miller encoding, a binary 1 is represented by a high-to-low or a low-to-high transition occurring in the middle of the bit window and a binary 0 causes no change to the signal level unless it is followed by another 0, in which case a transition to the other level will take place at the end of the first bit period, as shown in Fig.3.5. Miller-encoding thus guarantees a transition in every other bit. As compared with Manchester encoding, Miller encoding has a fewer number of transitions and larger pulse width, it therefore requires less bandwidth to transmit data. Timing information is also embedded generically in the data and can therefore be recovered conveniently.

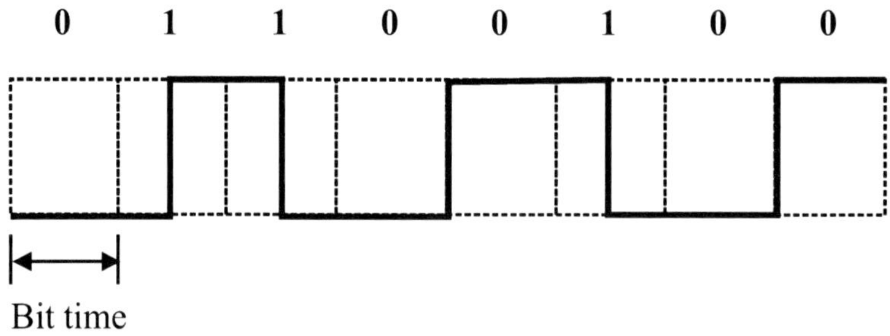

Figure 3.5. Baseband waveform of a Miller-encoded data stream. The minimum pulse width is one bit time T_b. The number of transitions is 5, and the normalized average value of the encoded bit stream is 3.5/8.

3.5 Miller-Modulated Sub-carrier Encoding

Since backscattering is used to transmit date from a passive RFID tag to its interrogator in EPC class-1 generation-2 UHF RFID protocols and the backscattered signal is quite weak, it is highly desirable to have the frequency of the backscattered signal be different from that of the incident signal from the interrogator so that the interrogator can pick up the backscattered signal with a less difficulty. Miller-modulated sub-carriers (MMS) can be used to achieve this. A Miller-modulated sub-carrier is obtained by multiplying the baseband Miller encoded data with a square wave whose frequency is M times that of the baseband data. Fig.3.6 shows the waveform of a Miller-modulated sub-carrier ($M = 2$) of the same digital bit stream used for illustrating the definition of other encoding schemes. It is seen that the Miller-modulated sub-carrier contains two sub-carrier cycles per bit. Note that the value of M is typically set by the base station when it initiates a communication request. It becomes evident from Fig.3.6 that the number of the transitions per bit is increased. As a result, it becomes easier for the reader to detect the backscattered data in the presence of noise. This, however, is at the expense of bandwidth. For the same bandwidth, the data rate of systems with Miller-modulated sub-carrier encoding is lower. Another key advantage of having more transitions per bit is that the spectrum of the response from the tag is displaced far away from the carrier frequency, making the detection of the backscattered signal by the reader more easier. Miller-modulated sub-carrier encoding with $M = 2, 4, 8$ is used to encode data backscattered from RFID tags to interrogators in EPC class-1 generation-2 UHF RFID protocols.

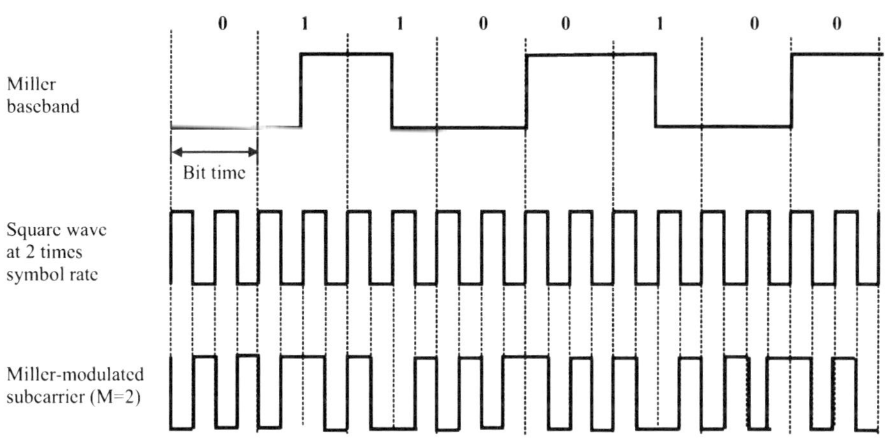

Figure 3.6. Baseband waveform of a Miller-modulated sub-carrier with $M = 2$. The minimum pulse width is $T_b/4$. The number of transitions is 26, and the normalized average value of the encoded bit stream is 0.5.

3.6 FM0 Encoding

FM0 encoding is bi-phase-space encoding. In FM0 encoding, a binary 1 is represented by a constant voltage occupying the entire bit window and a binary 0 is represented by a transition in the middle of the bit window, as shown in Fig.3.7. FM0 inverts the phase at the boundary of each symbol, as illustrated in Fig.3.8. As compared with Manchester encoding, FM0 encoding has a larger number of transitions and the same minimum pulse width. The bandwidth requirement of FM0 encoding is therefore high. In addition, since FM0 encoding guarantees to have a transition in every bit, timing information is embedded in the data intrinsically and can be recovered conveniently. FM0 encoding is used in tag-to-interrogator communications in EPC class-1 generation-2 UHF RFID protocols [53].

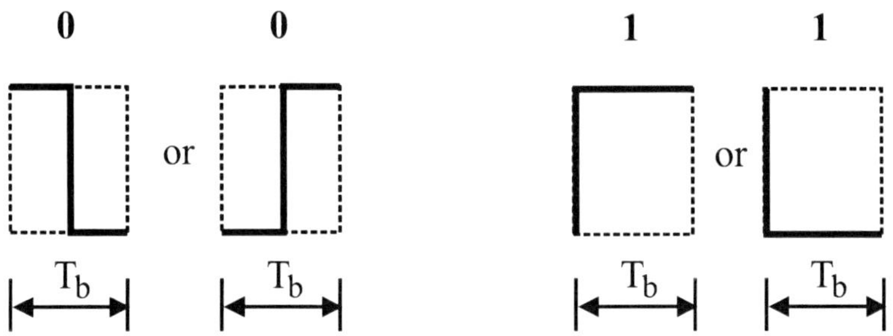

Figure 3.7. Baseband waveform of FM0 encoding.

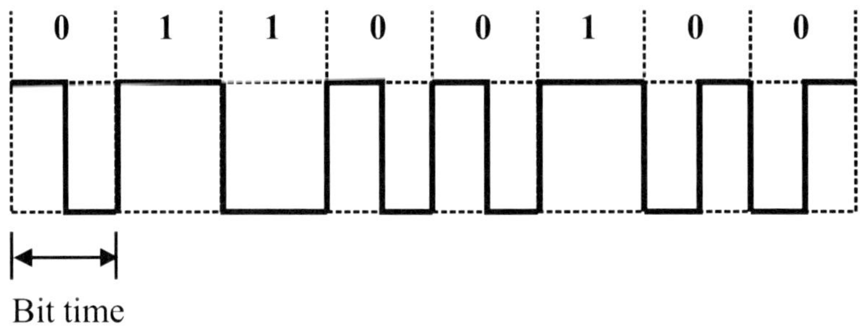

Figure 3.8. Baseband waveform of a FM0-encoded data stream. The minimum pulse width is $T_b/2$. The number of transitions is 12, and the normalized average value of the encoded bit stream is 4.5/8.

3.7 Pulse Interval Encoding

In pulse interval encoding (PIE), a binary 1 is represented by a long full-power pulse terminated with a short power-off pulse while a binary 0 is represented by a short full-power pulse terminated with the same short power-off pulse as that of binary 1, as shown in Fig.3.9. The length of the full-power pulse for binary 1 and that for binary 0 are specified in the figure.

PIE encoding is used in data communications from an interrogator to a passive tag in EPC class-1 generation-2 UHF RFID protocols. A distinct characteristic of PIE is that there is always a flow of power from the interrogator to the tag during data communications from the interrogator to the tag. Since a passive RFID tag harvests its operational power from the RF waves emitted by it interrogator, maximizing the duration of the full-power interval of both binary 1 and binary 0 is essential to maximize the amount of the power flowing from the interrogator to the tag. A drawback of PIE is that the data rate is data-dependent, as shown in 3.10. This is because the duration of binary 0 is shorter as compared with that of binary 1. As a result, a stream of 0s will be transmitted at a higher rate than a stream of 1s. Also, as compared with NRZ, PIE has narrow power-off pulses for the same average data rate. As a result, PIE-encoded data will require a larger bandwidth to transmit [54].

To obtain the minimum pulse width, the number of transitions, and the average value of the encoded bit stream shown in Fig.3.10, we assume that the PW of PIE is 0.5 Tari and the bit time of binary 1 is 2 Tari. Note that the bit time of binary 0 is one Tari. The minimum pulse width of the PIE-encoded bit stream in Fig.3.10 is PW=0.5 Tari. Assume that the average of the data rate of PIE-encoded bit stream is that same as that of others, we have

$$11Tari = 8T_b, \tag{3.1}$$

from which we have $Tari = 8/11T_b$. The normalized average value of the PIE-encoded bit stream is obtained from

$$\frac{5 \times \frac{1}{2} + 3 \times \frac{3}{4}}{11} = \frac{4.75}{11} \tag{3.2}$$

Table 3.1 compares the performance of the preceding data encoding schemes for the bit stream shown in Fig.3.2~Fig.3.10. The different characteristics of these data encoding schemes are evident.

3.8 Chapter Summary

Digital data encoding schemes encountered in wireless communications and their suitability for passive wireless microsystems have been examined. NRZ encoding has the characteristics of small bandwidth requirement, making it particularly suitable for high-speed data links over wire channels. This encoding scheme, however, is not particularly suitable for down-links in passive wireless microsystems due to the lack of power flow from base stations to passive wireless microsystems when a long train of 0s are encountered in the

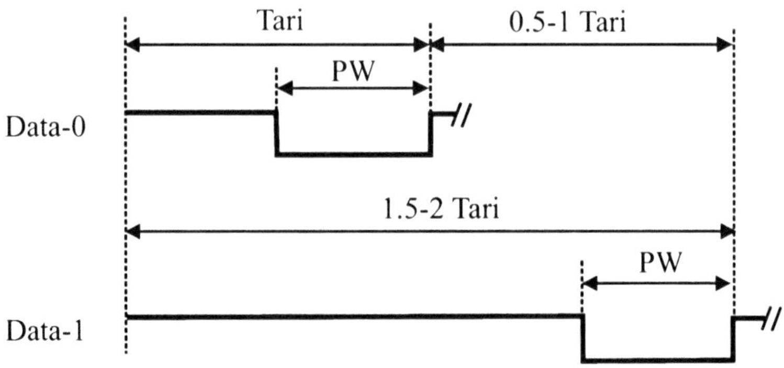

Figure 3.9. Baseband waveform of pulse interval encoding. Tari, an abbreviation of *Type A Reference Interval*, is the unit of time duration with its value varying from 6.35μs to 25μs. The width of the termination power-off pulse of binary 0 and binary 1 is between 0.265 Tari and 0.525 Tari.

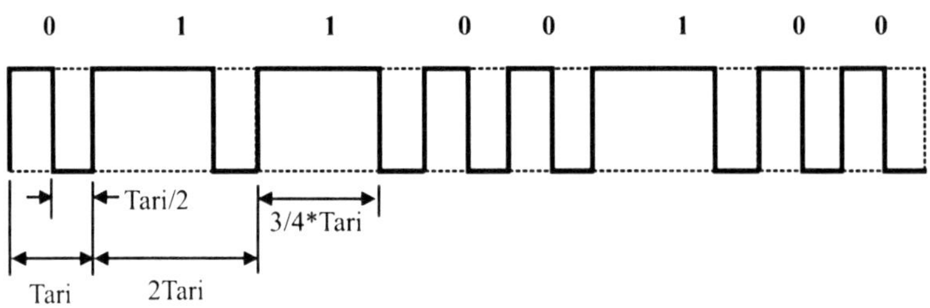

Figure 3.10. Baseband waveform of a PIE-encoded data stream. The minimum pulse width is $\frac{8}{11}T_b$. The number of transitions is 15, and the normalized average value of the encoded bit stream is 4.75/11.

data. It is also not suitable for up-links from passive wireless microsystems to base stations due to the lack of transitions in each data bit. RZ encoding requires more bandwidth as compared with NRZ due to the increased number of transitions. This encoding scheme contains timing information intrinsically. For the same reasons as those for NRZ, RZ is also not suitable for encoding data transmitted between passive wireless microsystems and their base stations.

Manchester encoding guarantees a transition in each data bit. As a result, a Manchester-encoded data stream not only has a zero DC component but is also self-clocking. A drawback of Manchester coding is the existence of a large number of transitions. More bandwidth is needed to transmit data. Miller-encoding guarantees to have a transition in every other bit. As compared with Manchester encoding, Miller encoding has a fewer number of transitions and larger pulse width, it therefore requires less bandwidth to transmit data. Miller-modulated sub carrier increases the number of transitions per bit by multiplying the baseband Miller encoded data with a square wave whose frequency is mul-

Table 3.1. Comparison of data encoding schemes.

Encoding	Min. pulse width	No. of transition	Normalized average
NRZ	T_b	5	1/2
RZ	$T_b/2$	6	1.5/8
Manchester	$T_b/2$	11	1/2
Miller	T_b	5	3.5/8
MMS	$T_b/4$	26	4.5/8
FM0	$T_b/2$	12	4.5/8
PIE	2/11 Tari	15	4.75/11

Legends :
MMS - Miller-Modulated Subcarrier.

tiple times that of the baseband data. It allows the reader to detect the data more reliably in the presence of noise, however, at the expense of bandwidth. Miller-modulated sub carrier encoding also allows the response from the tag to be displaced away from the carrier frequency, making the detection of the backscattered signal more easier. FM0 inverts the phase at the boundary of each symbol. A compared with Manchester encoding, FM0 encoding has a larger number of transitions and the same minimum pulse width. The bandwidth requirement of FM0 encoding is therefore high. Since FM0 encoding guarantees a transition in every data bit, timing information is embedded in data and can be recovered conveniently.

PIE is used in data communications from an interrogator to a passive tag in EPC class-1 generation-2 UHF RFID protocols. A distinct characteristic of PIE is that there is always a flow of power from the interrogator to the tag during data communications from the interrogator to the tag. A drawback of PIE is that the data rate is data-dependent. Also, as compared with NRZ, PIE has narrow power-off pulses for the same average data rate and requires more bandwidth to transmit data.

Chapter 4

MODULATORS AND DEMODULATORS

Data transmitted between passive wireless microsystems and their base stations must be modulated prior to their transmission. Modulation is a process of converting the information to be transmitted through a medium such as a wire line or a wireless channel to a proper form such that the information can be transferred over the medium with a high degree of fidelity. In digital communications, modulation is to convert a digital bit stream to a sinusoidal carrier with its parameters, such as amplitude, frequency, or phase varied in accordance with the bit value of the digital bit stream. If the amplitude of the analog carrier varies with the digital bit stream, the modulation is termed amplitude-shift-keying (ASK). If the frequency of the carrier is modulated by the digital bit stream, it is called frequency-shift-keying (FSK). If the phase of the carrier varies in accordance with the digital bit stream, it is referred to as phase-shift-keying (PSK). Complex digital modulation schemes, such as quadrature phase-shift-keying (QPSK), offset QPSK (OQPSK), minimum shift-keying (MSK), Gaussian MSK (QMSK), etc, are all built upon generic ASK, FSK, or PSK schemes to improve data rate and bit-error-rate (BER). Although a variety of modulation schemes are available for data communications over wireless channels, the limited power resource of passive wireless microsystems and constraints imposed on carrier frequency limit the choice of modulation schemes for these systems to only a few.

Demodulation is a process of extracting digital baseband data from a modulated analog waveform. Baseband data embedded in an ASK-modulated signal can be recovered (demodulated) by simply distinguishing the envelope corresponding to binary 1 and that corresponding to binary 0. To recover the baseband data embedded in a FSK-modulated signal, bandpass filters with passband frequencies set to the corresponding carrier frequencies will differentiate binary 1 and binary 0. To demodulate a PSK-modulated signal, a local

F. Yuan, *CMOS Circuits for Passive Wireless Microsystems*,
DOI 10.1007/978-1-4419-7680-2_4, © Springer Science+Business Media, LLC 2011

oscillator whose frequency is exactly the same as the carrier frequency is required for the coherent recovery of the baseband data. This chapter deals with digital modulation and demodulation for passive wireless microsystems. The finite power resource of passive wireless microsystems limits the primary modulation schemes to be ASK, FSK, and PSK. Our emphasis will be on the design of low-power demodulators as the configurations of modulators are typically simpler as compared with corresponding demodulators.

The chapter is organized as the follows : Section 4.1 examines three basic modulation schemes, namely amplitude-shift-keying, frequency-shift-keying, and phase-shift-keying. The pros and cons of these modulation schemes are studied and compared in detail. ASK modulators and demodulators for passive wireless microsystems are investigated in Section 4.2. A significant portion of this section is devoted to the design of ASK demodulators. A comparison of the performance of recently reported ASK demodulators for passive wireless microsystems is also provided. FSK modulators and demodulators are dealt with in Section 4.3. Section 4.4 is devoted to PSK modulators and demodulators. The chapter is summarized in Section 4.5.

4.1 Basic Modulation Schemes

4.1.1 Amplitude Shift Keying

In ASK, the amplitude of a sinusoidal carrier varies with the bit value of the digital bit stream to be transmitted, as illustrated graphically in Fig.4.1. The symbol of an ASK-modulated digital bit stream, denoted by $s(t)$, is depicted mathematically using

$$s(t) = A(t)\cos(\omega_c t), \tag{4.1}$$

where $A(t)$ and ω_c are the amplitude and frequency of the carrier, respectively. For analog ASK, $A(t)$ is a continuous function of time while for digital binary ASK, $A(t)$ is a piecewise-constant function of time with A_1 for a binary 1 and A_0 for a binary 0, as shown in Fig.4.1. The modulation index, also known as modulation depth, is defined as

$$m = \frac{A_1 - A_0}{A_1}. \tag{4.2}$$

Clearly the larger the modulation index, the larger the variation of the amplitude of the carrier, the smaller the chance that the ASK-modulated data will be corrupted by noise and disturbances, and the easier the baseband data can be recovered. If a binary 1 is transmitted by the carrier with full power while a binary 0 is transmitted by the carrier with zero power, this particular ASK is called On-Off Keying (OOK). OOK is the simplest form of binary ASK. It has a modulation index of 100% and provides the highest degree of immunity to disturbances and noise in all ASK schemes.

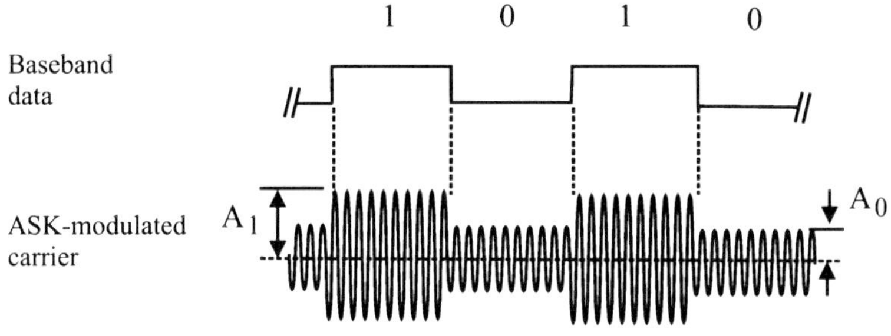

Figure 4.1. Waveform of a binary ASK-modulated bit stream.

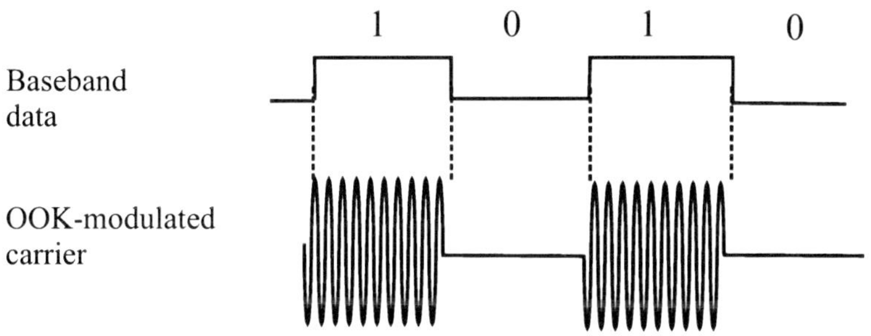

Figure 4.2. Waveform of an OOK-modulated bit stream.

ASK is susceptible to noise and disturbances. This is because noise and disturbances directly affect the amplitude of the received RF signal. The effect of noise and disturbances on ASK-modulated signals is particularly severe if the modulation index is small. When binary ASK is used to encode the data transmitted from a base station to a passive wireless microsystem, the amount of the power that flows from the base station to the microsystem fluctuates with baseband data. This is because passive wireless microsystems harvest their operational power from the RF waves emitted by the base station, as detailed in Chapter 2. The variation of the amplitude of the received RF signals, arising from the randomness of baseband data, results in the fluctuation of the total amount of power transmitted from the base station to the microsystem. The degree of this fluctuation will be maximized when OOK is used. This is because there will be no power transmitted from the base station to the microsystem when the transmitted data are 0. ASK also suffers from the drawback of a low data rate, mainly due to the difficulties encountered in extracting the envelope of ASK-modulated RF signals when the data rate is high, as to be seen in

Section 4.2. Finally, ASK is silicon-consuming especially when the carrier frequency is low, mainly due to the need for a low-pass filter with a large time constant to generate the dc component of the extracted envelope of the received RF signal. It should be noted that this drawback is largely eliminated with active ASK demodulators. This, however, is at the cost of an additional amount of power consumption, as to be seen shortly in Section 4.2. Finally, ASK is not particularly attractive for applications where only a low supply voltage is available. This is simply because the lower the supply voltage, the smaller the modulation index subsequently the higher the bit-error-rate, as evident from the constellation diagram shown in Fig.4.3.

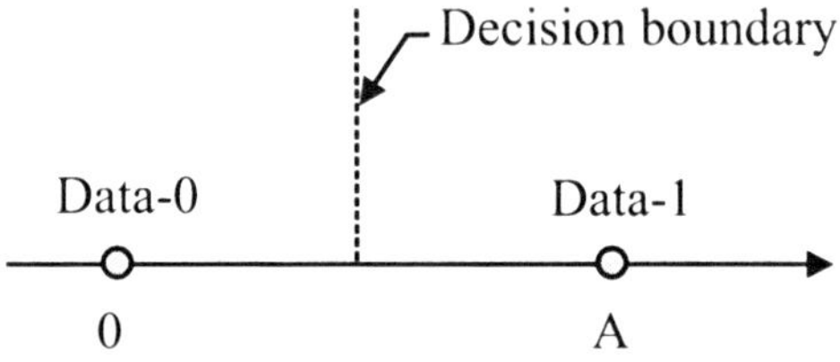

Figure 4.3. Constellation diagram of OOK. $v_{BASK} = A(t)\cos(\omega t)$ with $A(t) = A$ if data is a binary 1 and 0 otherwise.

Despite of the aforementioned drawbacks, binary ASK is perhaps the most widely used modulation scheme for passive wireless microsystems largely due to the simple configuration of ASK modulators and demodulators subsequently their low power consumption and the fact that the data rate of a large number of passive wireless microsystems, such as RFID tags and temperature/pressure sensors, is typically low.

4.1.2 Frequency Shift Keying

Unlike ASK , the frequency of the analog carrier of a FSK-modulated signal varies with the bit value of its digital bit stream, as shown in Fig.4.4 where the waveform of a binary FSK-modulated digital bit stream is shown. The symbol of a FSK-modulated bit stream can be depicted mathematically by

$$s(t) = A\cos\left[(\omega_c + \omega)\, t\right], \tag{4.3}$$

where ω varies with the digital bit stream. Often ω takes the value of 0 and ω_c for binary 1 and binary 0, or vice versa. FSK is a constant-envelope modulation scheme because the envelope of a FSK-modulated signal does not change with the baseband data. The constellation diagram of binary FSK is shown in Fig.4.5.

When FSK is used to encode the data transmitted from a base station to a passive wireless microsystem, the drawback of the variation of the power trans-

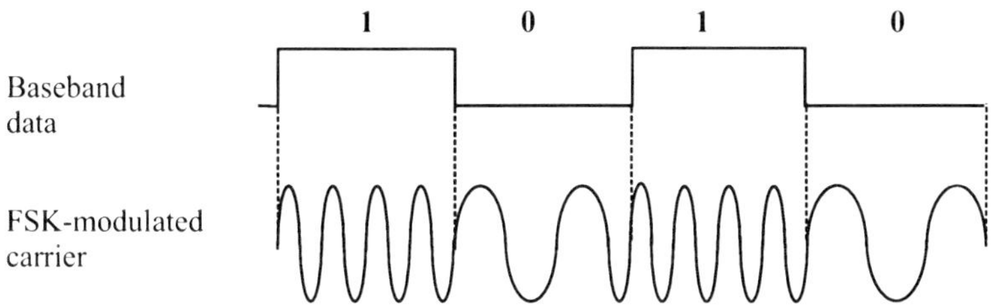

Figure 4.4. Waveform of a binary FSK-modulated bit stream.

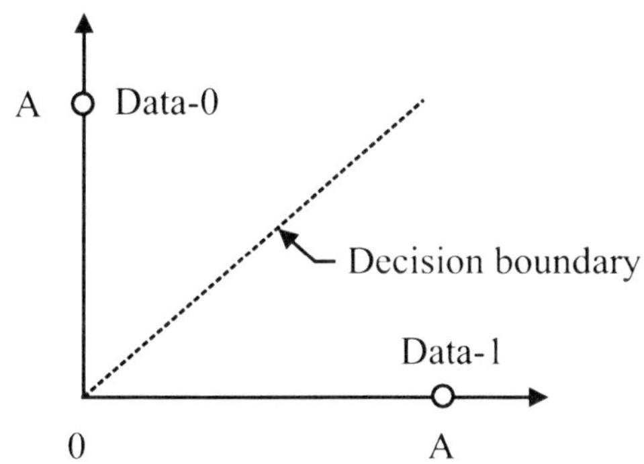

Figure 4.5. Constellation diagram of binary FSK. $v_{BFSK} = A_1(t)\cos(\omega_1 t) + A_0(t)\cos(\omega_0 t)$ with $A_1(t) = A$, $A_0(t) = 0$ for binary 1 and $A_1(t) = 0$, $A_0(t) = A$ otherwise.

mission from the base station to the passive wireless microsystem encountered in ASK is largely eliminated. Since digital data are represented by the frequency rather than the amplitude of the carrier, the effect of noise and disturbances on data transmission is also greatly reduced. Another unique characteristic of FSK is its ability to transmit data at a high data rate as compared with ASK, owing to the fact that the demodulation of a FSK-modulated data stream does not need to extract the envelope of the carrier using low-pass filters with large time constants. For example, the data rate of passive wireless microsystems with FSK modulation can be several Mbps with a carrier frequency of 13.56 MHz only. As compared with ASK demodulators, FSK demodulators are generally more complex and subsequently consume more power. Also, the carrier frequency of FSK passive wireless microsystems is low, typically 13.56 MHz. FSK has been primarily used in biomedical implants due to the need for a constant flow of power from base stations to implants, the need for a high data rate, and the high degree of the absorption of electromagnetic waves by living bodies at high frequencies. We will investigate FSK modulators and demodulators in detail in Section 4.3.

4.1.3 Phase Shift Keying

In PSK, the phase of an analog carrier is modulated by the digital bit stream to be transmitted. The symbol of a PSK-modulated bit stream is depicted mathematically by

$$s(t) = A\cos\left(\omega_c t + \phi\right), \tag{4.4}$$

where ϕ is modulated by the digital bit stream to be transmitted. Fig.4.7 shows the waveform of a binary PSK (BPSK) modulated bit stream. It uses only two distinct phases, typically separated by 180 degrees, to represent binary 1 and binary 0. BPSK is the most robust PSK and offers the highest degree of immunity to noise and disturbances. This is evident from the constellation diagram of binary PSK shown in Fig.4.6.

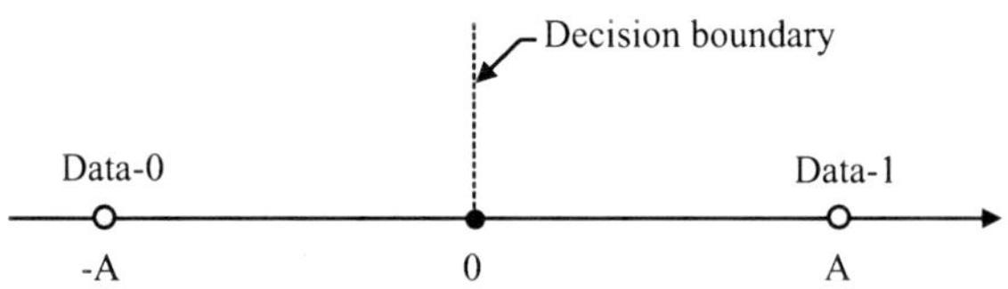

Figure 4.6. Constellation diagram of binary PSK.

Since there is no change in the envelope of the carrier when transmitting binary 1 and binary 0, BPSK is also a constant-envelope modulation scheme. The flow of power from a base station to a passive wireless microsystem with BPSK modulation is independent of baseband data.

One drawback of BPSK is that it is only able to modulate baseband data at the rate of 1 bit per symbol. As a result, BPSK is not particularly suitable for applications where data rates are high. This, however, is not of a critical concern for most passive wireless microsystems as the data rate of these systems is typically low. As compared with ASK and FSK, BPSK demodulators are more complex, often require phase-locked loops to recover baseband data (coherent detection). As a result, the power consumption of BPSK demodulators is generally high. Quadrature PSK (QPSK) doubles the data rate, however, at the cost of complex system configurations subsequently high power consumption. Although QPSK plays a dominant role in cellular wireless communications, its deployment in passive wireless microsystems is largely hindered by the limited power resource of these microsystems.

4.2 ASK Modulators and Demodulators

ASK demodulators are widely used in passive wireless microsystems due to the low data rate and the ease to implement ASK demodulators subsequently

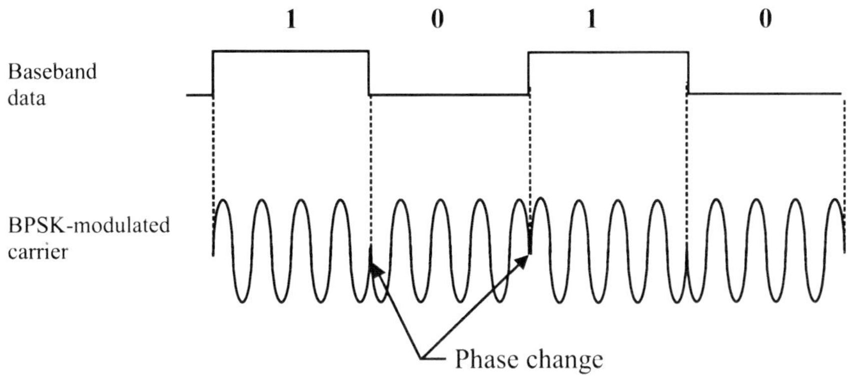

Figure 4.7. Waveform of a BPSK-modulated bit stream.

low power consumption. In this section, we focus on analog ASK demodulators. Digital ASK demodulators are purposely excluded in the presentation as the high power consumption of these ASK demodulators makes them less attractive for passive wireless microsystems.

4.2.1 ASK Modulators

An ASK-modulated signal can be generated conveniently using an oscillator whose frequency is the carrier frequency and an AND2 gate, as shown in Fig.4.8 [5]. The output of the AND2 gate is modulated by the digital bit stream. Other configurations of ASK modulators are also available, and are mostly based on the same principle.

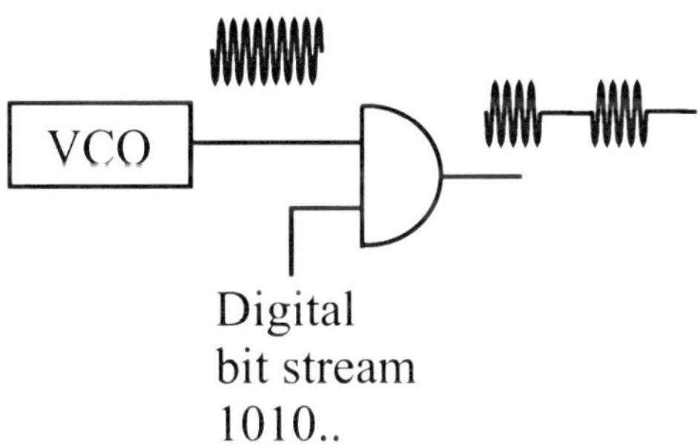

Figure 4.8. ASK modulator by Lee-Lee [5].

4.2.2 Classification of ASK Demodulators

ASK demodulators encountered in passive wireless microsystems can be loosely classified into voltage-mode, current-mode, and mixed-mode. In a voltage-mode ASK demodulator the information to be processed is represented by th nodal voltages of the modulator. A typical configuration of voltage-mode ASK demodulators consists of a voltage-mode envelope detector, an average detector, and a voltage comparator. The envelope detector filters out the high-frequency carrier and extracts the low-frequency envelope of the received RF signal. The function of the average detector is to generate the average voltage of the extracted envelope. The average voltage is then used as the reference with which the envelope will be compared. The voltage comparator compares the exacted envelope with its average and recovers the baseband data.

Current-mode ASK demodulators are those where the information to be processed is represented by branch currents of the modulators. These demodulators typically consist of a current-mode envelope detector where the envelope of the received RF signal is extracted, an average detector whose function is to produce the average of the extracted envelope, and a current comparator where the extracted envelope is compared with its average to recover baseband data. Clearly, current-mode ASK demodulators are the counterparts of voltage-mode ASK demodulators in the current domain. These demodulators are particularly attractive for applications where signals to be processed are currents and only a low supply voltage is available.

Mixed-mode ASK demodulators are those where both nodal voltages and branch currents are utilized to process information in demodulation in order to achieve optimal performance. The input stage of these demodulators is often a transconductor where an input voltage is converted to a current. The current is then fed to a current-mode envelope detector. The exacted current envelope can be either fed to a current comparator directly or a current-to-voltage converter followed by a voltage comparator to recover the data in the baseband. Mixed-mode ASK demodulators take the advantages of both voltage-mode and current-mode circuits and are capable of operating at a high data rate while providing a large signal swing.

4.2.3 Design Challenges of ASK Demodulators

Although conventional ASK demodulators have been around for a long time, the limited power resource and the fluctuating supply voltage of passive wireless microsystems impose stringent constraints on the design of ASK demodulators for these systems. In this section, we look into the design challenges of ASK demodulators for passive wireless microsystems.

Low and Fluctuating Supply Voltage

It was shown in the last chapter that the operational power of a passive wireless microsystem is harvested from the radio-frequency waves sent by its base station. The supply voltage of the microsystem drops once it is activated. If the storage capacitor of the passive wireless microsystem is charged during data transmission, for example, a RFID tag is charged during data transmission from the interrogator to the tag, the supply voltage of the microsystem will increase. As a result, the amount of power that the passive wireless microsystem harvests before and during data transmission is not constant and its supply voltage fluctuates during data transmission. A voltage regulator is typically required to stabilize the supply voltage. This is at the cost of additional power consumption. In addition, the supply voltage of these systems is typically low.

Carrier Suppression and Envelope Extraction

An ASK demodulator typically consists of a passive envelope detector, an RC-based average detector, and a voltage comparator, as shown in Fig.4.9. The low-pass network R_1C_1, together with the diode, form a rectifying envelope detector. During the positive cycle of the envelope, capacitor C_1 is charged and $v_1(t)$ rises with time. Ripples caused by the sinusoidal carrier also exist in $v_1(t)$. In the negative cycle of the envelope, the diode is off and C_1 is discharged through the path provided by R_1. Ripples also exist during this time interval due to the coupling of the parasitic capacitance of the diode.

The second low-pass network formed by R_2C_2 functions as an average detector that produces the average of $v_1(t)$, i.e. its dc component. In order for the demodulator to function properly, the choice of the value of the time constants $\tau_1 = R_1C_1$ and $\tau_2 = R_2C_2$ becomes critical, and is governed by the following considerations : τ_1 should be sufficiently large such that $v_1(t)$ will not contain a significant number of ripples and yet small enough such that $v_1(t)$ will be able to follow the envelope of the received RF signal. If the ripples are overly large, it will be difficult for the downstream comparator to correctly recover the data. On the other hand, if the time constant is overly large, $v_1(t)$ will not be able to follow the envelope of $v_{RF}(t)$, making the data recovery difficult. Clearly it will become more difficult to choose proper τ_1 when the data rate is high. This is the intrinsic drawback of passive RC-based envelope detectors.

The time constant of the average detector τ_2 must be sufficiently large such that the high-frequency components of $v_1(t)$ can be filtered out completely. Typically $\tau_2 = 10\tau_1$ [31]. This, however, is at the cost of a large silicon area. Alternative approaches that minimize silicon consumption without sacrificing performance, such as active envelope and average detectors, are highly desirable.

Comparators with Tunable Hysteresis

As pointed out earlier that the high-frequency components of the received RF signal will remain at the output of the envelope detector, as shown in Fig.4.9. These ripples will give rise to uncertainty at the threshold-crossing of the comparator, resulting in errors at the output of the comparator, as illus-

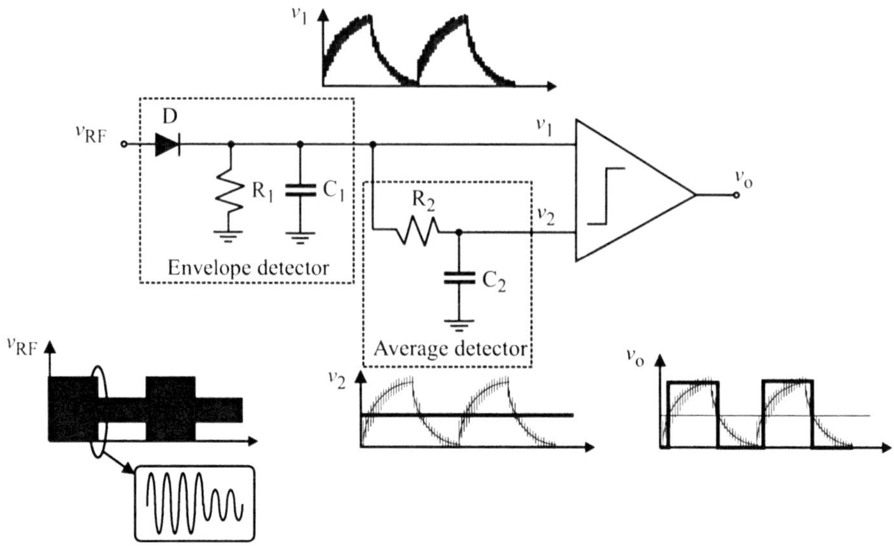

Figure 4.9. Configuration of passive ASK demodulators.

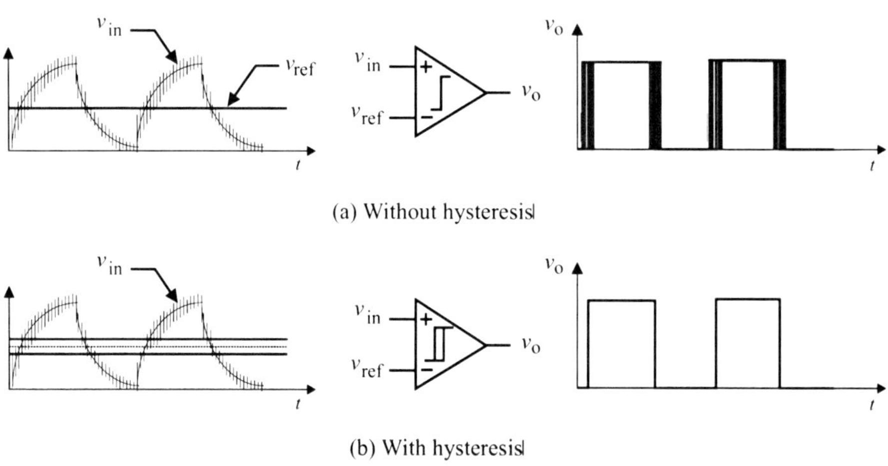

Figure 4.10. Comparators with and without hysteresis.

trated graphically in Fig.4.10(a). Comparators with hysteresis can effectively eliminate this uncertainty [55, 3, 56, 57]. This is illustrated in Fig.4.10(b).

Comparators with hysteresis are known as Schmitt triggers, attributed to American biophysicist Otto Herbert Schmitt (1913-1998) [59]. Schmitt triggers are bistable networks that have been widely used to enhance the immunity of circuits to external noise and disturbances. Schmitt triggers are traditionally

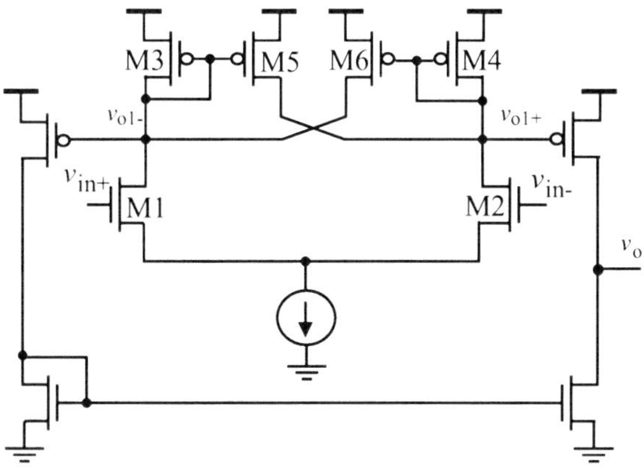

Figure 4.11. Simplified schematic of Schmitt trigger proposed by Allstot [58].

implemented using operational amplifiers with a resistive positive feedback. These Schmitt triggers suffer from the drawbacks of high power consumption due to the need for an operational amplifier. Perhaps the most widely cited single-ended Schmitt trigger was proposed by Dokic in 1984 [60]. The schematic of Dokic's Schmitt trigger is shown in Fig.4.12.

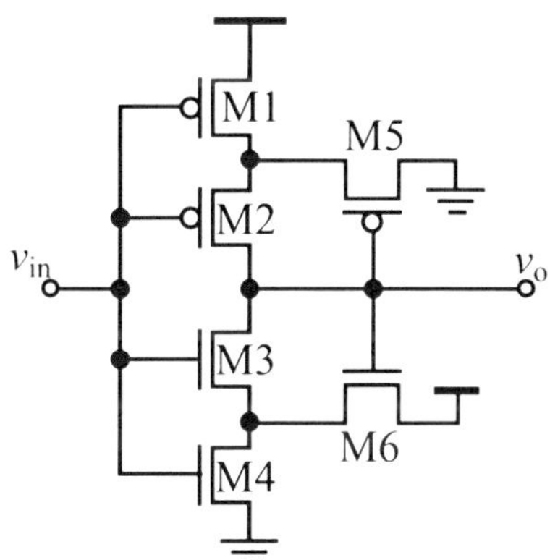

Figure 4.12. Schmitt trigger proposed by Dokic [60].

A detailed analysis of this Schmitt trigger, however, only became available ten years later by Filanovsky [61]. Dokic's version of Schmitt trigger stacks four transistors between the power and ground rails and is therefore not particularly attractive for applications where only a low supply voltage is available. Steyaert and Sansen modified Dokic's design by proposing a low-voltage Schmitt trigger with only two transistors piled between the power and ground rails, as shown in Fig.4.13 [62]. Although only one triggering voltage was implemented in

Steyaert and Sansen's circuit, their design can readily be extended to have two distinct triggering voltages, as demonstrated by Wang [63]. As seen in Fig.4.13, M3 and M4 provide an additional path for charging and discharging the capacitance at node A. It is this additional charging and discharging path that gives rise to hysteresis.

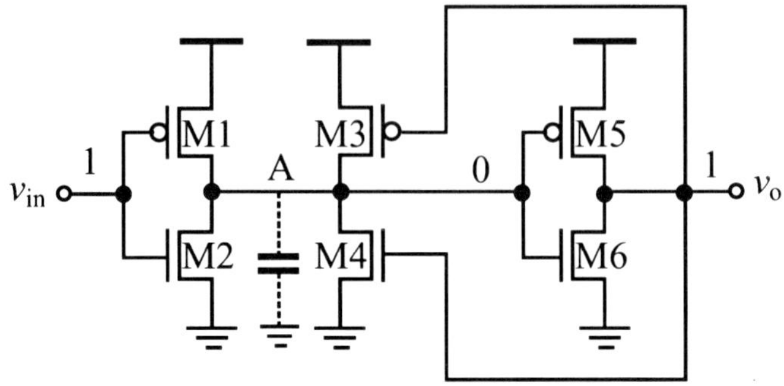

Figure 4.13. Schmitt trigger proposed by Steyaert and Sansen [62].

Schmitt trigger proposed by Kim *et al.* uses ten transistors to form the needed regenerative feedback [64]. A low-power CMOS Schmitt trigger was also proposed by Al-Sarawi [65]. Zhang *et al.* introduced an ultra low-voltage Schmitt trigger by using a body biasing technique to lower the threshold voltage of devices [66]. Schmitt trigger proposed by Pedroni employs two static inverters that have distinct threshold voltages such that different triggering voltages can be obtained [67]. A distinct characteristic of Pedroni's design is that it is an open-loop approach with no regenerative mechanism typically needed to create hysteresis. Sapawi *et al.* replaced the two cascaded pMOS transistors in Dokic's design with two pMOS transistors that are connected in parallel such that the modified Schmitt trigger can operate at a lower supply voltage [68].

A common drawback of the proceeding single-ended Schmitt triggers is that the hysteresis is set by device dimensions, process parameters, and supply voltage. These parameters vary with either process or operational conditions. As a result, the hysteresis of these Schmitt triggers can not be determined preciously. Schmitt triggers with tunable hysteresis are therefore highly desirable as it allows users to adjust the hysteresis to suite the need of a specific application. Comparators with tunable hysteresis are critically needed in applications where the level of noise and disturbances coupled to the triggering signals is not known a prior.

Wang modified Steyaert-Sansen's design by varying the current of the regenerative networks subsequently the intensity of the positive feedback such that the triggering voltages can be adjusted, as shown in Fig.4.14 [63]. As compared with the Schmitt trigger proposed by Steyaet and Sansen shown in Fig.4.13, the regenerative current in Wang's Schmitt trigger is now controlled by the biasing voltages V_p and V_n. By adjusting V_p and V_n, the regenerative current subsequently the hysteresis can be adjusted.

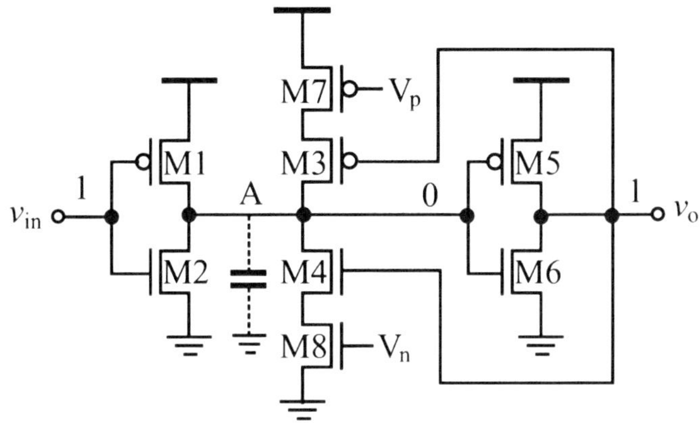

Figure 4.14. Schmitt trigger proposed by Wang [63].

Pfister added two additional transistors and a control voltage in the regenerative feedback network of Dokic's design so that the gain of the source-follower can be adjusted subsequently the triggering voltages and hysteresis of the Schmitt trigger [69]. Schmitt trigger with variable hysteresis proposed by Katyal *et al.* uses two two-input inverters to generate hysteresis [70].

The preceding Schmitt triggers are all single-ended. Surprisingly only a few differentially configured Schmitt triggers have been reported. Perhaps the most widely used differential Schmitt trigger is the one proposed by Allstot with its simplified schematic shown in Fig.4.11 [58]. Allstot's Schmitt trigger has been used in wireless transponders and sensors [3, 71]. In what follows we briefly examine the operation of this Schmitt trigger.

Transistors M5-M6 form the positive feedback to provide additional paths to charge the capacitor at the output nodes. When the positive feedback is absent, i.e. without M5 and M6, since

$$\begin{aligned} v_{in}^{+} &= (V_{ss} - V_T) + \sqrt{\frac{k_{3,4}}{k_{1,2}}}\left(V_{DD} - v_{o1}^{-} - V_T\right), \\ v_{in}^{-} &= (V_{ss} - V_T) + \sqrt{\frac{k_{3,4}}{k_{1,2}}}\left(V_{DD} - v_{o1}^{+} - V_T\right), \end{aligned} \quad (4.5)$$

we have

$$v_{in}^{+} = v_{in}^{-} + \sqrt{\frac{k_{3,4}}{k_{1,2}}}\left(v_{o1}^{+} - v_{o1}^{-}\right), \tag{4.6}$$

where

$$k_{1,2} = \frac{1}{2}\mu_n C_{ox}\left(\frac{W}{L}\right)_{1,2} \tag{4.7}$$

and

$$k_{3,4} = \frac{1}{2}\mu_p C_{ox}\left(\frac{W}{L}\right)_{3,4}. \tag{4.8}$$

If we assume the state transition of the Schmitt trigger takes place when $v_o^+ = v_o^-$, it becomes evident from (4.6) that this will occur when $v_{in}^+ = v_{in}^-$.

Now consider the case where positive feedback is present, i.e. M5 and M6 exist. Write KCL at v_{o1}^+ and v_{o1}^- nodes

$$\begin{aligned} k_1\left(v_{in}^{+} - V_{ss} - V_T\right)^2 &= k_3\left(V_{DD} - v_{o1}^{-} - V_T\right)^2 \\ &+ k_4 k_{46}\left(V_{DD} - v_{o1}^{+} - V_T\right)^2, \end{aligned} \tag{4.9}$$

$$\begin{aligned} k_2\left(v_{in}^{-} - V_{ss} - V_T\right)^2 &= k_4\left(V_{DD} - v_{o1}^{+} - V_T\right)^2 \\ &+ k_3 k_{35}\left(V_{DD} - v_{o1}^{-} - V_T\right)^2, \end{aligned} \tag{4.10}$$

where k_{46} and k_{35} are the aspect ratio of the current mirror M4-M6 and M3-M5, respectively. (4.9) and (4.10) can be written as

$$\begin{aligned} v_{in}^{+} &= \left(V_{ss} + V_T\right) \\ &+ \sqrt{\frac{k_3}{k_1}\left(V_{DD} - v_{o1}^{-} - V_T\right)^2 + \frac{k_4 k_{46}}{k_1}\left(V_{DD} - v_{o1}^{+} - V_T\right)^2}, \end{aligned} \tag{4.11}$$

$$\begin{aligned} v_{in}^{-} &= \left(V_{ss} + V_T\right) \\ &+ \sqrt{\frac{k_4}{k_2}\left(V_{DD} - v_{o1}^{+} - V_T\right)^2 + \frac{k_3 k_{35}}{k_2}\left(V_{DD} - v_{o1}^{-} - V_T\right)^2}, \end{aligned} \tag{4.12}$$

(4.11) can be modified to

$$v_{in}^{+} = (V_{ss} + V_T) + \sqrt{\frac{k_3}{k_1}}\left(V_{DD} - v_{o1}^{-} - V_T\right)\sqrt{1 + \frac{i_{D6}}{i_{D3}}}. \qquad (4.13)$$

where i_{D3} is the total current of transistor M3. The same notation is used for other transistors as well. Since

$$\frac{i_{D6}}{i_{D3}} \ll 1 \qquad (4.14)$$

and

$$\sqrt{1+x} \approx 1 + \frac{1}{2}x, \qquad (4.15)$$

and assume $i_{D5} = i_{D6} = i_{D5,6}$, $i_{D3} = i_{D4} = i_{D3,4}$, $k_3 = k_4 = k_{3,4}$, and $k_1 = k_2 = k_{1,2}$, (4.13) becomes

$$v_{in}^{+} \approx (V_{ss} + V_T) + \sqrt{\frac{k_3}{k_1}}\left(V_{DD} - v_{o1}^{-} - V_T\right)\left(1 + \frac{1}{2}\frac{i_{D6}}{i_{D3}}\right). \qquad (4.16)$$

In a very similar way, one can show that

$$v_{in}^{-} \approx (V_{ss} + V_T) + \sqrt{\frac{k_4}{k_2}}\left(V_{DD} - v_{o1}^{+} - V_T\right)\left(1 + \frac{1}{2}\frac{i_{D5}}{i_{D4}}\right). \qquad (4.17)$$

Subtracting (4.17) from (4.16) yields

$$v_{in}^{+} = \left[v_{in}^{-} + \sqrt{\frac{k_{3,4}}{k_{1,2}}}\left(v_{o1}^{+} - v_{o1}^{-}\right)\right] + \frac{i_{D5,6}}{i_{D3,4}}\sqrt{\frac{k_{3,4}}{k_{1,2}}}\left(v_{o1}^{+} - v_{o1}^{-}\right). \qquad (4.18)$$

Comparing (4.6) with (4.18), one observes that the second term on the right hand side of (4.18) quantifies the change of the triggering voltage and is the hysteresis generated by the regenerative feedback.

In [72], a current regenerative Schmitt trigger shown in Fig.4.15 was proposed. A distinct characteristic of this Schmitt trigger is that the critical nodes of the Schmitt trigger all have a low-impedance. This Schmitt trigger thus possesses the intrinsic characteristics of current-mode circuits. When the current feedback is absent, i.e. M5 and M6 do not exist, since $i_{D1} = i_{D4}$ and $i_{D2} = i_{D3}$ and

$$v_{in}^{+} = (V_{DD} - V_T) + \sqrt{\tfrac{i_{D4}}{k_1}},$$
$$v_{in}^{-} = (V_{DD} - V_T) + \sqrt{\tfrac{i_{D3}}{k_2}}, \tag{4.19}$$

we have

$$v_{in}^{+} = v_{in}^{-} + \left(\sqrt{\frac{i_{D4}}{k_1}} - \sqrt{\frac{i_{D3}}{k_2}}\right). \tag{4.20}$$

It is evident from (4.20) that if $i_{D3} = i_{D4}$ and $k_1 = k_2$, i.e. perfect device matching, $v_{in}^{+} = v_{in}^{-}$ will follow. This result reveals that the state transition of the circuit without the regenerative current feedback occurs only if the input voltages are equal.

When the current feedback is present,

$$i_{D4} + i_{D5} = k_1(v_{ss} - v_{in}^{+} - |V_{tp}|)^2, \tag{4.21}$$

we have

$$v_{in}^{+} = (v_{ss} - |V_{tp}|) - \sqrt{\frac{i_{D4}}{k_1}}\sqrt{1 + \frac{i_{D5}}{i_{D4}}}. \tag{4.22}$$

If the feedback current i_{D5} is set to be smaller than i_{D4}, we can make use of (4.11) to simplify (4.22) to

$$v_{in}^{+} \approx (v_{ss} - |V_{tp}|) - \sqrt{\frac{i_{D4}}{k_1}}\left(1 + \frac{1}{2}\frac{i_{D5}}{i_{D4}}\right). \tag{4.23}$$

Similarly, because

$$i_{D3} + i_{D6} = k_2(V_{ss} - v_{in}^{-} - |V_{tp}|)^2, \tag{4.24}$$

one can show that

$$v_{in}^{-} \approx (v_{ss} - |V_{tp}|) - \sqrt{\frac{i_{D3}}{k_2}}\left(1 + \frac{1}{2}\frac{i_{D6}}{i_{D3}}\right). \tag{4.25}$$

Subtract (4.25) from (4.22) yields

$$v_{in}^{+} = \left(v_{in}^{-} + \sqrt{\frac{i_{D3}}{k_2}} - \sqrt{\frac{i_{D4}}{k_1}} \right) + \frac{1}{2} \left(\frac{i_{D6}}{i_{D3}} \sqrt{\frac{i_{D3}}{k_2}} - \frac{i_{D5}}{i_{D4}} \sqrt{\frac{i_{D4}}{k_1}} \right). \quad (4.26)$$

A comparison of (4.20) and (4.26) reveals that the second term in (4.26) quantifies the shift of the triggering voltage obtained from the regenerative current feedback, and is the hysteresis of the Schmitt trigger.

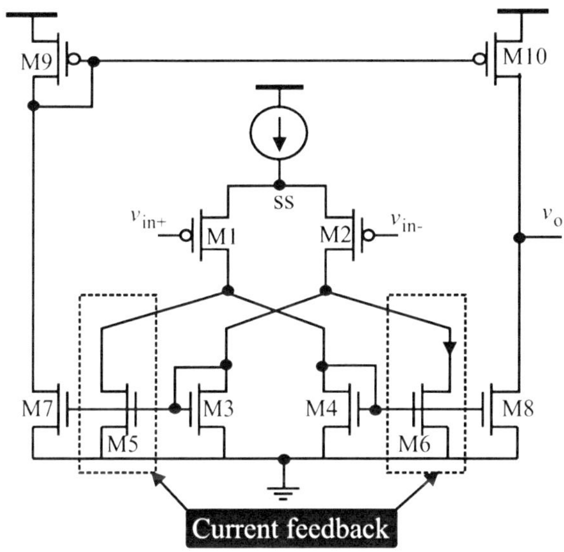

Figure 4.15. Simplified schematic of current regenerative Schmitt trigger proposed by Yuan [72].

Sensitivity of Comparators

The sensitivity of a comparator is the minimum voltage difference that the comparator can detect. It is set by the input offset voltage of the comparator. It is well known that the input offset voltage of a comparator is due to the mismatches of the devices of the comparator [73]. Although the input offset voltage of comparators can be compensated using digitally-trimmed capacitor arrays [74] or digitally-adjusted current arrays [75], capacitor array and current array-based approaches require digital tuning and are therefore not particularly attractive for passive wireless microsystems due to their limited power resource. Auto-zero techniques that use capacitors to sample and store the input offset voltage of the comparator in the sampling phase and subtract the stored input offset voltage from the input of the comparator in the evaluation phase are preferred [73].

Power Consumption

The power consumption of a passive ASK demodulator is dominated by the power consumption of its rectifying circuitry and the comparator. The power consumption of diode-based rectifying circuitry is due to the forward conduction of diodes. The power consumption of the comparator consists of both static and dynamic components. The former occurs when the logic state of the comparator remains unchanged while the latter takes place during the transition of the logic states of the comparator. Active ASK demodulators that use active devices to perform envelope extraction and average generation consume more power as compared with their passive counterparts.

To minimize the power consumption of ASK demodulators, the comparator should be switched off during the power harvesting phase where the power storage capacitor is charged and should be activated only in demodulation. Note that although the rectifying circuitry of the demodulator is activated during this phase, it is powered by the base station and its power consumption is not of a concern.

4.2.4 Voltage-Mode ASK Demodulators

A distinct characteristic of a near-field passive wireless microsystem is that the microsystem and its base station are coupled inductively, much like a transformer with the base station on the primary winding side and the passive wireless microsystem on the secondary winding side. As a result, the amplitude of the signal at the input of the microsystems is typically large enough such that diode-based envelope detectors can be used. To minimize the voltage loss across diodes, Schottky diodes that are formed by bringing metal into contact with a moderately doped n-type semiconductor and are characterized with a low forward conduction threshold voltage are usually used [76]. Because Schottky diodes are not available in standard CMOS processes, MOS-based diodes are usually used instead. MOS-based diodes suffer from a voltage loss of at least one threshold voltage. Native MOS transistors that have a nearly zero-threshold voltage have been used with the downside of a large channel resistance.

Fig.4.16 shows the simplified schematic of the ASK demodulator used for smart cards [77]. Transistor M1 serves as a voltage shifter that shifts the incoming voltage down by v_{GS1}. Transistors M2 and M3 are biased in triode and behave as both diodes and resistors when on with the resistance controlled by the width of the transistors. When the voltage of the drain of M1 exceeds V_T, M2 and M3 are on. Otherwise, they are off. M2 and the capacitance of M2 and M5 form a low-pass filter that extracts the envelope of the RF signal at the drain of M1. M4 is a MOS capacitor with a large capacitance. M3 and M4 thus form a low-pass with a large time constant and behave as an envelope detector, ensuring v_{in}^- is the average value of the envelope of the RF signal at the drain of M1. Transistors M5-M8 form a generic voltage amplifier with a single-ended output. The output is further amplified by the following common-source amplifier formed by M9 and M12 and finally restored to full swing by the static CMOS inverter.

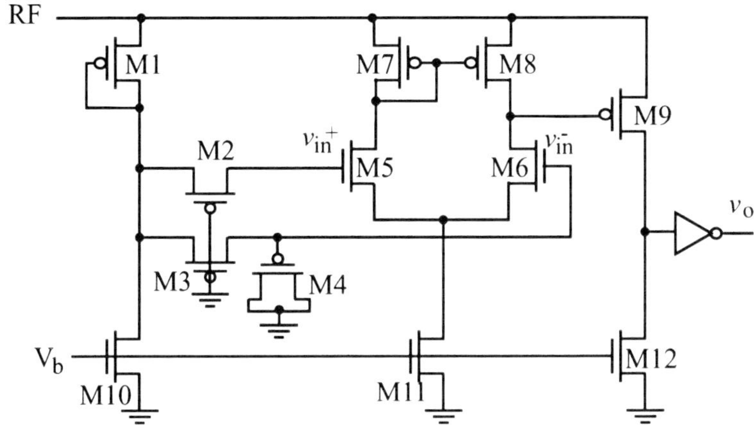

Figure 4.16. Simplified schematic of ASK demodulators proposed by Bouvier *et al.* for smart cards [77].

The discharge of the capacitors of both the envelope and average detectors is provided by M10 with its current controlled by V_b and the width of M10. Note that the "comparator" used in this design is perhaps the simplest with no hysteresis. The use of active devices in construction of both the envelope detector and the average detector enables the entire ASK demodulator to be implemented using digital CMOS processes. A similar configuration was used in [55, 78] for neural recording microprobes.

The ASK demodulator proposed by Yu and Najafi for retinal prosthetic devices is show in Fig.4.17 [55]. Allstot's Schmitt trigger is used. M1 and M2 behave as both diodes and resistors when on. Together with C_1 and C_2, they form two low-pass filters with distinct time constants, one for envelope extraction and the other for average extraction. The discharge of C_1 and C_2 is carried out by J_1 and J_2, respectively. A similar configuration was employed in [57] where a Dokic single-ended Schmitt trigger was employed in design of an ASK demodulator for cochlear prosthesis. Given the large input voltage, only a single diode was used for rectification and one RC network is used for envelope detection in this design. The same approach was used in [79] for biomedical implants.

The ASK demodulator proposed in [5] employs an inverting amplifier to boost the voltage swing of incoming RF signals, as shown in Fig.4.18. The OR2 gate reconstructs the modulated signal with an increased modulation index. The output of the OR2 gate is fed to a low-pass filter where high-

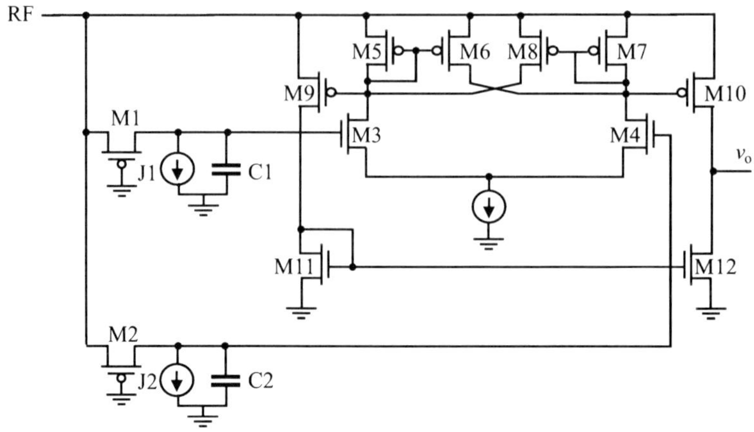

Figure 4.17. Simplified schematic of ASK demodulators proposed by Yu and Najafi for retinal prosthetic devices [55].

frequency components are eliminated. The following voltage comparator with a user-defined reference voltage recovers the baseband data.

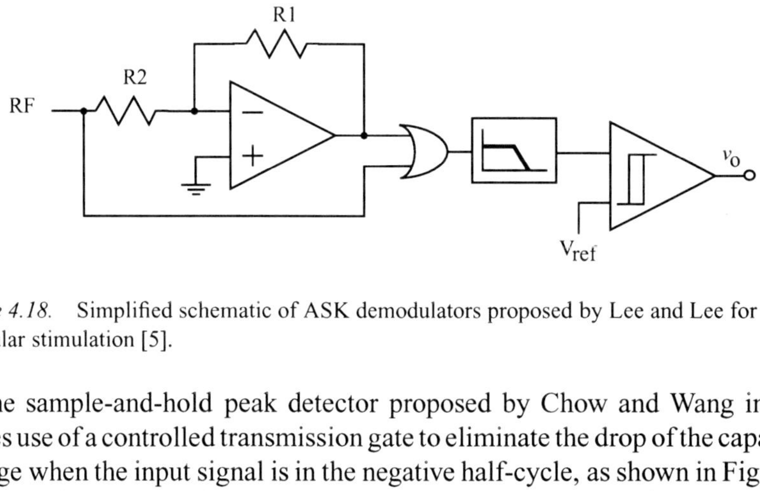

Figure 4.18. Simplified schematic of ASK demodulators proposed by Lee and Lee for neuromuscular stimulation [5].

The sample-and-hold peak detector proposed by Chow and Wang in [80] makes use of a controlled transmission gate to eliminate the drop of the capacitor voltage when the input signal is in the negative half-cycle, as shown in Fig.4.19. When the input signal is in its positive half-cycle, capacitor C is charged by the diode. At the same time, it is discharged by the current source. Since the discharge current is set to be small, the voltage of the capacitor will track the input signal with a small error. In the negative half-cycle of the input voltage, the transmission gate turns off, eliminating the discharge path of the capacitor. The voltage of the capacitor holds until the arrival of the next positive half-cycle of the input signal. It should be noted that when the amplitude of the input signal drops in the positive half-cycle, the diode will switch off, preventing v_C from tracking the input signal. To avoid this, the delay of the inverter must be made smaller as compared with that of the diode path such that when the amplitude of the input drops, the capacitor will be drained first, ensuring v_c is less than the amplitude of the input voltage such that v_c can track the input signal.

Several additional constraints affect the performance of the demodulator. For example, the voltage swing of the input signal must be large enough such that it will trigger the first inverter. Also, the threshold voltage of the inverter should be set to the common-mode voltage of the input. Clearly, if the received RF signal is small, this demodulator will not function properly.

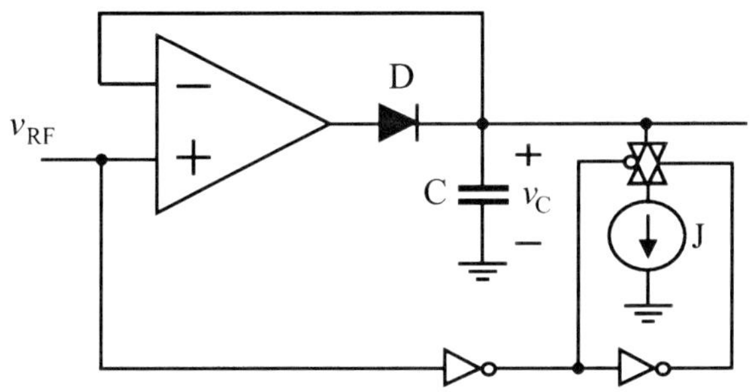

Figure 4.19. Simplified schematic of envelope detector proposed by Chow and Wang [80].

The ASK demodulator proposed by Yu and Bashirullah in [81] for biomedical implants makes use of an unbalanced source-coupled rectifier, which was initially proposed by Kimura in 1993 in [82], to perform rectification and amplification, as shown in Fig.4.20. The dimensions of M1-M4 are sized as $(W/L)_{1,4} = A(W/L)_{2,3}$, where A is the aspect ratio. Assume that the two differential pairs are matched such that nodes ss1 and ss2 have the same voltage V_{ss}. Let

$$\begin{aligned} v_{in}^{+} &= v_{in,cm} + \frac{v_{in}}{2}, \\ v_{in}^{-} &= v_{in,cm} - \frac{v_{in}}{2}, \end{aligned} \tag{4.27}$$

where $v_{in,cm}$ is the common-mode input voltage and v_{in} is the differential-mode input. It can be shown that the currents induced by the incoming RF signal v_{in} are given by

$$\begin{aligned} i_{d1} &= Ak_{2,3}(V_{in,cm} - V_{ss} - V_T)v_{in} + \frac{Ak_{2,3}}{4}v_{in}^2, \\ i_{d2} &= -k_{2,3}(V_{in,cm} - V_{ss} - V_T)v_{in} + \frac{k_{2,3}}{4}v_{in}^2, \\ i_{d3} &= k_{2,3}(V_{in,cm} - V_{ss} - V_T)v_{in} + \frac{k_{2,3}}{4}v_{in}^2, \\ i_{d4} &= -Ak_{2,3}(V_{in,cm} - V_{ss} - V_T)v_{in} + \frac{Ak_{2,3}}{4}v_{in}^2, \end{aligned} \quad (4.28)$$

where

$$k_{2,3} = \frac{1}{2}\mu_n C_{ox}\left(\frac{W}{L}\right)_{2,3}, \quad (4.29)$$

$i_{d1} \sim i_{d4}$ are the ac currents of M1~M4, respectively. Assume $(W/L)_5 = (W/L)_6$. Because

$$i_{d7} = (i_{d2} + i_{d3}) - (i_{d1} + i_{d4}), \quad (4.30)$$

substituting (4.28) into (4.30) yields

$$i_{d7} = \frac{k_{2,3}}{2}(1 - A)v_{in}^2. \quad (4.31)$$

The rectification operation on the input voltage is evident from (4.31). If the two differential pairs are balanced, i.e. $A = 1$, no rectification will be obtained. Also observed is that the output current can be increased by either increasing the aspect ratio A or by adjusting the current gain of the current mirror M7-M8.

The ASK demodulator proposed in [83] for implantable neural interfacing chips minimizes silicon consumption by eliminating the capacitors required for envelope detection in conventional ASK demodulators. The demodulator is therefore termed C-less ASK demodulator. The simplified schematic of this C-less ASK demodulator is shown in Fig.4.21. The rectifying portion of the demodulator is essentially a bandgap reference [73] with M5 functioning as the start-up transistor. During the negative half cycle of v_{in}, the demodulator is off and v_{o1} remains unchanged. During the positive half cycle of v_{in}, if all transistors are in saturation, $v_{o1} = v_{DS4} + R_s i_{D4}$.

There are a number of constraints that affect the application of the C-less ASK demodulator : (i) To start the demodulator, $v_{in,min} = 3V_T$ is required.

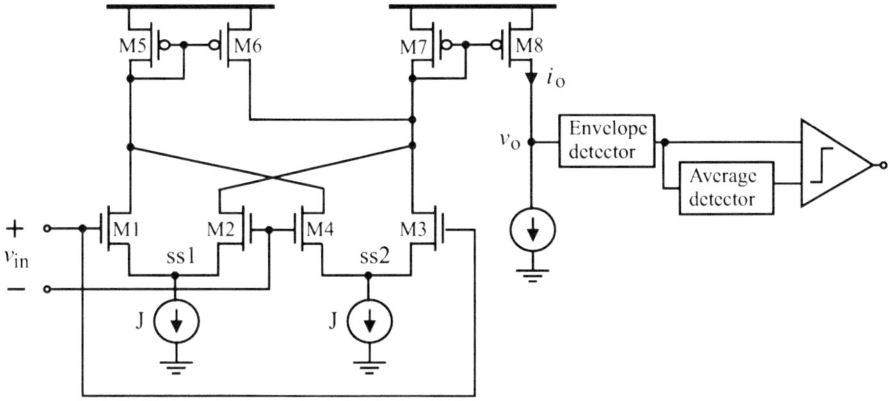

Figure 4.20. Simplified schematic of ASK demodulator proposed by Yu and Bashirullah for implantable electronics [81].

Note that we have assumed $V_{tn} = |V_{tp}| = V_T$. This sets the lower bound of the peak voltage of the incoming modulated signal. (ii) Since $v_{o1,max} = v_{in,max} - V_T$, the amplitude of v_{in} must be sufficiently large. As a result, this ASK demodulator is not particularly suitable for far-field applications where the received RF signal is weak. (iii) The threshold voltage of data recovery is set by the hysteresis voltages of the Schmitt trigger, which are the functions of the supply voltage. As a result, a sufficient safe margin must be kept from the threshold voltages of the Schmitt trigger to avoid errors. Note that conventional ASK demodulators employ a large shunt capacitor to exact the dc value of the modulated signal and use this dc voltage as the reference for data recovery. A key advantage of doing so is that the reference voltage varies with the strength of the incoming signal. This is, however, at the cost of increased silicon consumption due to the need for a large capacitor. (iv) v_{o1} must fall into the operating range of the downstream Schmitt trigger in order to function correctly. This is achieved by varying the size of the transistors and the value of R. (v) Since all transistors of the rectifying circuitry conduct during most of the positive half cycles of incoming data, the static power consumption of the demodulator can not be neglected.

The ASK demodulator proposed by Lee *et al.* in [84] improves the design of the preceding C-less ASK demodulator by further eliminating the resistor, making it a truly all-MOS ASK demodulator, as shown in Fig.4.22. The improved voltage swing of v_{o1} removes the Schmitt trigger used in the C-less ASK demodulator, further reducing its silicon consumption. Transistor M8 operates in the triode and behaves as a resistor. To start the demodulator, $v_{in,min} = 4V_T$ is required. This sets the lower bound of the peak voltage of the incoming modulated signals.

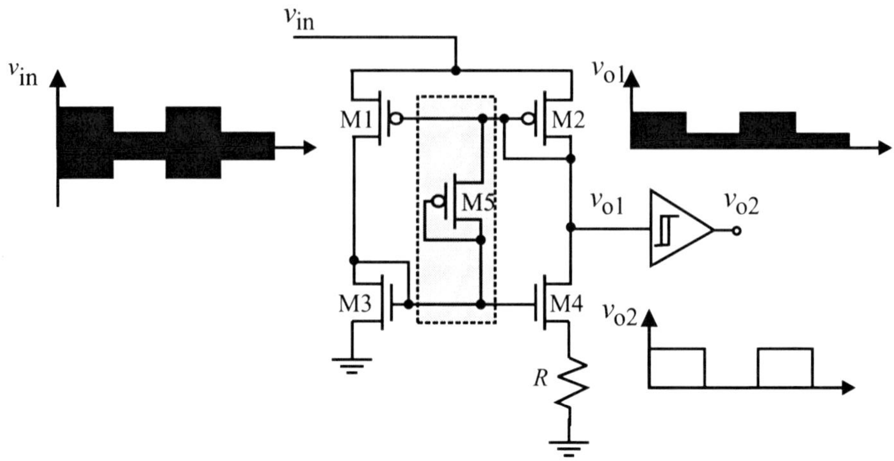

Figure 4.21. Simplified schematic of C-less ASK demodulator proposed by Wang *et al.* for implantable neural interfacing chips [83].

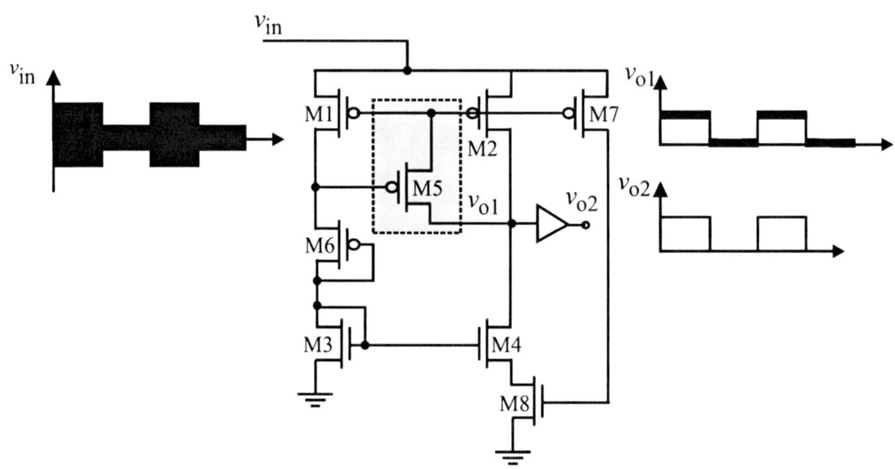

Figure 4.22. Simplified schematic of all-MOS ASK demodulator proposed by Lee *et al.* [84].

The amplitude of the input voltage of ASK demodulators for far-field passive microsystems is typically much smaller as compared with that of ASK demodulators for near-field counterparts. As a result, a voltage multiplier is required, as shown in Fig.4.23 [31, 56]. To minimize the voltage loss across the diodes so as to increase RF-DC conversion efficiency, rectifying diodes in passive wireless microsystems are typically implemented using Schottky diodes. Since Schottky is not available in standard CMOS technologies, the MOSFET realization of rectifying diodes is often used [56, 85]. This is at the cost of the voltage loss of one threshold voltage across each MOSFET diode. To minimize the voltage loss across MOSFET diodes, native MOSFETs whose threshold voltage is nearly zero can be used for both power harvesting and ASK demodulation [34]. The main disadvantages of native transistors is their larger size due to additional doping mask. Typical minimal size of the native NMOS gate is 2-3 times longer and wider than standard transistor. In addition, these transistors suffer from a low transconductance. Cost is also increased due to the need for additional doping operations.

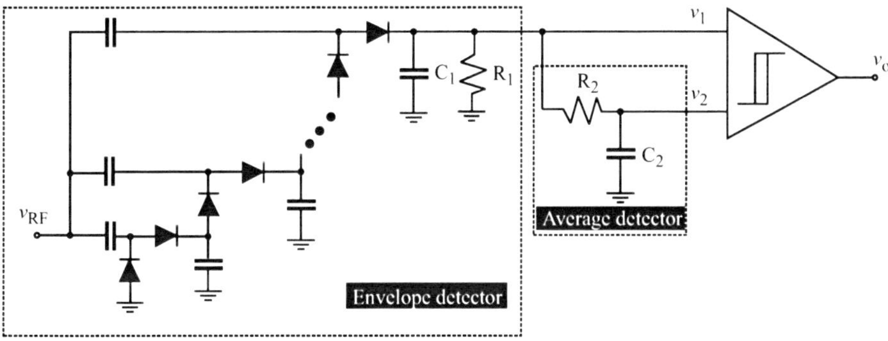

Figure 4.23. Configuration of passive ASK demodulators proposed by Karthaus and Fischer [34] for far-field microsystems.

4.2.5 Current-Mode ASK Demodulators

As pointed out earlier that current-mode ASK demodulators typically consist of a current-mode envelope detector, an average detector, and a current comparator. They are the counterparts of voltage-mode ASK demodulators in the current domain. The current-mode ASK modulator proposed by Djemouai and Sawan in [86, 87] consists of a current edge-detector, two current comparators, and an output buffer. The current edge detector detects both the rising and falling edges of incoming current. The outputs of the current edge detector are fed to two current comparators that generate corresponding logic outputs. Fig.4.24 shows the typical approaches for the detection of both the rising and the falling edges of a square wave. If only the rising edge or the falling edge are to be detected, the approaches given in Fig.4.25 can be used.

A current-mode implementation of edge detection is shown in Fig.4.26. The input current i_{in} is mirrored to i_{D6} and i_{D7}. It is also mirrored to i_{D2} after delay τ. Assume that the gain of all current mirrors is unity, we have $i_{D6,7} = i_{in}$ and $i_{D2,3,4} = i_{in}u(t-\tau)$, where $u(t)$ is a unit step function. The two output currents of the current-edge detector are given by : $i_{src} = i_{D4} - i_{D6}$ and $i_{sink} = i_{D7} - i_{D5}$.

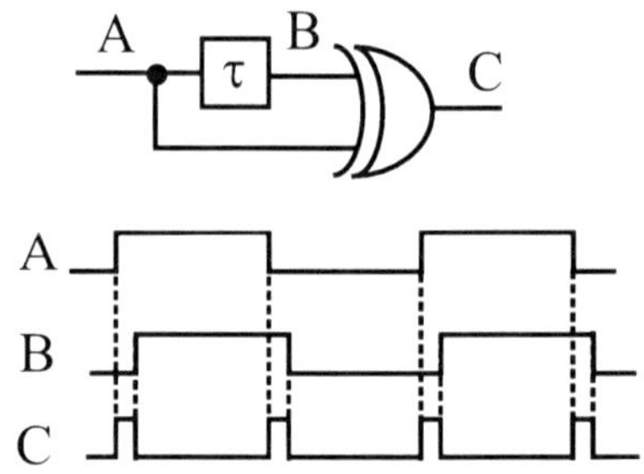

Figure 4.24. Detection of both rising and falling edges.

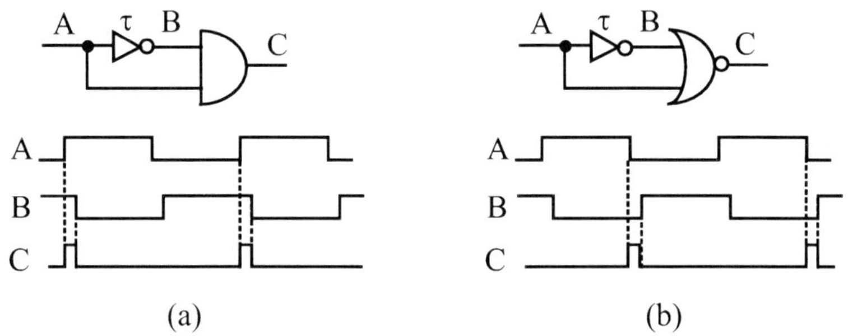

Figure 4.25. (a) Detection of rising edge. (b) Detection of falling edge.

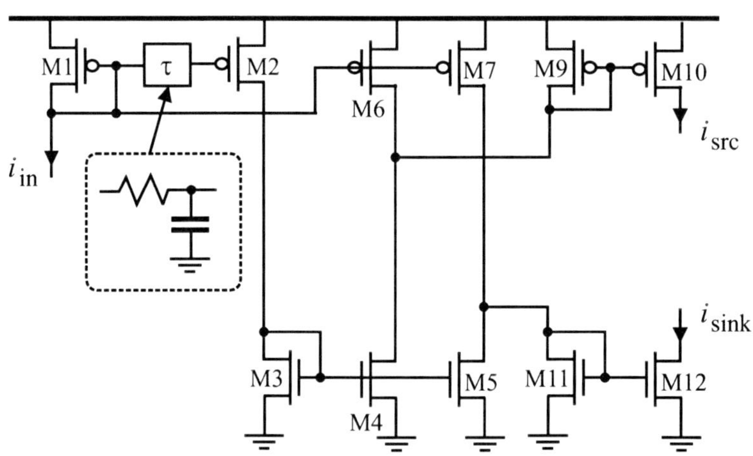

Figure 4.26. Simplified schematic of current edge detector proposed by Djemouai and Sawan [86, 87].

i_{src} and i_{sink} are fed to two downstream current comparators with their schematics shown in Fig.4.27 where i_{src} and i_{sink} are compared with the reference current i_{ref}. The outputs of the current comparators are then fed to a DFF where a full-swing output voltage is generated.

A drawback of this design is that the reference current of the current comparators must be properly set whereas in the preceding passive voltage-mode ASK demodulators, the reference voltage is generated by an averaging circuitry automatically.

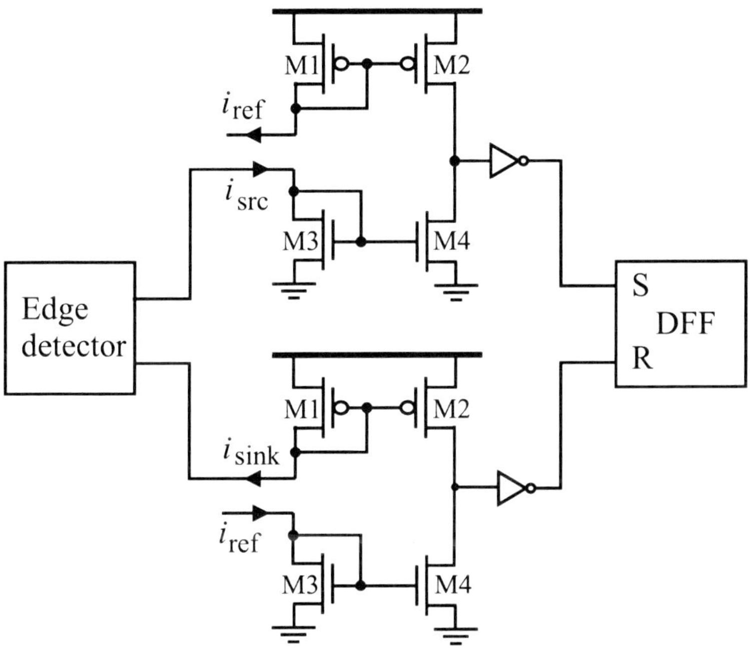

Figure 4.27. Simplified schematic of current comparator proposed by Djemouai and Sawan [86, 87].

The ASK demodulator proposed by Gudnason is shown in Fig.4.28 [88]. The incoming RF signal is a current generated from a passive RC network. It is rectified by the current squarer proposed by Oliaei and Loumeau [89, 90]. The negative feedback formed by M5-M6 lowers the input impedance of the current squarer. A 3rd-order $G_m - C$ low-pass is employed to extract the envelope of the output of the squarer. The level detector employs three transconductors to eliminate the long synchronization / start sequence required to establish a proper voltage reference for data recovery.

4.2.6 Mixed-Mode ASK Demodulators

Mixed-mode ASK demodulators are those where both nodal voltages and branch currents are employed to represent and process information in demodulation to achieve optimal performance. The input stage of these demodulators

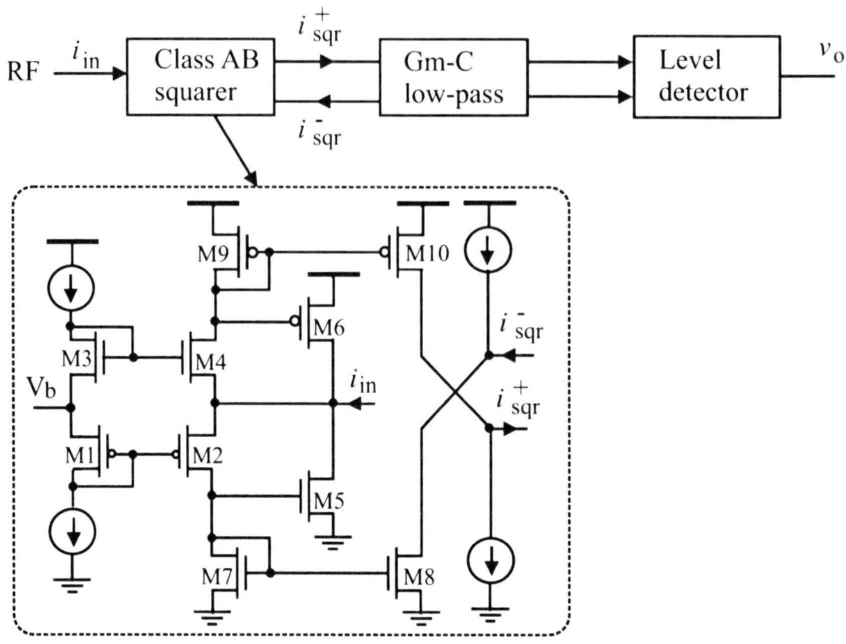

Figure 4.28. Simplified schematic of ASK demodulator proposed by Gudnason [88].

is typically a transconductor that converts an input voltage to a current. The current is then fed to a current-mode envelope detector. The extracted current envelope can be either fed to a current comparator directly or a current-to-voltage converter followed by a voltage comparator to recover the baseband data.

The ASK demodulator proposed by Harjani *et al.* in [91] and shown in Fig.4.29 is mixed-mode configured. It is based on the use of rectifying current subtracters. The input differential voltage is converted to two pairs of currents. If $v_{in}^+ > v_{in}^-$, $i_{d1,d3} < i_{d2,d4}$ follows. As a result, $i_{d3}-i_{d2} < 0$ and $i_{d4}-i_{d1} > 0$. It follows that

$$i_o = (i_{d3} - i_{d2}) - (i_{d4} - i_{d1}) < 0, \tag{4.32}$$

and $v_o > 0$. Similarly, one can show that when $v_{in}^+ < v_{in}^-$, $i_o > 0$, and $v_o < 0$. The polarity of the output voltage is set by the direction of i_o.

ASK demodulators can also be constructed using the envelope detector proposed by Alegre *et al.*, as shown in Fig.4.30 [92, 93]. The current-mode precision full-wave rectifier is derived from the current squarer reported by Bult and Wallinga in [94] and Chang and Liu in [95] and formalized in a full-wave rectification form by Khucharoensin and Kasemsuwan in [96].

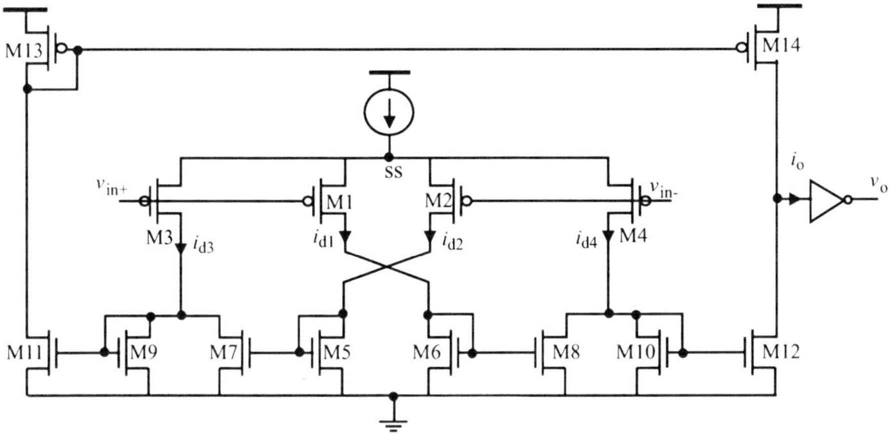

Figure 4.29. Simplified schematic of envelope detector proposed by Harjani *et al.* [91].

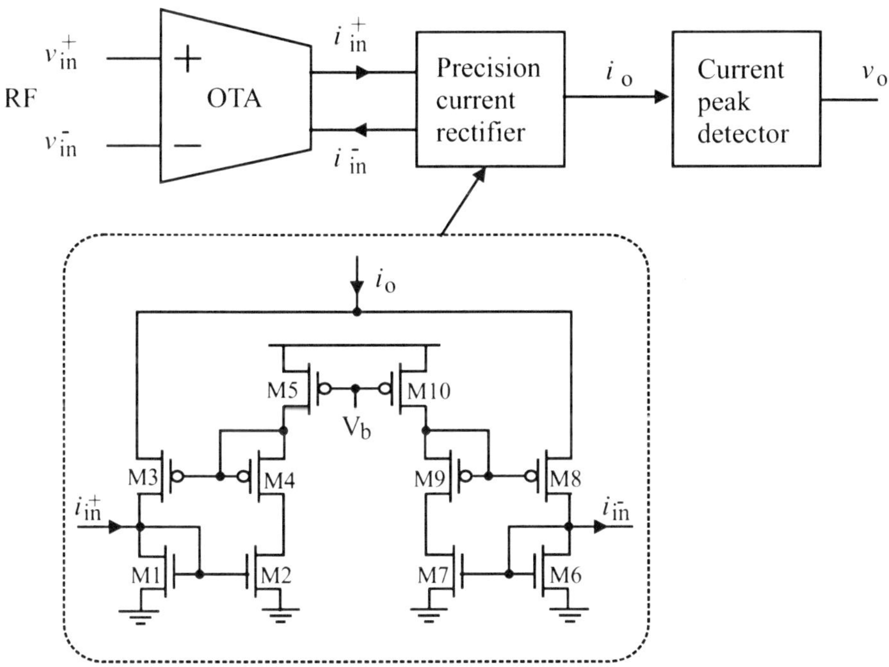

Figure 4.30. Simplified schematic of envelope detector proposed by Alegre *et al.*[92, 93].

To appreciate the elegance of the design, we need to understand the principle of the current squarer proposed by Bult and Wallinga [94] with its simplified schematic shown in Fig.4.31(a). Note that a similar circuit was reported in [97]. Assume that all transistors are identical and operate in the saturation. From

$$\begin{aligned} i_{D2} &= k_{2,3}(v_{GS2} - V_T)^2, \\ i_{D3} &= k_{2,3}(v_{GS3} - V_T)^2, \end{aligned} \tag{4.33}$$

and

$$V_b = v_{GS2} + v_{GS3}, \tag{4.34}$$

where

$$k_{2,3} = \frac{1}{2}\mu_n C_{ox}\left(\frac{W}{L}\right)_{2,3}, \tag{4.35}$$

and note that

$$\begin{aligned} i_{D2} + i_{D3} &= k_{2,3}\left(v_{GS2}^2 - 2v_{GS2}V_T + v_{GS3}^2 - 2v_{GS3}V_T + 2V_T^2\right), \\ i_{D2} - i_{D3} &= k_{2,3}\left(V_b - 2V_T\right)\left(v_{GS2} - v_{GS3}\right), \end{aligned} \tag{4.36}$$

we arrive at

$$i_{D2} + i_{D3} = \frac{k_{2,3}}{2}(V_b - 2V_T)^2 + \frac{(i_{D2} - i_{D3})^2}{2k_{2,3}(V_b - 2V_T)^2}. \tag{4.37}$$

Since

$$i_o = i_{D2} + i_{D3}, \tag{4.38}$$

and

$$i_{in} = i_{D2} - i_{D3}, \tag{4.39}$$

we have

$$i_o = \frac{k_{2,3}}{2}(V_b - 2V_T)^2 + \frac{i_{in}^2}{2k_{2,3}(V_b - 2V_T)^2}. \tag{4.40}$$

It is evident from (4.40) that the output current is the square of the input current.

The current-mode full-wave rectifier proposed by Khucharonesin and Kasemsuwan in [96] shown in Fig.4.31(b) is evolved from Bult-Wallinga current squarer. Transistors M1-M8 are carefully biased on the edge of conduction such that when $i_{in}^+ = i_{in}^- = 0$, $i_o = 0$. When i_{in}^+ flows into the rectifier and i_{in}^- flows out of the rectifier, M3 turns off / M1 turns on, and M6 turns off / M8 turns on. As a result, $i_o = i_{in}^-$. Similarly when i_{in}^+ flows out of the rectifier and i_{in}^- flows into the rectifier, $i_o = i_{in}^+$. As a result, i_o is the rectified version of the input current. A drawback of this current rectifier is its slow response when the input current is small. This is because when i_{in}^+ flows into the rectifier and i_{in}^- flows out of the rectifier, $C_{gs1,2}$ must be sufficiently charged in order to drive M1-M2 into conduction while $C_{gs6,7}$ must be sufficiently discharged in order to drive M6-M7 into cut-off.

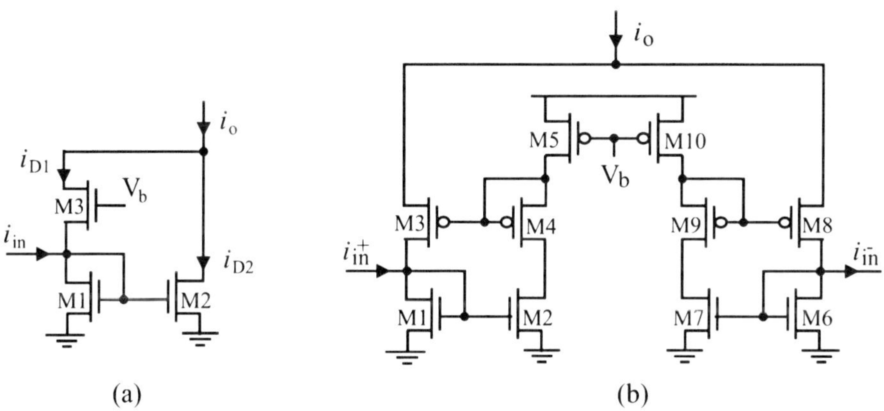

Figure 4.31. (a) Simplified schematic of current squarer proposed by Bult and Wallinga [94]. (b) Simplified schematic of precision full-wave current rectifier proposed by Khucharonesin and Kasemsuwan [96].

The output current of the rectifier is fed to a current peak detector that extracts the envelope of the current. The early work on current peak detectors includes those reported in [98, 99] and shown in Fig.4.32(a). The operation of the current peak detector can be briefly depicted as the follows : When i_{in} rises, capacitor $C \approx C_{gs1} + C_{gs2}$ is charged by M3 and i_o follows i_{in} accordingly. When i_{in} drops, v_{in} will drop. If the discharge current source i_d is absent, v_c will be held unchanged. The same holds for i_o as well. To track the decreasing peak of i_{in}, i_d is needed to discharge C and lower v_c such that i_o will follow the peak of i_{in}. A similar design was used in [92, 93] with the exception that the basic current mirror in Fig.4.32(a) was replaced with its cascode counterpart, as shown in Fig.4.32(b), to boost the output impedance of the current peak detector.

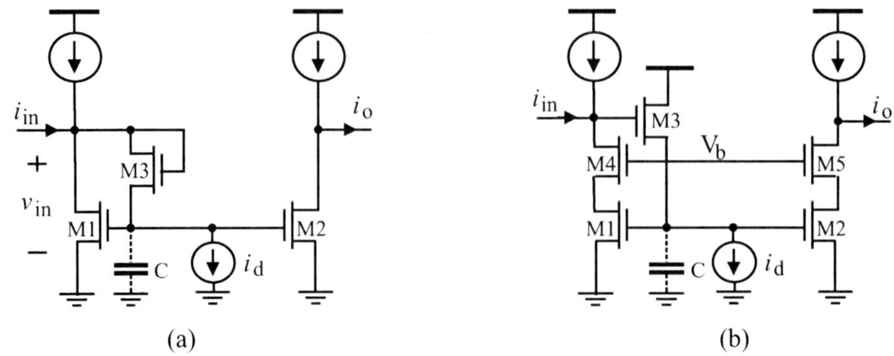

Figure 4.32. (a) Simplified schematic of current peak detector by proposed Ismail and Tiez [98]. (b) Simplified schematic of current peak detector proposed by Sarpeshkar *et al.* [100].

A similar design was employed in [101] for wideband envelope detectors, as shown in Fig.4.33. Active inductors were employed to improve the bandwidth of the operational transconductor amplifier. The differential output from the transconductor amplifier is fed to the full-wave current rectifier whose output is then fed to the downstream peak detector with its operation the same as that given in Fig.4.32.

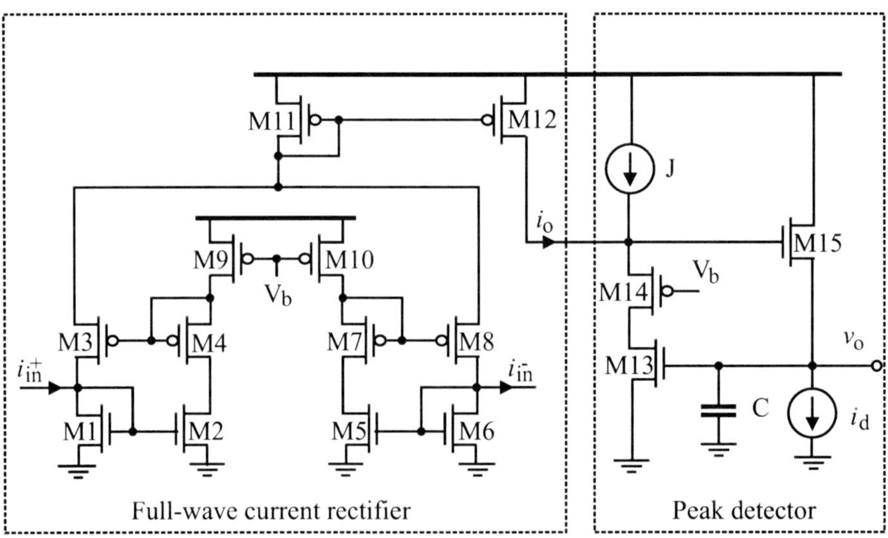

Figure 4.33. Simplified schematic of envelope detector proposed by Zhou *et al.* [101].

The ASK demodulator proposed in [8, 9] for medical endoscopy applications employs a transconductor to convert the differential input voltage v_{in}

to four currents $i_{d1} \sim i_{d4}$, as shown in Fig.4.34. Rectification is performed in the current domain. The differential output current of the rectifier is fed to a current-mode Schmitt trigger evolved from Allstot voltage-mode Schmitt trigger examined earlier.

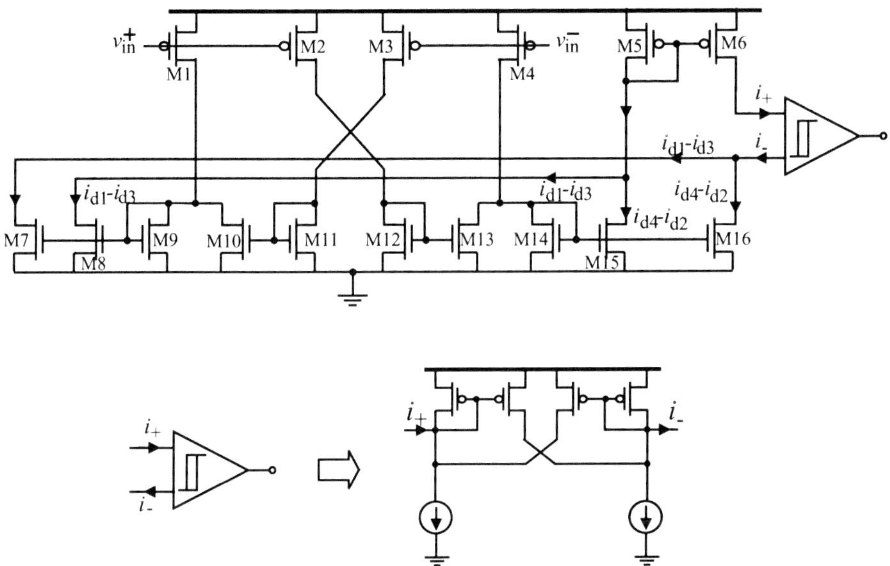

Figure 4.34. Simplified schematic of ASK demodulator proposed by Chi *et al.* for medical endoscopy applications [8, 9].

The envelope detector proposed in [102, 103] for hearing-aids applications is another mixed-mode ASK demodulator. As shown in Fig.4.35, it consists of a transconductor with a wide dynamic range to perform V/I conversion, and a current-mode peak detector. To improve the dynamic range of the transconductor, the differential inputs are well-inputs, i.e. from the substrate. Transistors M1 and M2 are biased near the edge of conduction such that when $i_{in} > 0$, M1 is off and M2 conducts. and when $i_{in} < 0$, M2 is off and M1 conducts. To speed up the current rectification and minimize the distortion when the input current i_{in} is low, an auxiliary amplifier is employed to amplify V_A such that a small variation of v_A will result in a large voltage change at the output of the amplifier, speeding up the turn-on and turn-on processes of M1 and M2. Also, the current feedback provided by the transconductor G_f is employed to minimize the effect of the offset current of the input transconductor G_{in}. A care must be taken in choosing V_{ref1} and V_{ref2}. V_{ref1} should be set to the dc biasing value of v_A. V_{ref2} should be set to $V_{DD}/2$, assuming $V_C = V_{DD}/2$ in the dc steady state.

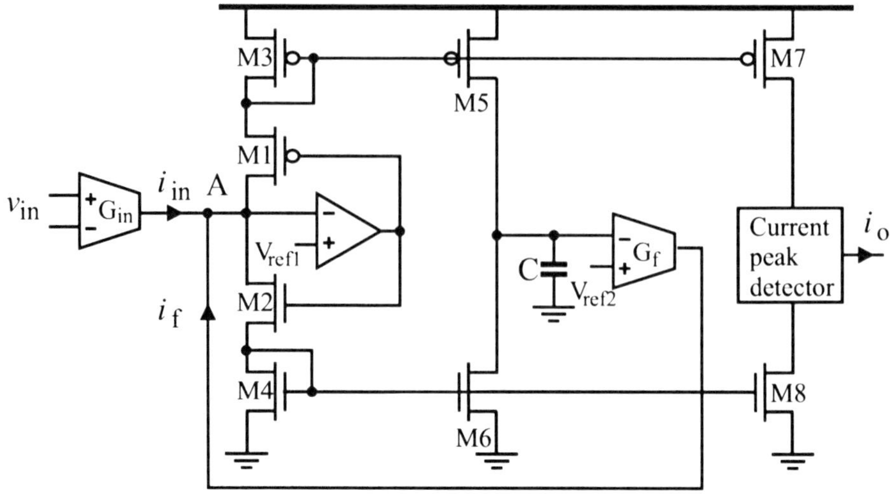

Figure 4.35. Simplified schematic of envelope detector proposed by Baker *et al.* for hearing-aids [102, 103].

4.2.7 Performance Comparison of ASK Demodulators

Table 4.1 compares the performance of recently published ASK demodulators for passive wireless microsystems. As can be seen that the data rate of ASK demodulators is generally low. The carrier frequency ranges from low MHz to GHz.

4.3 FSK Modulators and Demodulator

It was pointed out earlier in Section 4.1 that FSK is a constant-envelope modulation scheme. When FSK is used to encode the data transmitted from a base station to a passive wireless microsystem, the drawback of fluctuating power transmission encountered in ASK is eliminated [108]. Since baseband data are represented by the frequency rather than the amplitude of the carrier, the effect of noise and disturbances on data transmission is greatly reduced. Another unique characteristic of FSK is its ability to transmit data at a high data rate as compared with ASK, owing to the fact that the demodulation of a FSK-modulated signal does not need to extract the envelope of the carrier using a low-pass filter with a large time constant. This is particularly important for biomedical implants as not only the data rate of these microsystems such as cochlear implants and visual prosthesis implants is high, the carrier frequency of these implants is typically 13.56 MHz in ISM bands due to the high degree of the absorption of electromagnetic waves by living bodies at high frequencies.

Table 4.1. Performance comparison of ASK demodulators for passive wireless microsystems.

Ref.	Tech.	Carrier freq.	Data rate
Bouvier *et al.*[77](97)	0.5μm	4.9152 MHz	9.6 Kbps
Liu *et al.* [3](00)	1.2 mm	1-10 MHz	25-250 Kbps
Harjani [91](00)	0.5μm	10 MHz	1 Mbps
Yu & Najafi [55](01)	1.5μm	4 MHz	–
Chow & Wang [80](02)	0.5μm	860 KHz	–
Yu & Najafi [78](03)	0.8μm	4 MHz	60 Kbps
Djemouai & Sawan [86](04)	0.18μm	250 MHz	–
Naghmouchi *et al.* [57](04)	0.35μm	20 MHz	1 Kbps
Wang [83](04)	0.35μm	2 MHz	10 Kbps
Lee & Lee [5](05)	0.35μm	2 MHz	1 Mbps
Lin *et al.*[104](05)	0.18μm	0.1-1.9 GHz	2 Mbps
Cho [85](05)	0.25μm	860-960 MHz	–
Hmida *et al.*[79](06)	0.35μm	13.56 MHz	1 Mbps
Cinco-Galicia/ Sandoval-Ibarra [105](06)	0.5μm	915 MHz	250 Kb/s
Alegre [92, 93](06/08)	0.35μm	10 MHz	–
Pongsawat-Thanachayanout [31](06)	0.35μm	900 MHz	160 Kbps
Hsia *et al.* [56](06)	0.35μm	915 MHz	–
Yu & Bashirallah [81](06)	0.6μm	1 MHz	4-18 Kbps
Tran *et al.* [106](07)	0.35μm	900 MHz	2.2 MHz
Chi *et al.*[8, 9](07)	0.25μm	1.83 GHz	256 Kbps
Lee *et al.* [107, 84](07/08)	0.35μm	2 MHz	20 Kb/s
Zhou *et al.* [101](08)	0.18μm	100 Hz-1.6 GHz	10 Kb/s

It was shown in Chapter 2 that an inductive link in ASK should have a high quality factor in order to boost the voltage of the received RF signal so as to maximize the power efficiency of the downstream voltage multiplier. In FSK, the inductive link must transfer an sufficient amount of power at the two frequencies of the carrier. These two frequencies are typically far apart. For instance, the carrier frequency corresponding to binary 1 is usually twice that corresponding to binary 0 or vice versa [109–112]. This requires that the quality factor of the inductive link be rather moderate such that the coupling factor of the inductive link at both frequencies of the carrier can be equally large. FSK is particularly suitable for passive wireless microsystems that use planar coils for wireless links with their base stations. This is because the quality factor of these coils is typically low. A downside of using coupling coils with a low quality factor is the reduced efficiency of power transmission from the base stations to the implants, mainly due to the reduced power efficiency of the following voltage multipliers when the voltage at the input of the multipliers is low [113]. Finally, FSK is also suitable for low-voltage applications, owing to the independence of the baseband data on the amplitude of the FSK-modulated signal. ASK, on the contrary, becomes less attractive with the continuous reduction of the supply voltage.

4.3.1 FSK Modulators

Since the function of a digital FSK modulator is to represent binary 1 and binary 0 with two sinusoids of the same amplitude but different frequencies, a FSK modulator can be constructed using two oscillators that oscillate at different frequencies and a multiplexer with its inputs from the two oscillators and its output selected by the modulating digital bit stream, as shown in Fig.4.36 [114, 111]. A drawback of this configuration is its high power consumption, arising from the need for two oscillators.

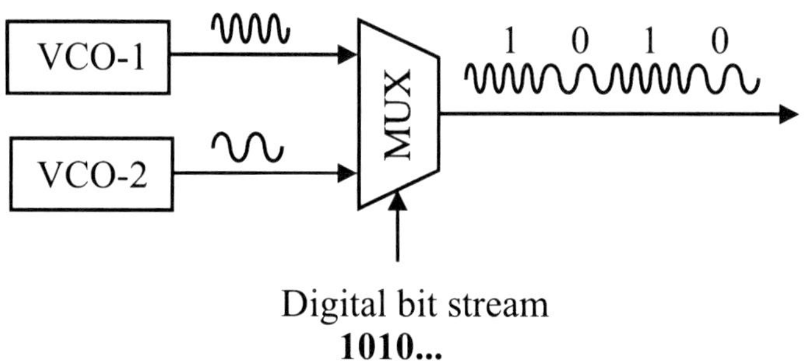

Figure 4.36. Binary FSK-modulator using two oscillators and a multiplexer.

To minimize the power consumption, a FSK modulator can also be constructed using a single oscillator whose frequency is controlled by the modulating digital bit stream. An example is the FSK modulator using a Colpitts oscillator with the capacitors of the oscillator controlled by the modulating digital bit stream, as shown in Fig.4.37 [115]. The frequency of the oscillator is varied by connecting/disconnecting capacitor C_5 in parallel with $C_3 \sim C_4$.

A FSK-modulated signal can be demodulated using a phase-locked loop that locks onto the two frequencies of the carrier [116, 117]. This, however, is at the cost of high silicon and power consumption. FSK-demodulated signals can also be recovered using differentiators and mixers [118, 119]. The need for mixers in these approaches makes the reduction of the power consumption of the modulators a rather difficult goal to achieve.

It was shown earlier that FSK has been used in biomedical implants to take the advantages of the high data rate and the continuous flow of radio-frequency power from base stations to biomedical implants. The high degree of the

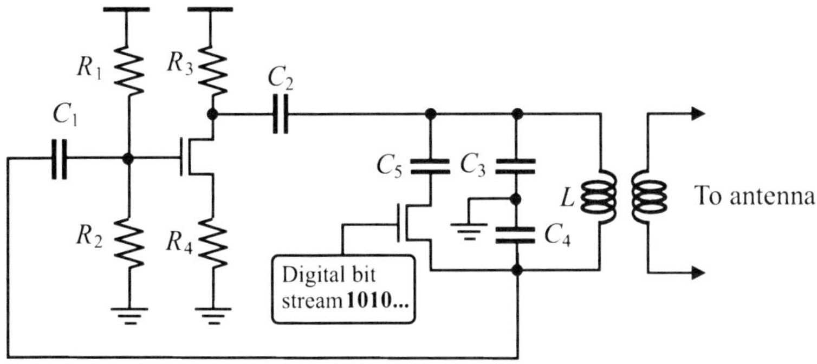

Figure 4.37. Simplified schematic of binary FSK-modulator proposed by Chuah *et al.* [115].

absorption of electromagnetic waves by living bodies at high frequencies and the low self-resonant frequency of planar coils that link biomedical implants and their base stations set the upper bound of the carrier frequency of biomedical implants. 13.56 MHz in ISM bands is often the choice of the carrier frequency for biomedical implants. The low carrier frequency significantly relaxes design constraints and opens a door for innovations in design of ultra-low power FSK demodulators for biomedical implants. In what follows we will study several recently published FSK demodulators for biomedical implants.

4.3.2 Ghovanloo-Najafi FSK Demodulator

Ghovanloo and Najafi proposed a low-power FSK demodulator for biomedical implants with the configuration of the data recovery portion of the demodulator shown in Fig.4.38 [109, 110]. The FSK-modulated signal is converted to a full-swing square wave using a clock recovery block, as shown in Fig.4.39. This allows us to use digital circuits to perform demodulation conveniently The clock recovery circuit can be implemented using either a Schmitt trigger [120] or simply a pair of CMOS inverters that are connected in series [111]. Since two sinusoids of different frequencies are used to represent binary 1 and binary 0, the value of the baseband data can be determined by measuring the period of the sinusoids. To facilitate measurement, the two sinusoids for binary 1 and binary 0 are converted to a square wave with a data-dependent period by the clock recovery circuit. The period of the square wave is measured by using a local oscillator whose period is constant and much smaller as compared with that of the two sinusoids. A counter is used to record the number of the oscillation cycles of the local oscillator during binary 0 and binary 1 intervals. The frequency of the local oscillator is chosen such that the most significant bit of the counter is set to 1 when a binary 1 is encountered and 0 otherwise. Specifically, the counter starts to record the number of the oscillation cycles of the oscillator at the rising edge of $\widehat{RF}$. The counting continues until $\widehat{RF}$ becomes zero.

Since the pulse width is determined by the number of the oscillation cycles of the local oscillator, the frequency of the oscillator should be sufficiently high in order to yield an accurate measurement of the period of the square waves for binary 0 and binary 1. This, however, is at the cost of a high level of power consumption as the higher the frequency of the oscillator, the higher its power consumption. Since our objective is to distinguish binary 1 and binary 0 rather than to accurately measure the two frequencies of the carrier, and the ratio of the frequency of the sinusoid representing binary 1 to that representing binary 0 is typically 2 or vice versa, it is advantageous to set the frequency of the local oscillator low to minimize power consumption. This, of course, is on the condition that binary 1 and binary 0 can be distinguished reliably.

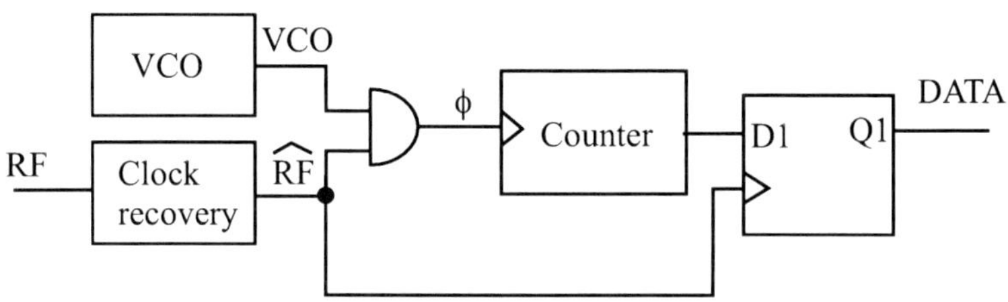

Figure 4.38. Simplified configuration of the data recovery portion of binary FSK-demodulator proposed by Ghovanloo and Najafi [109, 110].

4.3.3 Jung FSK Demodulator

The configuration of the FSK demodulator proposed by Jung *et al.* for biomedical implants is shown in Fig.4.40 [120]. A binary 1 is represented by a sinusoid of period T_1 and a binary 0 is represented by a sinusoid of period T_0 with $T_1 = 2T_0$. The clock recovery block is simply a Schmitt trigger that converts the incoming FSK-modulated sinusoid to a full-swing square wave for convenient signal processing, as shown in Fig.4.41. The square wave is delayed by τ_d and used to sample the square wave. The sampling instant is controlled by adjusting τ_d. The amount of time delay τ_d is critical to the proper operation of the FSK demodulator. In the case shown in the figure, $\frac{T_0}{2} < \tau_d < T_0$ must be satisfied. The constraint on the time delay τ_d ensures that during the interval of binary 0, D_1 is 0 at the rising edge of ϕ_1. Similarly, during the interval of binary 1, D_1 is 1 at the rising edge of ϕ_1. Note that sampling binary 0 must take place when $D_1 = 0$, the maximum timing margin is therefore $T_0/2$. When taking the timing jitter of the received RF signal into consideration, the timing margin will be smaller. It becomes evident that this method becomes less reliable once the frequency of the carrier is high.

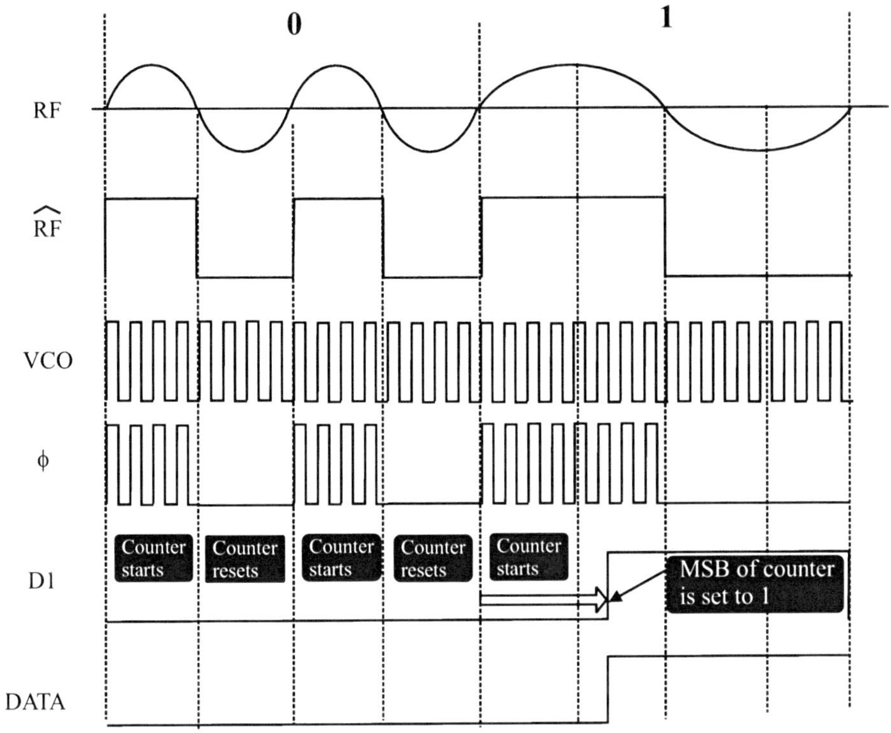

Figure 4.39. Waveforms of binary FSK-demodulator proposed by Ghovanloo and Najafi.

As compared with Ghovanloo-Najafi FSK demodulator studied earlier, Jung FSK demodulator recovers the baseband data by sampling the incoming data at a specific time instant rather than measuring the period of the two sinusoids of the carrier. The elimination of the high-frequency oscillator used in Ghovanloo-Najafi FSK demodulator greatly lowers the power consumption of Jung FSK demodulator. The price paid is the reduced degree of reliability.

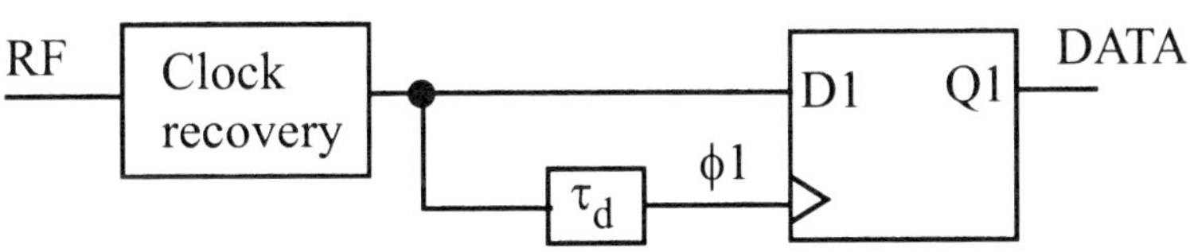

Figure 4.40. Simplified configuration of data recovery portion of binary FSK demodulator proposed by Jung *et al.* [120].

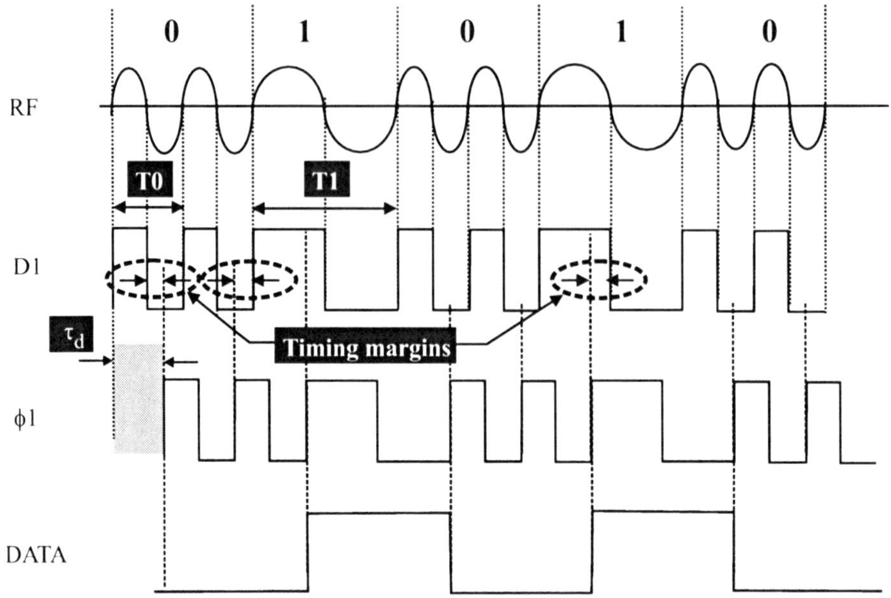

Figure 4.41. Waveforms of binary FSK demodulator proposed by Jung *et al.*

4.3.4 Weng FSK Demodulator

The data recovery portion of the FSK demodulator proposed by Weng *et al.* for biomedical implants is shown in Fig.4.42 [111]. Similar to Jung FSK demodulator, a binary 1 is represented by a sinusoid of period T_1 and a binary 0 is represented by a sinusoid of period T_0 with $T_1 = 2T_0$. The inverter chain converts the received FSK-modulated sinusoid wave to a full-wing square wave. The two frequencies used to represent binary 1 and binary 0 are 5 MHz and 10 MHz, respectively. A constant reference signal REF of frequency 10 MHz is generated locally and used to demodulate the FSK-modulated signal. The XOR2 gate generates a positive pulse with pulse width $T_1/2$ each time a binary 1 is received.

The square wave generated from the received FSK-modulated signal is delayed by τ_d. The amount of time delay τ_d is chosen with the constraint $\frac{T_0}{2} < \tau_d < T_0$ to ensure that the rising edge of ϕ samples $D = 1$ when a binary 1 is detected and samples $D = 0$ when a binary 0 is detected, as seen from Fig.4.43. As compared with Jung FSK demodulator studied earlier, Weng FSK demodulator requires the generation of a reference clock and a XOR2 gate, both consume an additional amount of power. Also, $\widehat{RF}$ and $\widehat{REF}$ must have the same frequency and their phases must also align up perfectly. Otherwise, there will be a large number of ripples in the output of the XOR2, resulting in a failure in data recovery.

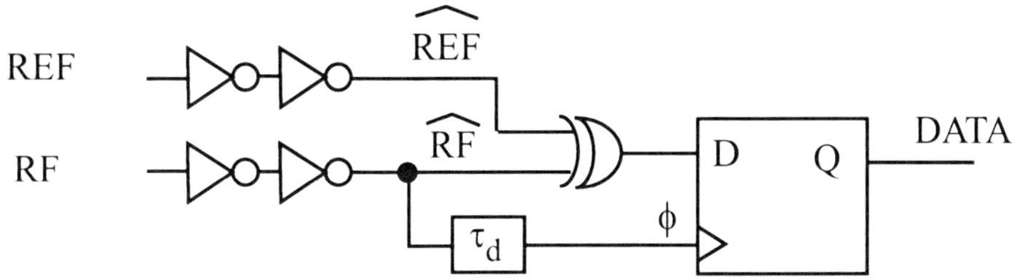

Figure 4.42. Simplified configuration of data recovery portion of binary FSK-demodulator proposed by Weng *et al.* [111].

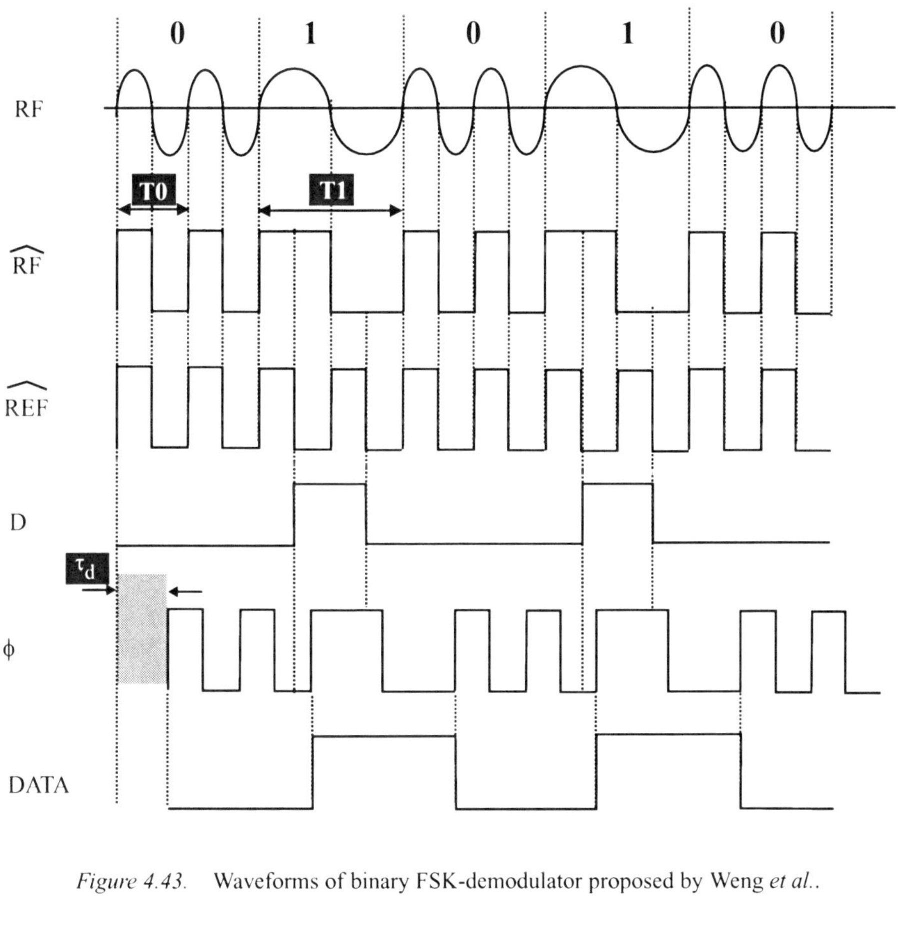

Figure 4.43. Waveforms of binary FSK-demodulator proposed by Weng *et al.*.

4.3.5 Performance Comparison of FSK Demodulators

Having investigated the pros and cons of various FSK demodulators for passive wireless microsystems, let us now compare the performance of these demodulators quantitatively in terms of carrier frequency, data rate, and power consumption. Table 4.2 tabulates the performance indicators of some recently reported FSK demodulators for biomedical implants. It is seen that the carrier frequency of FSK-modulated signals is typically in low MHz ranges. This is mainly due to the low self-resonant frequency of planar coupling coils and the high absorption of electromagnetic waves by living bodies at high frequencies. The data rate is typically in low Mbps ranges. Innovative designs are needed to increase the data rate and at the same time to lower the power consumption of FSK demodulators.

Table 4.2. Performance comparison of binary FSK demodulators for passive wireless microsystems.

Ref.	Tech.	Carrier [MHz]	Data rate [Mbps]	Power [mW]
Jung *et al.* [120](07)	Discrete	4.17/6.25	2.083	58
Ghovanloo-Najafi [109](04)	1.5μm	5/10	2.5	0.38
Weng *et al.* [111](07)	0.18μm	5/10	5	0.022
Hwang-Lin [112](09)	0.35μm	6.78/13.56	1	0.96

4.4 PSK Modulators and Demodulators

As pointed out earlier that the two key characteristics of a PSK-modulated carrier are the constant amplitude and constant frequency of the carrier. The constant amplitude of the carrier ensures that the flow of the power from a base station to a passive wireless microsystem is independent of baseband data. The constant frequency of the carrier allows the transmission of power from a base station to a passive wireless microsystem to be maximized by optimizing the quality factor of the coupling coils between the base station and the passive wireless microsystem. As a result, the drawback associated with FSK, in particular, a low power transmission efficiency arising from the use of low-Q coupling coils to accommodate the two distinct frequencies of the carrier is eliminated [113, 121]. As stated earlier in Section 4.1, BPSK uses only two distinct phases, typically separated by 180 degrees, to represent binary 1 and binary 0. It is the most robust PSK as it offers the highest degree of immunity to noise and disturbances among all PSK modulation schemes. The configuration of BPSK demodulators is in general more complex as phase-locked loops are usually required for coherent demodulation. This is accompanied with a high level of power consumption. A key challenge encountered in design of BPSK demodulators for passive wireless microsystems is how to minimize the power consumption of these systems without sacrificing the performance.

4.4.1 PSK Modulators

A PSK modulator can be constructed by modulating the phase of the carrier by the digital bit stream to be transmitted while keeping the amplitude of the carrier unchanged. When PSK modulation is used in backscattering, the imaginary part of the impedance seen by the antenna should be modulated by the modulating digital bit stream. Fig.4.44 shows the simplified schematic of the PSK modulator by Karthaus and Fischer [34]. The voltage swing of the digital bit stream is first boosted to full swing using a pair of static inverters. The voltages at nodes A and B are controlled by the incoming data. For example, if a binary 1 appears at the input, the voltage at node A will drop while that at node B will rise. The transient behavior of the voltages at nodes A and B is determined by the current J, which sets the speed at which the capacitors at nodes A and B are charged and discharged. If a binary 0 appears at the input, the voltage at node A will rise while that at node B will drop. The preceding observation shows that the polarity of the voltage across C_2 has a one-to-one relation with the bit value of the incoming data. To modulating the capacitance, C_2 is implemented using a MOS varactor. The change of the polarity of the voltage across C_2 modulates the capacitance of the varactor between its maximum and minimum values. Capacitors C_1 and C_3 are poly capacitors whose capacitance is constant. Together with C_2, they form the total capacitance seen by the antenna.

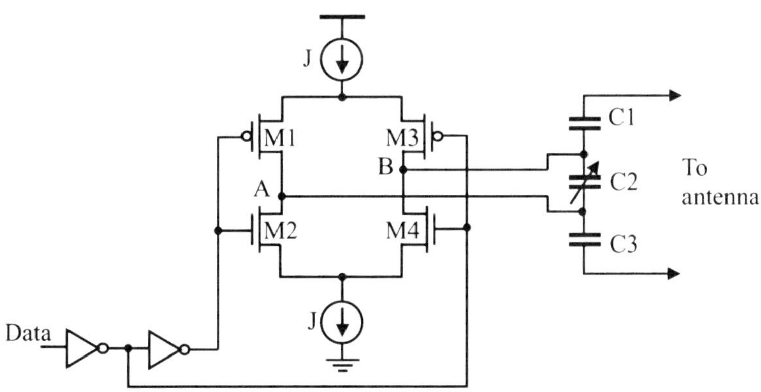

Figure 4.44. Backscatter BPSK modulator proposed by Karthaus and Fischer [34].

4.4.2 Coherent BPSK Demodulators

To demodulate a PSK-modulated signal, coherent detection using a Costas loop is perhaps the most widely used technique [122]. Initially proposed by John Costas (1923-2008) for optimizing amplitude modulation (AM) systems, a Costas loop consists of two loops, an in-phase (I) loop and a quadrature-phase (Q) loop, as shown in Fig.4.45. The synchronous detectors function as multipliers that detect the phase difference between the incoming AM signal and the output of the local oscillator. The frequency of the oscillator is the same as the carrier frequency. If the phase difference between the incoming AM signal and the output of the local oscillator is zero, the in-phase audio amplifier will contain the demodulated signal while the output of the quadrature audio amplifier will diminish. If the phase of the local oscillator slightly drifts from its desired value, i.e. the phase of the incoming AM signal, the output of the quadrature-phase audio amplifier will emerge from zero. The Q-branch thus provides an effective mean to detect whether a lock state is established or not. By combining the in-phase and quadrature-phase signals in the discriminator, a dc control voltage can be generated. This voltage is then used to adjust the frequency of the local oscillator in such a way that the phase error can be tuned to zero.

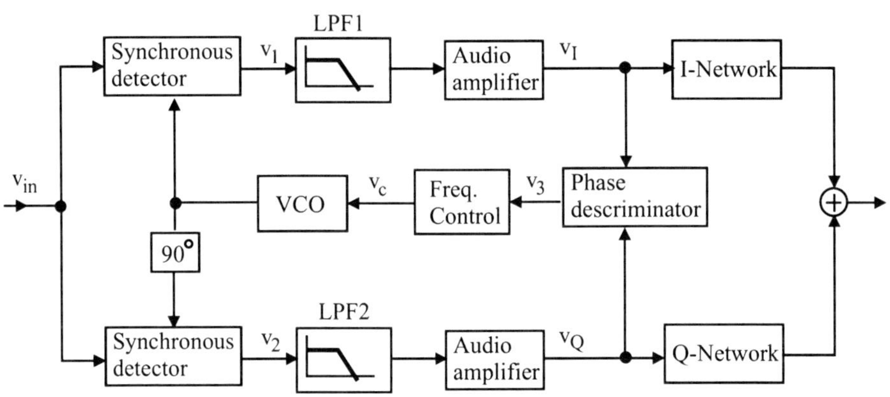

Figure 4.45. Costas loop proposed by Costas for AM systems [122].

A Costas loop can be used to demodulate a BPSK-modulated signal, as illustrated in Fig.4.46. Without losing generality, let the incoming BPSK-modulated signal be

$$v_{in} = A(t)\sin(\omega t + \phi_1), \tag{4.41}$$

where $A(t)$ is the baseband data and only takes the value of either +1 or -1, ω is the carrier frequency, and ϕ_1 is the phase of the carrier. The frequency of the local oscillator is set to be the same as the carrier frequency (coherent detection). The two low-pass filters, one in the I-branch and the other in the Q-branch immediately after the multipliers, extract the low-frequency components of the output of the multipliers. Making use of the identities of trigonometric functions

$$\begin{aligned} \sin\phi_1\sin\phi_2 &= \frac{1}{2}\left[\cos(\phi_1-\phi_2)-\cos(\phi_1+\phi_2)\right], \\ \sin\phi_1\cos\phi_2 &= \frac{1}{2}\left[\sin(\phi_1-\phi_2)+\sin(\phi_1+\phi_2)\right], \end{aligned} \tag{4.42}$$

we obtain the output of multipliers MX1 and MX2

$$\begin{aligned} v_1(t) &= A(t)\sin(\omega t+\phi_1)\sin(\omega t+\phi_2) \\ &= \frac{A(t)}{2}\left[\cos(\phi_1-\phi_2)-\cos(2\omega t+\phi_1+\phi_2)\right], \end{aligned} \tag{4.43}$$

$$\begin{aligned} v_2(t) &= A(t)\sin(\omega t+\phi_1)\cos(\omega t+\phi_2) \\ &= \frac{A(t)}{2}\left[\sin(\phi_1-\phi_2)+\sin(2\omega t+\phi_1+\phi_2)\right]. \end{aligned} \tag{4.44}$$

The high-frequency components of $v_1(t)$ and $v_2(t)$ will be filtered out by the low-pass filters and only the low-frequency terms will make their way to the output of the low-pass filters

$$\begin{aligned} v_I(t) &= \frac{A(t)}{2}\cos(\phi_1-\phi_2), \\ v_Q(t) &= \frac{A(t)}{2}\sin(\phi_1-\phi_2). \end{aligned} \tag{4.45}$$

The extracted low-frequency components are further multiplied each other with the help of MX3 to generate the control voltage of the voltage-controlled oscillator

$$v_c(t) = \frac{A^2(t)}{8}\sin\left[2(\phi_1-\phi_2)\right]. \tag{4.46}$$

Since $A(t) = +1$ or -1, $A^2(t) = 1$ follows. As a result, $v_c(t)$ is only a function of the phase difference $\phi_1 - \phi_2$. The loop filter LPF3 is needed to filter out any high-frequency component that might exist at the output of MX3. A close examination of the Costas loop reveals that the lower portion of the Costas loop containing the quadrature branch is essentially a phase-locked loop that locks onto the carrier. If there is a phase difference, i.e. $\phi_1 \neq \phi_2$, a corresponding non-zero control voltage v_c will be generated and the frequency of the oscillator will be adjusted accordingly. In the lock state where $\phi_1 = \phi_2$, the control voltage vanishes and the frequency of the oscillator remains unchanged. The quadrature branch of the Costas loop thus tracks the frequency and phase of the carrier while the in-phase branch of the Costas loop yields the baseband data $v_I(t) = A(t)/2$.

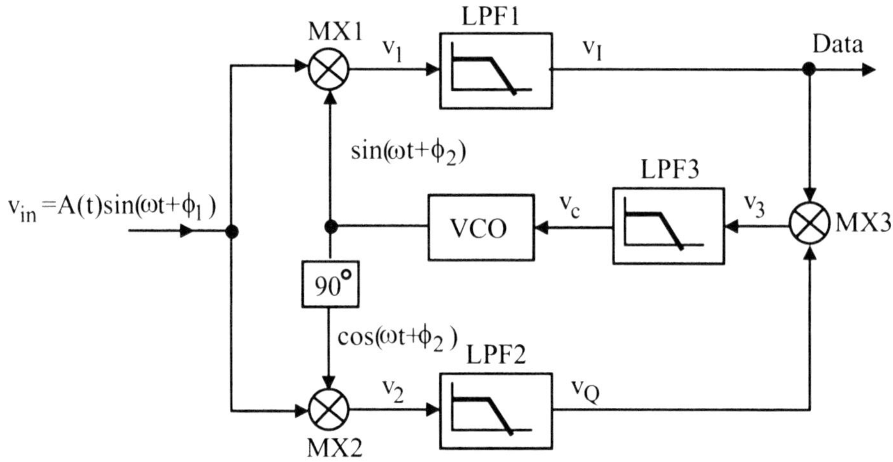

Figure 4.46. BPSK demodulation using a Costas loop.

A number of issues must be addressed when using Costas-loop based BPSK demodulators for passive wireless microsystems :

- The frequency of the local oscillator must be the same as the carrier frequency for coherent detection. The effect of process variation and the change of the environment in which the local oscillator resides such as temperature, however, will cause the frequency of the oscillator to deviate from the carrier frequency. Note that the conventional Costas loop of Fig.4.46 does not have the ability to detect subsequently correct the frequency difference between the local oscillator and the carrier.

- The acquisition range of a Costas loop is determined by the linear range of its multipliers. If the amplitude of the input of the multipliers is large, the effect of the nonlinearities of the multipliers will create low-frequency components that will pass through the low-pass filters and affect the performance of the

Costas loop. The small linear range of multipliers will limit the acquisition range of the Costas loop [123].

- The need for multiple multipliers makes the minimization of the power consumption of Costas loops a rather challenging task.

- Since the carrier frequency of biomedical implants is typically 13.56 MHz, the highest frequency at the output of the multipliers in the I-branch and Q-branch is 13.56 MHz $\times$ 2 = 27.12 MHz. To extract the low-frequency components of the output of the multipliers, the cutoff frequency of the low-pass filters must be sufficiently low. These low-pass filters sometimes are realized using off-chip components [124]. When implemented on-chip, their silicon consumption could be significant.

- The control voltage given in (4.46) shows that in the vicinity of the lock state, the sensitivity of v_c to $\phi_1 - \phi_2$ is rather low. This is because $\sin(\phi_1 - \phi_2) \approx \phi_1 - \phi_2$ when $\phi_1 \approx \phi_2$, resulting in a long locking process to achieve $\phi_1 = \phi_2$. The constant in the denominator of (4.46) further lowers the sensitivity.

To improve the performance of Costas loop based BPSK demodulators, Hu and Sawan showed that the preceding conventional Costas loop can be converted to a hard-limited Costas loop by employing a Schmitt trigger to convert the incoming BPSK-modulated RF signal to a square wave and another Schmitt trigger in the quadrature branch, as shown in Fig.4.47 [113, 121]. The purpose of using a Schmitt trigger to convert the BPSK-modulated RF signal to a square wave is to allow the use of digital phase detectors, such as XOR2 phase detectors, to replace conventional multipliers in both the in-phase and quadrature-phase branches of the Costas loop such that not only the power consumption can be reduced, the acquisition range can also be increased. The loop gain of the overall system will also become independent of the amplitude of the input. The insertion of the Schmitt trigger in the quadrature-branch of the Costas loop relaxes the requirement of multiplier MX3.

Costas-based BPSK demodulators including those using hard-limited Costas loops require that the frequency of the local oscillator be identical to that of the carrier. As pointed out earlier that the effect of process variation and the change of the environment in which passive wireless microsystems reside will cause the frequency of the local oscillator of these microsystems to deviate from the carrier frequency. It is therefore highly desirable to have the Costas loop to have the ability to detect the difference between the frequency of the local oscillator and that of the carrier, and the difference between the phase of the local oscillator and that of the carrier such that both the frequency and phase of the local oscillator can be adjusted to establish a lock state.

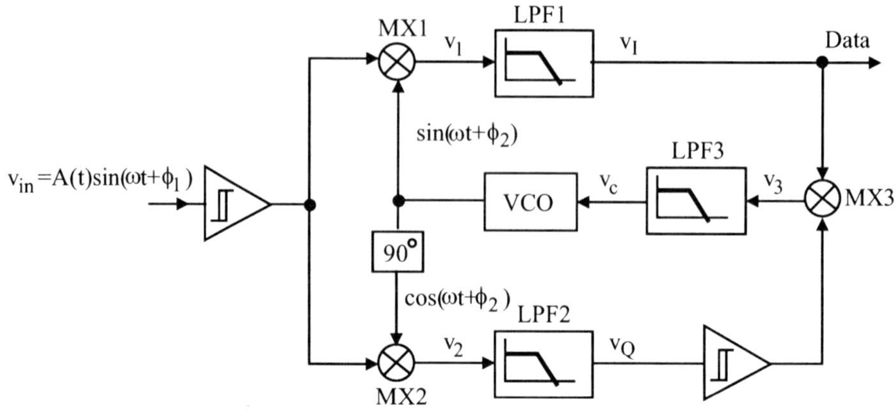

Figure 4.47. BPSK demodulation using a hard-limited Costas loop [113, 121].

Luo and Sonkusale showed that the XOR2 multiplier in the I-branch of a Costas loop can be replaced with a DFF phase/frequency detector (PFD) to give the Costas loop the ability to detect both frequency and phase differences between the local oscillator and the carrier so as to increase the acquisition range and speed up the locking process of the Costas loop [124, 125, 123]. The configuration of Lou-Sonkusale BPSK demodulator is shown in Fig.4.48. The phase/frequency detector is a conventional DFF phase/frequency detector while MX2 and MX3 are XOR2 phase detectors. If the frequency of the local oscillator differs from the carrier frequency, the PFD and the charge pump (CP) will generate a current that will either charge or discharge the loop filter LPF1, depending upon the polarity of the difference between the frequency of the oscillator and that of the carrier. Readers are referred to [52] for an exhaustive treatment of CMOS phase/frequency detectors and charge pumps. The generated control voltage v_c will adjust the frequency of the oscillator such that both the frequency and phase differences between the carrier and the local oscillator will become zero. Once the frequency of the local oscillator becomes identical to that of the carrier, we have

$$\begin{aligned} v_2(t) &= A(t)\sin(\omega t+\phi_1)\sin(\omega t+\phi_2) \\ &= \frac{A(t)}{2}\left[\cos(\phi_1-\phi_2)-\cos(2\omega t+\phi_1+\phi_2)\right]. \end{aligned} \tag{4.47}$$

With the help of LPF2, we arrive at

$$v_D(t) = \frac{A(t)}{2}\cos(\phi_1-\phi_2). \tag{4.48}$$

The trigger-and-hold block exacts data transition information. Specifically, for each transition in the baseband data, the trigger-and-hold block is triggered and holds the data for a short period of time during which a lock state of the Costas loop can be established. In the lock state, since

$$\begin{aligned} v_3(t) &= \frac{A(t)}{2}\sin(\omega t+\phi_2)\cos(\phi_1-\phi_2)|_{\phi_1=\phi_2} \\ &= \frac{A(t)}{2}\sin(\omega t+\phi_1), \end{aligned} \tag{4.49}$$

it becomes evident that the I-branch of the Costas loop replicates the incoming BPSK-modulated input $v_{in}(t)$ while the Q-branch of the Costas loop yields the baseband data.

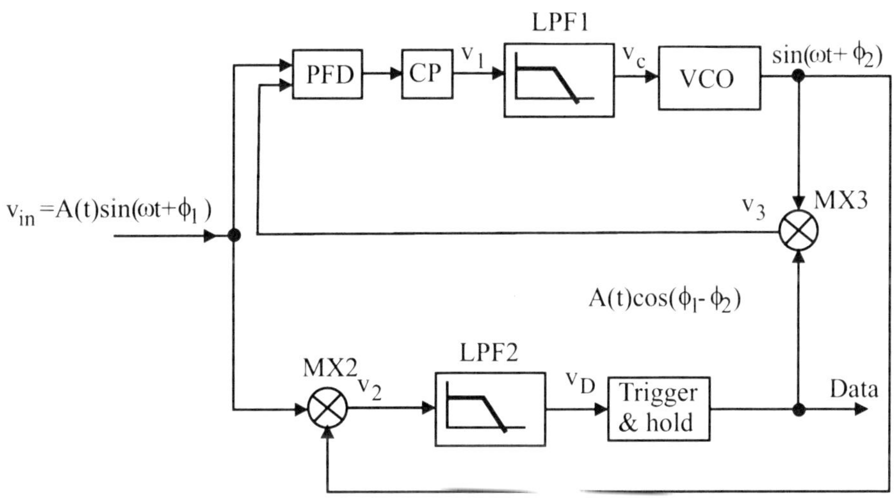

Figure 4.48. BPSK demodulation using modified Costas loop proposed by Luo and Sonkusale [124, 125].

4.4.3 Non-Coherent BPSK Demodulators

The preceding BPSK demodulators fall into the category of coherent demodulators as a local oscillator whose frequency is the same as the carrier frequency is required for demodulation. The power consumption of coherent BPSK modulators is generally high due to the need for a Costas loop. To lower the power consumption, non-coherent BPSK demodulators have emerged for biomedical implants. Non-coherent BPSK demodulators recover baseband data from a BPSK-modulated carrier without using a Costas loop.

The configuration of the non-coherent BPSK demodulator proposed by Asgarian and Sodagar is shown in Fig.4.49 [126, 127]. The carrier frequency is 8 MHz, the same as that of the baseband data. The LC tank is tuned to the carrier frequency such that it picks up and amplifies the incoming BPSK-modulated signal. The comparator acts as a 1-bit analog-to-digital converter that converts the incoming BPSK-modulated analog signal to a full-swing square wave for the purpose of convenience in signal handling and processing.

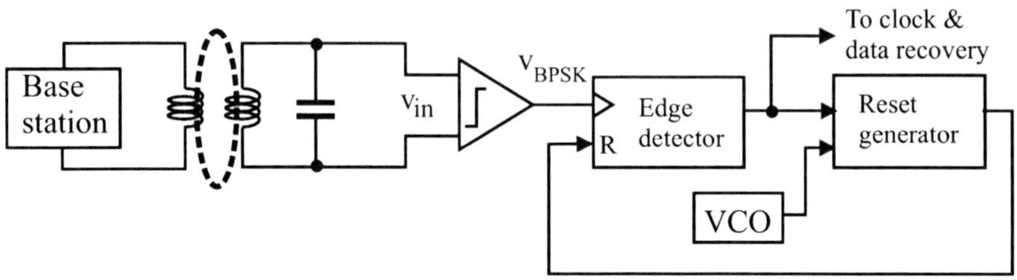

Figure 4.49. Simplified configuration of non-coherent BPSK demodulator proposed by Asgarian and Sodagar [126, 127].

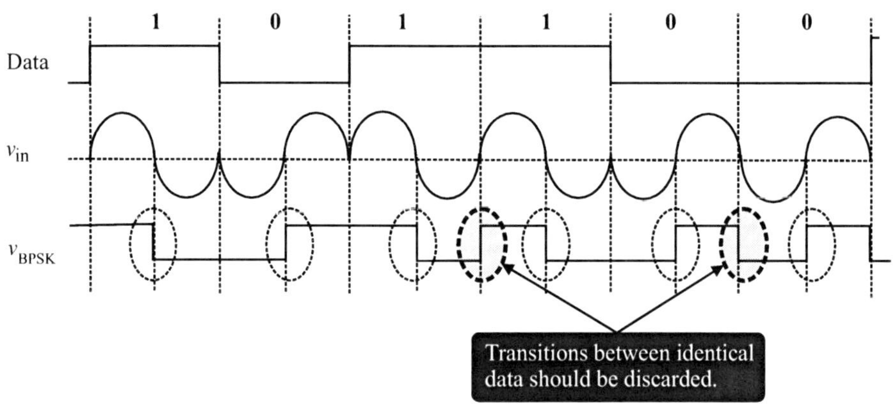

Figure 4.50. Waveforms of non-coherent BPSK demodulator proposed by Asgarian *et al.*.

Since the frequency of the carrier is the same as that of the baseband data, as shown in Fig.4.50, binary 1 is represented by a high-to-low transition and binary 0 is represented by a low-to-high transition. An edge detector can be employed to detect the arrival of a rising/falling edge so as to determine the bit value of the data. Note that the distance between neighboring transitions is one bit time. If the distance is half the bit time, the transition must be discarded, as shown in Fig.4.50.

The transition detector can be implemented using two DFFs with a reset control, as illustrated in Fig.4.51. To correctly detect high-to-low and low-to-high transitions, the two DFFs must be reset before the arrival of a transition edge. Specifically, the reset signal should be asserted in the time interval bounded by the end of each symbol and the beginning of the transition edge of the next symbol. To generate the reset signal, a 3-bit counter driven by an oscillator is employed. The counter starts counting immediately after a transition edge is detected. The frequency of the oscillator is set in such a way that the most significant bit of the counter becomes 1 when $T_b/2 < t < T_b$ with $t = 0$ at the start of the counting. Mathematically, $3T_{vco} > 0.5T_b$ and $4T_{vco} < T_b$ must be satisfied, where T_b is the bit time of the baseband data and T_{vco} is the period of the oscillator. It follows that $4f_b < f_{vco} < 6f_b$, where $f_b = 1/T_b$ and $f_{vco} = 1/T_{vco}$. The waveforms of the edge detector are shown in Fig.4.52.

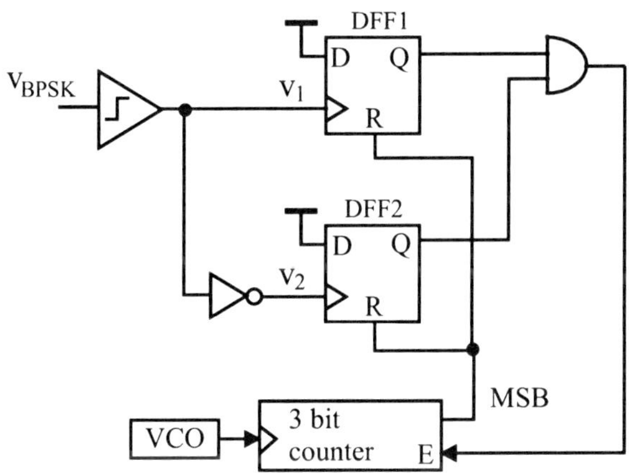

Figure 4.51. Edge detector using two DFFs and reset generator.

A drawback of this approach is the need for an oscillator whose frequency is approximately 5 times that of the carrier. The power consumption of the oscillator is significant. Gong *et al.* proposed an elegant way to generate the reset signal of the DFFs of the edge detector without using an oscillator and a counter, as shown in Fig.4.53 [128]. The output of the two DFFs are ANDed and used to charge capacitor C. A single-ended Schmitt trigger is used to boost the voltage of the capacitor to full swing. The output of the Schmitt trigger is used to reset the DFFs and sample the received BPSK-modulated signal. The waveforms of the critical nodes of Gong PSK demodulator are sketched in Fig.4.54. The elimination of the oscillator in Gong PSK demodulator greatly lowers the power consumption.

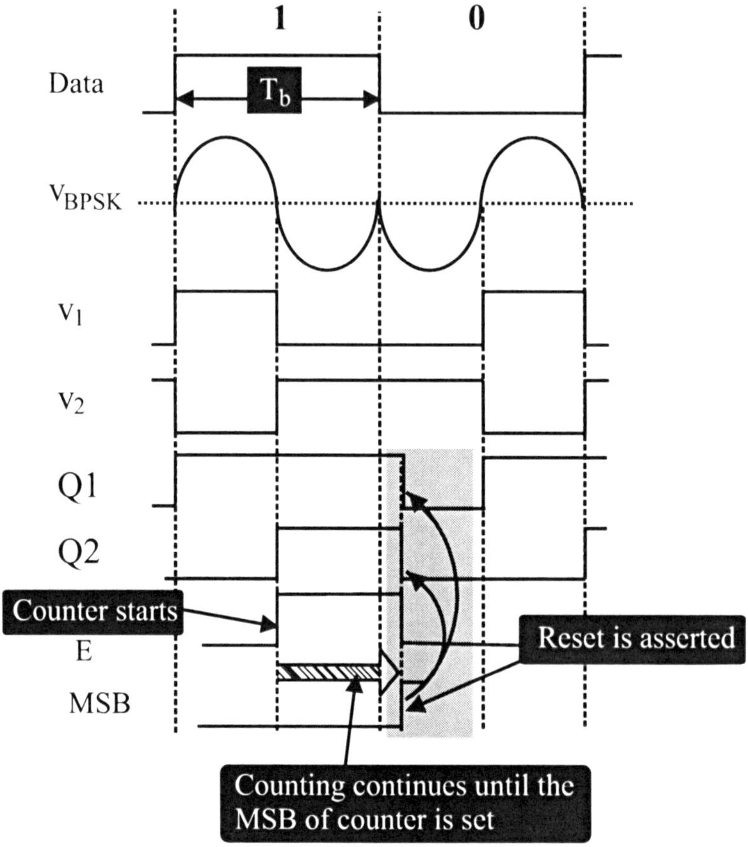

Figure 4.52. Waveform of edge detector using two DFFs.

4.4.4 Performance Comparison of PSK Demodulators

Table 4.3 tabulates the key parameters of some recently reported PSK demodulators for biomedical implants. It is seen that the carrier frequency of PSK-modulated signals is typically in low MHz ranges, the same as FSK-modulated signals. This is mainly due to the absorption of electromagnetic waves by living bodies at high frequencies and the low self-resonant frequency of planar coils between base stations and biomedical implants. The data rate is typically in low Mbps ranges. The power consumption of coherent PSK demodulators is high as compared with that of non-coherent PSK demodulators. The power consumption of Gong PSK demodulator is lower as compared with that of Asgarian-Sodagar PSK demodulator due to the elimination of the high-frequency oscillator.

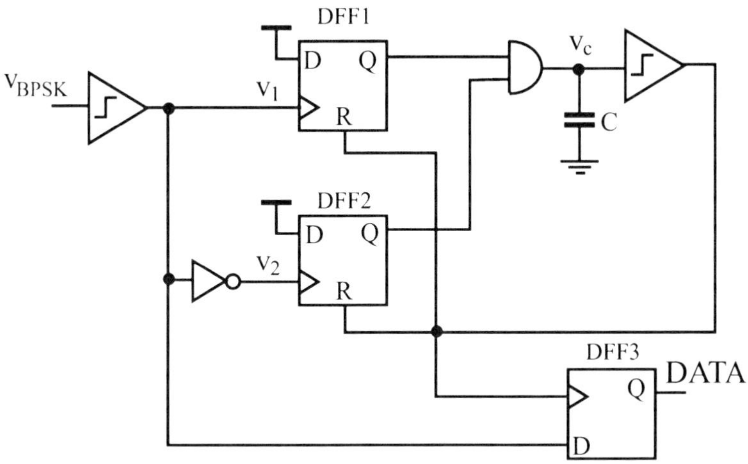

Figure 4.53. Non-coherent BPSK demodulator proposed by Gong *et al.* [128]

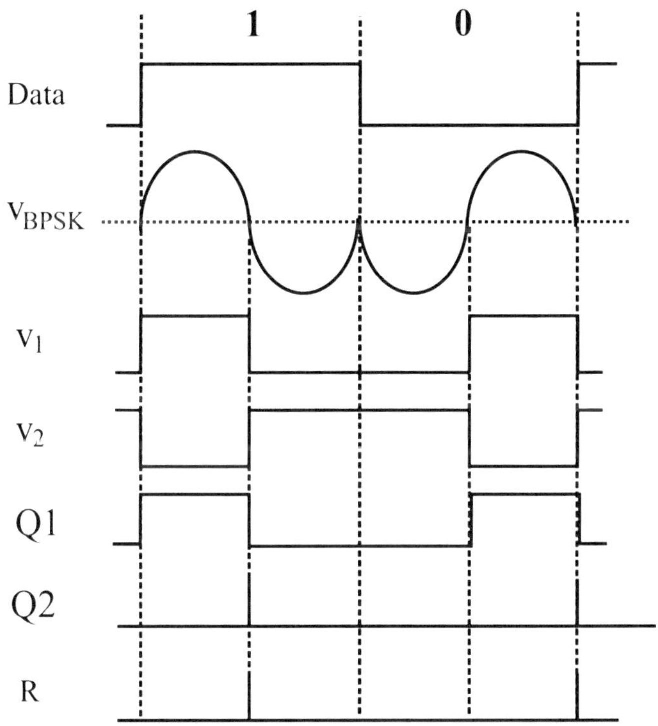

Figure 4.54. Waveforms of non-coherent BPSK demodulator proposed by Gong *et al.*.

4.5 Chapter Summary

Modulation schemes and the design of modulators and demodulators for passive wireless microsystems have been studied. We have shown that ASK is

Table 4.3. Performance comparison of BPSK demodulators for passive wireless microsystems.

Ref.	Tech.	PSK	Det.	Carrier [MHz]	Data rate [Mbps]	Power [mW]
Hu & Sawan [121](05)	0.18μm	BPSK	Coh.	10	1.12	0.61
Deng *et al.*[129](06)	0.18μm	QPSK	Coh.	13.56	4	0.75
Lu-Sawan [130](08)	0.18μm	OQPSK	Coh.	13.56	8	0.68
Gong *et al.* [128](08)	0.18μm	BPSK	Ncoh	4	0.8	0.059
Asgarian & Sodagar [126](09)	0.18μm	BPSK	Ncoh.	8	8	0.148
Asgarian & Sadagar [127](09)	0.18μm	BPSK	Ncoh.	10	10	0.232

Legends :
1. Coh : coherent.
2. Ncoh : non-coherent.

more susceptible to noise and disturbances as compared with FSK and PSK. The effect of noise and disturbances is particularly severe if modulation index is small. To improve BER, ASK with a large modulation index is preferred. This is at the cost of the increased fluctuation of power transmission from base stations to passive wireless microsystems. In addition, ASK suffers from the drawback of a low data rate, mainly due to the difficulties encountered in extracting the envelope of ASK-modulated RF signals at high data rates. Moreover, ASK is silicon-consuming especially when carrier frequency is low, mainly due to the need for low-pass filters with large time constants to perform envelope extraction. Finally, ASK is also not particularly attractive for applications where only a low supply voltage is available. Despite of the aforementioned drawbacks, ASK demodulators enjoy the advantage of simple configurations subsequently low power consumption, and are the most widely used in passive wireless microsystems.

ASK demodulators encountered in passive wireless microsystems can be loosely classified into voltage-mode, current-mode, and mixed-mode. Current-mode ASK demodulators are particularly attractive for applications where signals to be processed are currents and the supply voltage is low. Mixed-mode ASK demodulators take the advantages of both voltage-mode and current-mode circuits and are capable of operating at high frequencies while providing a large signal swing.

Unlike ASK, FSK is a constant envelope modulation scheme. The amount of the power transferred from base stations to passive wireless microsystems with FSK modulation is independent of baseband data. Also, the effect of noise and disturbances on FSK-modulated data is much smaller as compared with that on ASK-modulated data. FSK is capable of transmitting data at a higher rate because FSK demodulation does not need to extract the envelope of the carrier using low-pass filters with large time constants. FSK has been primarily used in biomedical implants due to the need for a continuous flow of power from base stations to implants, a high data rate, and the high degree of the absorption of electromagnetic waves by living bodies at high frequencies. Since the carrier of a FSK-modulated signal contains two frequencies that are typically far apart, the quality factor of the coils coupling base stations and biomedical implants must be moderate such that the coupling factors of the coils at the two FSK carrier frequencies are equally high. This, however, is at the expense of a reduced power transmission efficiency. FSK is also particularly suitable for low-voltage applications. A FSK-modulated signal can be demodulated using a phase-locked loop. This is at the cost of high silicon and power consumption. FSK-demodulated signals can also be recovered using differentiators and mixers. The need for mixers increases the power consumption, making it less attractive for passive wireless microsystems. Ghovanloo-Najafi FSK demodulator recovers baseband data by measuring the periods of the square waves corresponding to binary 1 and binary 0 using an oscillator whose frequency is much higher than that of the carrier frequencies. The need for a high-frequency oscillator, however, increases the overall power consumption. The FSK demodulator proposed by Jung *et al.* and that proposed by Weng recover the data by using a sampling technique. By properly controlling the sampling time, the data can be recovered correctly. The absence of a high-frequency oscillator in these demodulators lowers the power consumption. A drawback of this approach is the small timing margins for sampling, especially when carrier frequencies are high.

The two unique characteristics of a PSK-modulated carrier are the constant amplitude and frequency of the carrier. The former ensures that the flow of the power from a base station to a passive wireless microsystem is independent of baseband data while the latter allows the transmission of power from a base station to a passive wireless microsystem to be maximized by optimizing the quality factor of the coupling coils at the carrier frequency. BPSK is the most robust PSK and offers the highest degree of immunity to noise and disturbances. BPSK-modulated data can be recovered using either coherent detection and non-coherent detection. The former requires the use of a Costas loop that contains a local oscillator whose frequency is identical to the carrier frequency. The performance of Costas-based BPSK demodulators is severely affected by the frequency drift of the local oscillator caused by PVT and the small acquisition range. BPSK demodulators based on a hard-limited Costas loop increase the acquisition range. When phase/frequency detectors are employed in Costas loops, the frequency of the local oscillator can be tuned to the carrier frequency with a large acquisition range. A major drawback of

Costas-loop BPSK demodulators is the high power consumption. To reduce the power consumption of BPSK demodulators, techniques based on non-coherent detection have been proposed. The configuration of these BPSK demodulators is significantly simpler as compared with Costas-based BPSK demodulators subsequently consume a less amount of power. The non-coherent BPSK demodulator proposed by Asgarian and Sodagar recovers baseband data by distinguishing the type of transitions (low-to-high and high-to-low) for binary 1 and binary 0 of baseband data. The method requires that the frequency of the carrier be the same as that of the baseband data. The need for an oscillator whose frequency is approximately 5 times that of the carrier for edge detection results in a high level of power consumption. The non-coherent BPSK demodulator proposed by Gong *et al.* detects the type of transitions without using a high-frequency oscillator. As a result, its power consumption is greatly reduced.

Chapter 5

LOW-POWER PRECISION VOLTAGE REFERENCES

Temperature-insensitive precision voltage references are critically needed for passive wireless microsystems. Although bandgap voltage references are widely studied and numerous designs are available, the rapid reduction of supply voltages, the need for precision voltage references, and the low-power consumption requirement of passive wireless microsystems impose stringent constraints on the design of voltage references. This chapter deals with the principles and design of CMOS voltage references. The chapter starts with a brief examination of the figure-of-merits that characterize the performance of voltage references in Section 5.1. It is followed by an in-depth investigation of the temperature-dependent characteristics of semiconductors and MOS devices in Section 5.2. First-order voltage references are studied in Section 5.3 while high-order voltage references are dealt with in Section 5.4. The performance of recently reported first-order voltage reference and that of high-order voltage references such as temperature coefficient, power consumption, power supply rejection ratio, and minimum supply voltage are compared. Section 5.5 focuses on the design of ultra low-power voltage references where devices operate in weak inversion. These voltage references are of particular importance to passive wireless microsystems. The chapter is concluded in Section 5.6.

5.1 Characterization of Voltage References

There are a number of metrics that are often used to quantify the performance of a voltage reference. These metrics provide a quantitative measure of the effect of temperature variation and supply voltage fluctuation on the output of the voltage reference. In this section, we examine some of these metrics, specifically temperature coefficient, power supply rejection ratio, and the minimum supply voltage.

F. Yuan, *CMOS Circuits for Passive Wireless Microsystems*,
DOI 10.1007/978-1-4419-7680-2_5,

5.1.1 Temperature Coefficient

The temperature coefficient (TC) of a voltage reference quantifies the effect of temperature on the output voltage of the reference. Two temperature coefficients, namely normalized temperature coefficient, also known as fractional temperature coefficient, and normalized average temperature coefficient, also known as effective temperature coefficient, are often used. The fractional temperature coefficient is defined as

$$\mathrm{TC}_f = \frac{1}{V_{ref}(T)} \frac{\partial V_{ref}(T)}{\partial T}, \tag{5.1}$$

where V_{ref} is the output voltage of the reference and T is temperature. The fractional temperature coefficient quantifies the dependence of V_{ref} on temperature at a specific temperature. Since V_{ref} usually varies with temperature in a nonlinear fashion, to effectively quantify the dependence of V_{ref} on temperature over a specific temperature range, the normalized average temperature coefficient defined as

$$\mathrm{TC}_{eff} = \frac{1}{V_{ref,avg}(T_{max} - T_{min})} \int_{T_{min}}^{T_{max}} V_{ref}(T)dT \tag{5.2}$$

should be employed. Here $V_{ref,avg}$ is the average value of V_{ref} over the specified temperature range, T_{min} and T_{max} are the minimum and maximum temperature of the temperature range, respectively. Since we usually do not know the analytical expression of $V_{ref}(T)$, the following approximation is used to estimate the normalized average temperature coefficient

$$\mathrm{TC}_{eff} \approx \frac{1}{V_{ref,avg}} \left(\frac{V_{ref,max} - V_{ref,min}}{T_{max} - T_{min}} \right), \tag{5.3}$$

where $V_{ref,min}$ and $V_{ref,max}$ are the minimum and maximum output voltages of the reference over the temperature range, respectively. Note $V_{ref,avg}$ in (5.3) is often replaced with the desired reference voltage.

Both the fractional temperature coefficient and effective temperature coefficient have the unit ppm/°C (ppm is the abbreviation of *parts per million*). Often, the maximum variation of the reference voltage over a specific temperature range is used to quantify the effect of temperature on reference voltages in engineering reports and scientific publications. It can be easily converted to ppm/°C. For example, if the maximum variation of the output voltage of a voltage reference over $-20\sim100$°C is 5 mV and the average reference voltage over the temperature range is 1.2 V, the effective temperature coefficient of the reference voltage is calculated from

$$\frac{5 \times 10^{-3}\text{mV}}{1.2\text{V} \times 120°\text{C}} = 3.47 \times 10^{-5}/°C = 34.7 \text{ ppm}/°\text{C}.$$

5.1.2 Power Supply Rejection Ratio

Power supply rejection ratio (PSRR) is a measure of how well a voltage reference can reject the voltage ripples coming from the power supply. It is defined as

$$\text{PSRR} = 20 \log \left(\frac{\partial V_{ref}}{\partial V_{DD}} \right) \quad (\text{dB}). \tag{5.4}$$

PSRR is a small-signal figure-of-merit and varies with frequency. Since references are typically designed at dc, PSRR at dc is often used to compare the performance of voltage references. At high frequencies, because both intrinsic and parasitic capacitances of MOSFETs provide additional paths from the power supply to the output of the references, PSRR drops with frequencies.

5.1.3 Minimum Supply Voltage

The output voltage of a voltage reference should be independent of the supply voltage ideally. Although a lower supply voltage is always desirable for a given voltage reference, there exists a minimum supply voltage below which V_{ref} is no longer constant. This minimum supply voltage is an indicator of the low-voltage capability of the voltage reference, and is widely used as a figure-of-merit to quantify the performance of the voltage reference. Since $V_{ref}(T)$ is temperature-dependent, the minimum supply voltage is typically measured at room temperature. It is obtained by sweeping V_{DD} from 0 to the nominal supply voltage of the given technology while recording V_{ref}, as illustrated in Fig.5.1.

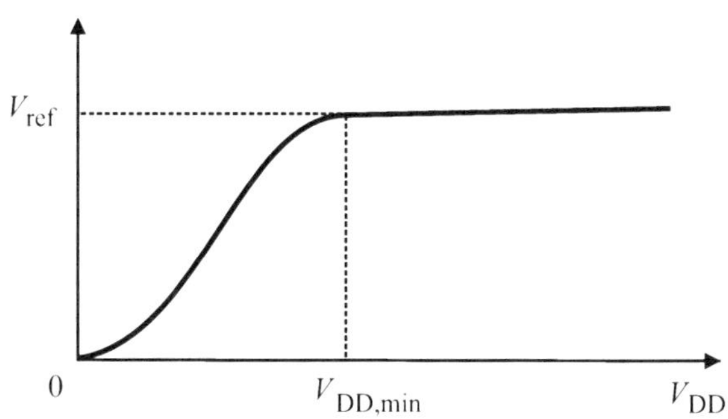

Figure 5.1. Dependence of V_{ref} on supply voltage.

5.2 Temperature Characteristics of MOS Devices

The characteristics of semiconductor devices, such as the mobility of minority charge carriers, the bandgap voltage of silicon, the forward basing voltage of pn-junctions, the threshold voltage of MOSFETs, the gate-source voltage of MOSFETs in weak inversion, and the resistance of poly and diffusion resistors are temperature-dependent. Some of them increase with temperature while others decrease with temperature. The temperature-dependent characteristics of semiconductor devices form the foundation of temperature-compensated voltage references. In this section, we investigate the temperature characteristics of MOS devices in detail. The utilization of these properties in design of temperature-compensated voltage or current references will be dealt with in Sections 5.3, 5.4, and 5.5.

5.2.1 Base-Emitter Voltage of BJTs

The collect current I_C of a bipolar junction transistor (BJT) is related to its base-emitter voltage V_{BE} by

$$I_C \approx I_S e^{V_{BE}/V_t}, \tag{5.5}$$

where I_S is the saturation current of the base-emitter pn-junction,

$$V_t = \frac{kT}{q} \tag{5.6}$$

is the thermal voltage, $k = 1.38 \times 10^{-23}$ V.C/°K is Boltsmann's constant, $q \approx 1.6 \times 10^{-16}$ C is the charge of an electron, and T is the absolute temperature in degrees Kelvin. $V_t \approx 25.9$ mV at room temperature (300°K). The saturation current I_S is given by [131]

$$I_S = \frac{qA_e D_n n_i^2}{W_b N_A}, \tag{5.7}$$

where A_e is the cross-sectional area of the emitter, D_n is the effective diffusion constant of electrons, n_i is the concentration of intrinsic carriers in silicon ($n_i \approx 1.5 \times 10^{10}/\text{cm}^3$ at 300°K [131]), W_b is the width of the base, and N_A the doping of the acceptors in the base. The relation between the effective diffusion constant and the mobility of electrons is characterized by Einstein relation [132]

$$D_n = V_t \mu_n. \tag{5.8}$$

The mobility of electrons is temperature-dependent with its dependence depicted by [132]

$$\mu_n(T) = \mu_n(T_o)\left(\frac{T}{T_o}\right)^m, \tag{5.9}$$

where $\mu_n(T)$ and $\mu_n(T_o)$ are the mobility of electrons at temperature T and reference temperature T_o, respectively. $m\approx -1.2 \sim -2.0$, often $m = -1.5$ is used. T_o is often set to room temperature as the performance of integrated circuits (ICs) is typically optimized at room temperature. The concentration of the intrinsic carriers in silicon is also a function of temperature [73]

$$n_i^2 = c_1 T^3 e^{-\frac{E_g}{kT}}, \tag{5.10}$$

where c_1 is a constant,

$$E_g = qV_G \tag{5.11}$$

is the bandgap energy of silicon, and V_G is the bandgap voltage of silicon. $E_g\approx 1.12$ eV (electron volt) at 300°K. Note 1 eV is the amount of the kinetic energy gained by an electron when accelerated through an electric field of a potential difference of one volt, i.e.

$$1\text{eV} = 1\text{V}\times 1\text{C}\approx 1.6\times 10^{-19}\text{Joule}. \tag{5.12}$$

Re-write (5.7) by utilizing (5.9) and (5.10)

$$I_S = c_2 T^{m+4} e^{-\frac{E_g}{kT}}, \tag{5.13}$$

where

$$c_2 = \frac{c_1 k A_e \mu_n(T_o)}{T_o^m W_b N_A}. \tag{5.14}$$

Eq.(5.13) shows that I_S is a strong function of temperature. The degree of the dependence of I_S on temperature can be quantified by differentiating I_S with respect to T

$$\frac{\partial I_S}{\partial T} = c_2(m+4)T^{m+3}e^{-\frac{E_g}{kT}} + c_2 T^{m+4} e^{-\frac{E_g}{kT}}\frac{E_g}{kT^2}. \tag{5.15}$$

We will make use of (5.15) shortly. Let us first solve (5.5) for V_{BE}

$$V_{BE} = V_t \ln\left(\frac{I_C}{I_S}\right). \tag{5.16}$$

Differentiate (5.16) with respect to T

$$\frac{\partial V_{BE}}{\partial T} = \frac{\partial V_t}{\partial T}\ln\left(\frac{I_C}{I_S}\right) - \frac{V_t}{I_S}\frac{\partial I_S}{\partial T}. \tag{5.17}$$

Substitute (5.15) into (5.17)

$$\frac{\partial V_{BE}}{\partial T} = \frac{1}{T}\left[V_{BE} - (m+4)V_t - \frac{E_g}{q}\right]. \tag{5.18}$$

To find out the typical value of $\frac{\partial V_{BE}}{\partial T}$ at room temperature (300°K), we use $V_{BE} = 750$ mV and $m = -1.5$. Since

$$\frac{E_g}{q} \approx 1.12\text{V}, \tag{5.19}$$

we obtain from (5.18) that

$$\frac{\partial V_{BE}}{\partial T} \approx -1.45\text{mV}/^{\circ}\text{K}. \tag{5.20}$$

The base-emitter voltage of bipolar junction transistors has a negative temperature coefficient. The temperature dependence of V_{BE} of BJTs on temperature with a negative temperature coefficient is one of the fundamental principles upon which temperature-insensitive voltage references are based.

So far, we have learned that the base-emitter voltage of bipolar junction transistors decreases with temperature. We, however, do not know whether the dependence of V_{BE} on temperature is linear or nonlinear. Such information is critical if one wants to generate a correction voltage that offsets the effect of temperature on V_{BE} such that a temperature-independent voltage can be generated. In what follows we investigate the exact relation between V_{BE} and temperature. We follow the approach by Brugler [133] and Tsividis [134].

Writing (5.16) at a reference temperature T_o, typically room temperature, and an arbitrary temperature T and subtracting the results give

$$V_{BE}(T) = \frac{T}{T_o}V_{BE}(T_o) + \frac{kT}{q}\ln\left[\frac{I_C(T)I_S(T_o)}{I_C(T_o)I_S(T)}\right], \tag{5.21}$$

where $V_{BE}(T)$ and $V_{BE}(T_o)$ denote V_{BE} at T and T_o, respectively. Similar conventions apply to I_C and I_S as well. Making use of the expression of I_S derived earlier, i.e.

$$\begin{aligned} I_S &= \frac{qA_e n_i^2 D}{W_b N_A} \\ &= \frac{qA_e c_1 T^3 e^{-\frac{qV_G}{kT}} \frac{kT}{q}\mu_n}{W_b N_A}, \end{aligned} \tag{5.22}$$

we have

$$\frac{I_S(T_o)}{I_S(T)} = \left(\frac{T}{T_o}\right)^4 e^{-\frac{q}{k}\left[\frac{V_G(T_o)}{T_o} - \frac{V_G(T)}{T}\right]} \frac{\mu_n(T_o)}{\mu_n(T)}. \tag{5.23}$$

Making use of (5.23), we can write (5.21) as

$$\begin{aligned} V_{BE}(T) &= V_G(T) + \frac{T}{T_o}\left[V_{BE}(T_o) - V_G(T_o)\right] + \frac{4kT}{q}\ln\left(\frac{T_o}{T}\right) \\ &+ \frac{kT}{q}\ln\left[\frac{\mu_n(T_o)}{\mu_n(T)}\right] + \frac{kT}{q}\ln\left[\frac{I_C(T)}{I_C(T_o)}\right], \end{aligned} \tag{5.24}$$

where $V_G(T)$ and $V_G(T_o)$ are the bandgap voltage of silicon at T and T_o, respectively. Substitute (5.9) into (5.24)

$$\begin{aligned} V_{BE}(T) &= V_G(T) + \frac{T}{T_o}\left[B_{BE}(T_o) - V_G(T_o)\right] \\ &- (m+4)\frac{kT}{q}\ln\left(\frac{T}{T_o}\right) + \frac{kT}{q}\ln\left[\frac{I_C(T)}{I_C(T_o)}\right]. \end{aligned} \tag{5.25}$$

If we assume that the dependence of the collector current on temperature is given by

$$I_C(T) = FT^\delta, \tag{5.26}$$

where F and δ are constants [134, 131], we can write (5.25) as

$$\begin{aligned} V_{BE}(T) = V_G(T) &+ \frac{T}{T_o}\left[V_{BE}(T_o) - V_G(T_o)\right] \\ &- (m+4-\delta)\frac{kT}{q}\ln\left(\frac{T}{T_o}\right). \end{aligned} \tag{5.27}$$

If the collector current of the BJT is independent of temperature, i.e. $\delta = 0$, (5.27) becomes

$$V_{BE}(T) = V_G(T) + \frac{T}{T_o}\left[V_{BE}(T_o) - V_G(T_o)\right] - (m+4)\frac{kT}{q}\ln\left(\frac{T}{T_o}\right). \tag{5.28}$$

If the collector current of the BJT is proportional to temperature, i.e. $\delta = 1$, we have

$$V_{BE}(T) = V_G(T) + \frac{T}{T_o}\left[V_{BE}(T_o) - V_G(T_o)\right] - (m+3)\frac{kT}{q}\ln\left(\frac{T}{T_o}\right). \tag{5.29}$$

Fig.5.2 plots (5.29) with $V_G(T)$ modeled using Tsividis's bandgap voltage model, specifically

$$V_G(T) = 1.20595 - 2.7325\times10^{-4}T \tag{5.30}$$

with $300°\text{K} \leq T \leq 400°\text{K}$ [134], and $T_o = 300°\text{K}$, $m = -1.5$, $V_{BE}(T_o) = 0.7$ V. The nonlinear portion is multiplied by 100 so that it can be seen clearly in the figure. As can be seen from the figure that V_{BE} drops with temperature approximately linearly. The nonlinear portion of V_{BE} is significantly smaller as compared with that of the linear portion of V_{BE}.
Eq.(5.27) can also be written as

$$\begin{aligned} V_{BE}(T) &= V_G(T) \\ &+ \left[\frac{V_{BE}(T_o) - V_G(T_o)}{T_o} + (m+4-\delta)\frac{k}{q}\ln(T_o)\right]T \\ &- \left[(m+4-\delta)\frac{k}{q}\right]T\ln(T). \end{aligned} \tag{5.31}$$

The only unknown function in (5.31) is the bandgap voltage of silicon $V_G(T)$.

Numerous studies confirm that the bandgap voltage of silicon decreases when temperature rises. The dependence of $V_G(T)$ on temperature is often quantified using Varshni's formula [135]

$$V_G(T) = V_G(0) - \frac{\alpha T^2}{T+\beta}, \tag{5.32}$$

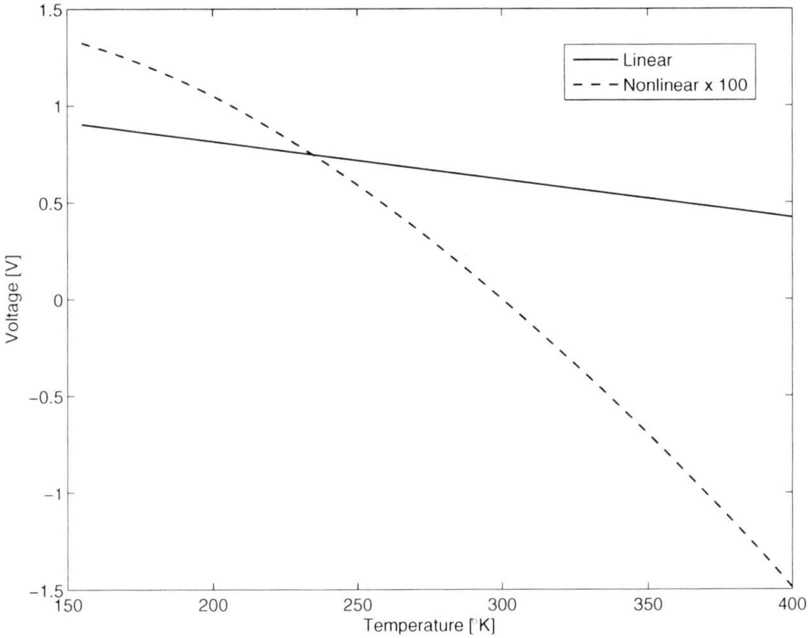

Figure 5.2. Dependence of V_{BE} on temperature when collector current is proportional to temperature. The nonlinear portion is multiplied by 100 so that it can be seen clearly.

where $V_G(0) = 1.170$ V is the bandgap voltage of silicon extrapolated to 0 °K, $\alpha = 4.73\times10^{-4}$ V/°K, and $\beta = 636$°K [134]. As pointed out by Tsividis in [134] that the values of the parameters in Varshni's formula were chosen to best fit to the measurement results in the temperature range $0 \sim 400$°K. They are by no means the best values for ICs, which normally operate in the temperature range $200 \sim 400$°K.

Bludau and Onton showed that the dependence of the bandgap voltage of silicon on temperature can be modeled using the following 2nd-order model [136]

$$V_G(T) = a - bT - cT^2, \tag{5.33}$$

where a, b, and c are empirical fitting constants with their values tabulated in Table 5.1. Tsividis modified the values of a, b, and c by noting the lack of accurate measurement at high temperature and the fact that most of the nonlinear variation of the bandgap voltage only takes place below 300°K. The modified values of a, b, and c are given in Table 5.2.

Table 5.1. Coefficients of Bludau-Onton model of bandgap voltage of silicon.

Temp. range	a	b	c
190∼300°K	1.17V	1.059×10^{-5}V/°K	-6.05×10^{-7}V/°K
150∼400°K	1.1785V	-9.025×10^{-4}V/°K	-3.05×10^{-7}V/°K

Table 5.2. Coefficients of Tsividis model of bandgap voltage of silicon.

Temp. range	a	b	c
150∼300°K	1.1785V	9.025×10^{-5}V/°K	3.05×10^{-7}V/°K
300∼400°K	1.20595V	2.7325×10^{-4}V/°K	0

Difficulties are encountered when applying the preceding 2nd-order models of the bandgap voltage of silicon as two sets of parameters must be used. Lin and Salama showed that the expressions of Bludau-Onton model can be replaced with the following single expression

$$V_G(T) = k_1 + k_2T + k_3T\ln(T) \tag{5.34}$$

that is valid over 150°K≤T≤400°K with $k_1 = 1.1774$ V, $k_2 = 3.042\times10^{-4}$ V/°K, and $k_3 = -8.459\times10^{-5}$ V/°K [137]. A key advantage of Lin-Salama model is that it allows us to combine $V_G(T)$ with $V_{BE}(T)$ as both have the same expression such that only a single correction voltage needs to be generated. Fig.5.3 provides a numerical comparison of these models. It is observed from the figure that the difference between these model is rather small.

Having obtained the mathematical approximation of the bandgap voltage silicon, we now show that V_{BE} can be represented by the following polynomial series of temperature

$$V_{BE} = a_0 + a_1T + a_2T^2 + ... \tag{5.35}$$

We first replace $T\ln(T)$ in (5.31) with its Taylor series expansion at reference temperature T_o

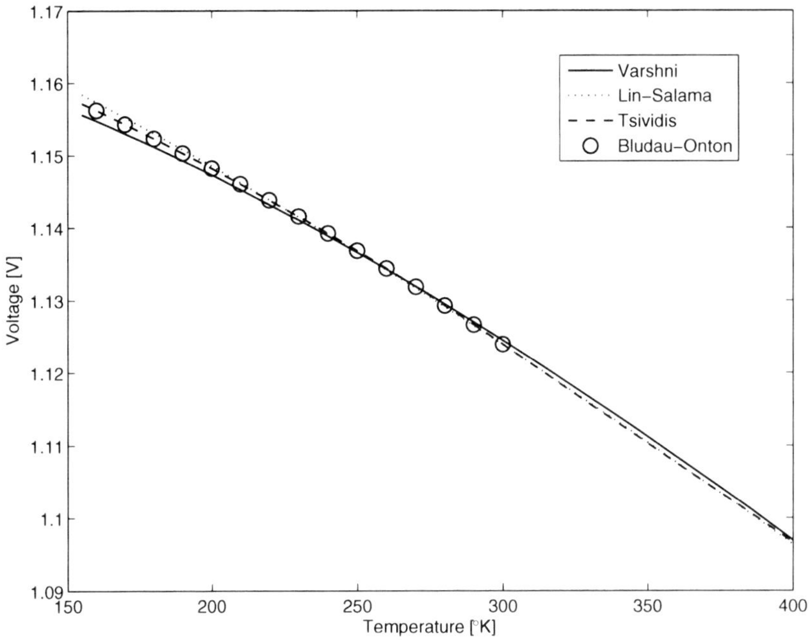

Figure 5.3. Comparison of models of bandgap voltage of silicon.

$$T\ln(T) = T_o\ln(T_o) + [1 + \ln(T_o)](T - T_o) + \frac{1}{T_o}(T - T_o)^2 + \ldots \quad (5.36)$$

Secondly, we use Lin-Salama model for the bandgap voltage of silicon and replace its nonlinear term with its Taylor series expansion at T_o

$$\begin{aligned} V_G(T) &= k_1 + k_2T \\ &+ k_3\left[T_o\ln(T_o) + [1 + \ln(T_o)](T - T_o) + \frac{1}{T_o}(T - T_o)^2 + \ldots\right]. \end{aligned} \quad (5.37)$$

Substituting (5.27) and (5.37) into (5.31) will yield (5.35).

5.2.2 Threshold Voltage of MOSFETs

The threshold voltage of MOS transistors is given by

$$V_T = V_{To} + \gamma\left(\sqrt{V_{SB} + 2\phi_F} - \sqrt{2\phi_F}\right), \quad (5.38)$$

where V_{SB} is the source-substrate voltage, ϕ_F is Fermi potential calculated from

$$\phi_F = V_t \ln\left(\frac{N_A}{n_i}\right) \tag{5.39}$$

with N_A the doping of the substrate, $\phi_F \approx 0.3$ V for a typical p-type silicon substrate [138], γ is the body effect constant calculated from

$$\gamma = \frac{\sqrt{2q\epsilon_{si}N_A}}{C_{ox}}, \tag{5.40}$$

with

$$C_{ox} = \frac{\epsilon_{ox}}{t_{ox}} \tag{5.41}$$

the gate capacitance per unit area, $\epsilon_{ox} = 3.5\times10^{-13}$ F/cm the dielectric constant of oxide, $\epsilon_{si} = 1.05\times10^{-12}$ F/cm the dielectric constant of silicon, and t_{ox} the thickness of the gate oxide. V_{To}, the threshold voltage when $V_{SB} = 0$, is a function of a number of parameters including Fermi potential [132, 138, 131]. Since ϕ_F is a strong function of temperature, the threshold voltage also varies with temperature. $V_T(T)$ decreases with temperature almost linearly and is typically depicted using the following first-order model [132]

$$V_T(T) = V_T(T_o) + \alpha_{V_T}(T - T_o), \tag{5.42}$$

where $V_T(T)$ and $V_T(T_o)$ are the threshold voltage at T and T_o, respectively, and α_{V_T} is the temperature coefficient of the threshold voltage. The typical value of α_{V_T} is in the range of -0.5 mV/°C $\sim$ -4 mV/°C with -2.4 mV/°C perhaps the most frequently used [131, 139, 140].

5.2.3 Gate-Source Voltage of MOSFETs in Weak Inversion

The small channel current of MOSFETs in weak inversion is the backbone of low power design. This unique characteristic of MOSFETs has been utilized recently in low-noise amplifiers [141–143], oscillators [144], and receivers [145], to name a few. A number of in-depth studies on the behavior of MOSFETs in weak inversion, in particular, the channel current, are available [131, 146, 132, 147]. In this section, we utilize the widely used EKV model, developed by C. Enz, F. Krummenacher and E. Vittoz [148], to investigate the temperature dependence of the gate-source voltage of MOSFETs in weak inversion.

The channel current of an nMOS transistor in weak inversion is given by [146]

$$I_D = I_{D0} e^{\frac{V_{GS}-V_T}{nV_t}} \left(1 - e^{-\frac{V_{DS}}{V_t}}\right), \tag{5.43}$$

where

$$I_{D0} = 2n\mu_n C_{ox} S V_t^2, \tag{5.44}$$

$$n = 1 + \frac{C_{ox}}{C_{js}}, \tag{5.45}$$

$$S = \frac{W}{L} \tag{5.46}$$

is the aspect ratio, C_{ox} is the gate-oxide capacitance per unit area, C_{js} is the capacitance of the depletion region under the channel as shown in Fig.5.4 graphically, and V_T is the threshold voltage of the transistor. Often $n = 1.5$ is used [149]. It is seen that the channel current of MOSFETs in weak inversion rises with V_{GS} in an exponential fashion. This bears a strong resemblance to $I_C \sim V_{BE}$ relation of BJTs.
When $V_{DS} \leq 3V_t \approx 78$ mV,

$$1 - e^{-\frac{V_{DS}}{V_t}} \approx 1,$$

(5.43) is simplified to

$$I_D \approx I_{D0} e^{\frac{V_{GS}-V_T}{nV_t}}. \tag{5.47}$$

I_D in this case becomes independent of V_{DS}, a distinct characteristic of MOSFETs in weak inversion. Another unique characteristic of MOSFETs in weak inversion is the dependence of the transconductance on the channel current

$$g_m = \frac{\partial I_D}{\partial V_{GS}} = \frac{I_D}{nV_t}. \tag{5.48}$$

Note the transconductance in strong inversion is given by

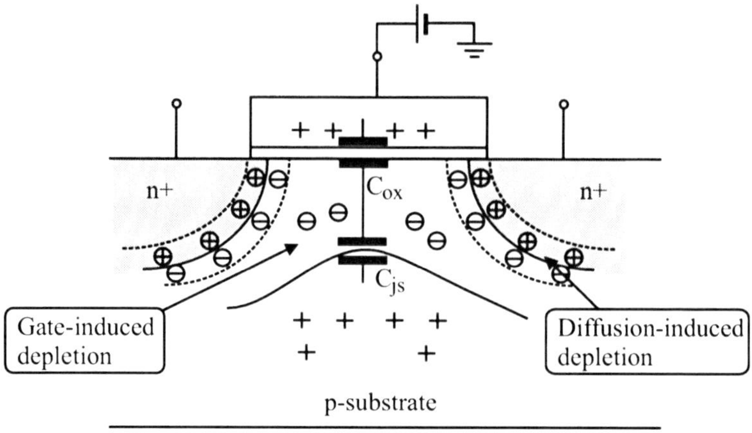

Figure 5.4. Cross-sectional view of MOSFETs.

$$g_m \approx \frac{2I_D}{V_{GS} - V_T}. \tag{5.49}$$

The boundary between weak and strong inversion regions is set by equating (5.48) and (5.49)

$$V_{GS} = V_T + 2nV_t. \tag{5.50}$$

Devices are therefore in weak inversion if $V_{GS} \ll V_T + 2nV_t$ and in strong inversion otherwise. Note when V_{GS} is in the neighborhood of $V_T + 2nV_t$, the device is said in moderate inversion. For this reason, it is strongly recommended to establish $V_{GS} \ll V_T + 2nV_t$ to ensure that devices do operate in weak inversion. It should also be noted that the small V_{GS} of devices in weak inversion enables us to lower the minimum V_{DD}.

Having briefly reviewed the characteristics of MOSFETs in weak inversion, we now turn our attention to the effect of temperature on the gate-source voltage V_{GS} of MOSFETs in weak inversion. The temperature dependence of the channel current of MOSFETs in weak inversion is evident from (5.44) and (5.47), specifically, μ_n, V_t, and V_T are all functions of temperature. Writing (5.47) at T and T_o, and dividing the results yield

$$\frac{I_D(T)}{I_D(T_o)} = \frac{\mu_n(T)}{\mu_n(T_o)} \frac{T^2}{T_o^2} e^{\frac{q}{nk}\left[\frac{V_{GS}(T) - V_T(T)}{T} - \frac{V_{GS}(T_o) - V_T(T_o)}{T_o}\right]}. \tag{5.51}$$

Solving (5.51) for $V_{GS}(T)$ and making use of the temperature-dependence of the mobility given in (5.9) yield

$$\begin{aligned} V_{GS}(T) = & \frac{T}{T_o}V_{GS}(T_o) + \left(1 - \frac{T}{T_o}\right)\left[V_T(T_o) - \alpha_{V_T}T_o\right] \\ + & \frac{nkT}{q}\ln\left[\frac{I_D(T)}{I_D(T_o)}\left(\frac{T_o}{T}\right)^{\alpha}\right], \end{aligned} \tag{5.52}$$

where $\alpha = m + 2$. If the temperature dependence of I_D is given by

$$I_D = FT^{\delta}, \tag{5.53}$$

(5.52) becomes

$$\begin{aligned} V_{GS}(T) = & \left[V_T(T_o) + \alpha_{V_T}T_o\right] + \left[\frac{V_{GS}(T_o)}{T_o} + \alpha_{V_T} - \frac{V_T(T_o)}{T_o}\right]T \\ - & \frac{nkT}{q}(\delta - \alpha)\ln\left(\frac{T}{T_o}\right). \end{aligned} \tag{5.54}$$

Grouping the like terms of (5.54) yields

$$\begin{aligned} V_{GS}(T) = & V_T(T_o) + \alpha_{V_T}T_o \\ + & \left[\frac{V_{GS}(T_o)}{T_o} + \alpha_{V_T} - \frac{V_T(T_o)}{T_o} - \frac{nk(\delta - \alpha)}{q}\ln(T_o)\right]T \\ + & \frac{nk}{q}(\delta - \alpha)T\ln(T). \end{aligned} \tag{5.55}$$

It is evident from (5.55) that the effect of temperature on V_{GS} of MOSFETs in weak inversion bears a strong resemblance to that of V_{BE} of BJTs. Eq.(5.55) can be written in the following general form

$$V_{GS}(T) = a_0 + a_1T + a_2T\ln(T), \tag{5.56}$$

with a_0, a_1, and a_2 given by

$$\begin{aligned} a_0 &= V_T(T_o) + \alpha_{V_T}T_o, \\ a_1 &= \frac{V_{GS}(T_o)}{T_o} + \alpha_{V_T} - \frac{V_T(T_o)}{T_o} - \frac{nk}{q}(\delta - \alpha)\ln(T_o), \\ a_2 &= \frac{nk}{q}(\delta - \alpha). \end{aligned} \tag{5.57}$$

The polarity of a_1 at room temperature (300°K) is determined as the follows : Use $m = -1.5$, we have $\alpha = m + 2 = 0.5$. Also, the value of α_{V_T} is in the range -0.5 mV/°C∼-4 mV/°C. If $\delta = 0$, i.e. the channel current is constant, we have $a_1 < 0$. If $\delta = 1$, i.e. the channel current is PTAT,

$$\frac{nk}{q}(\delta - \alpha)\ln(T_o) \approx 3.69 \times 10^{-4}\text{mV}, \tag{5.58}$$

which is much smaller as compared with α_{V_T}. Clearly $a_1 < 0$ in this case. The polarity of the nonlinear term depends upon the value of δ. If $\delta = 0$, $a_2 > 0$. Otherwise, $a_2 < 0$. Since the value of the nonlinear term is much smaller compared with that of the linear term, V_{GS} of MOSFETs in weak inversion has a negative temperature coefficient. This is very similar to the temperature characteristics of V_{BE} of BJTs.

5.2.4 Resistance of Diffusion and Poly Resistors

Poly resistors and diffusion resistors are the most widely used resistors in CMOS technologies. Both silicide and non-silicide types of these resistors are usually available. The former has a low sheet resistance with a large variance while the latter offers a large sheet resistance with a small variance. For example, the sheet resistance of silicide poly resistors is approximately 8 Ω and that of non-sicilide poly resistors is approximately 3 kΩ in a typical 0.18μm CMOS technology. Diffusion resistors consist of both n+ diffusion and n-well resistors. The former offer a very low sheet resistance while the latter provide a large sheet resistance. p+ and p-well resistors also exist in triple-well processes. Diffusion resistors have a large junction capacitance between the diffusion area and the substrate, and their resistance varies with the voltage of their terminals in a nonlinear fashion [73]. The resistance of both poly and diffusion resistors is temperature-dependent and is typically quantified using the following second-order model [150]

$$R(T) \approx R(T_o)\left[1 + \alpha_1(T - T_o) + \alpha_2(T - T_o)^2\right], \tag{5.59}$$

where $R(T)$ and $R(T_o)$ are the resistance at temperature T and reference temperature T_o, respectively, and α_1 and α_2 are the first-order and second-order temperature coefficients, respectively. α_1 and α_2 have the units ppm/°C and ppm/°C^2, respectively. The value and polarity of α_1 and α_2 are process-dependent. The temperature coefficients of diffusion resistors are typically several times those of poly resistors. For example, α_1 of n-well resistors in a 0.25μm CMOS process is 3600 ppm/°C while that of high-resistive poly resistors is -620 ppm/°C [151]. The temperature coefficients of high-resistive poly resistors and p-diffusion resistors of a 0.6 μm CMOS technology are -1050 ppm/°C and 1650 ppm/°C, respectively [152].

5.2.5 PTAT Voltage / Current Generators

It was shown in the preceding sections that the base-emitter voltage of BJTs varies with temperature with a negative temperature coefficient. To create a temperature-independent reference voltage, a correction voltage with a positive temperature coefficient is clearly needed in order to offset the effect of temperature on V_{BE}. In this section, we investigate the circuits that generate a proportional-to-absolute-temperature (PTAT) voltage / current.

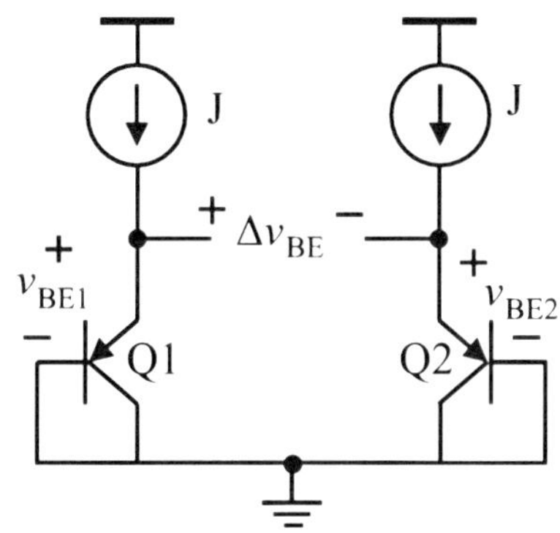

Figure 5.5. Circuit that generates a proportional-to-absolute-temperature voltage.

Consider the circuit shown in Fig.5.5. Let the ratio of the cross-sectional area of the emitter of Q2, denoted by A_{e2}, to that of Q1, denoted by A_{e1}, be n, i.e. $A_{e2} = nA_{e1}$. The collector current of Q1 is the same as that of Q2. Because

$$\Delta V_{BE} = V_{BE1} - V_{BE2} = V_t \ln\left(\frac{I_{S1}}{I_{S2}}\right) = \frac{kT}{q}\ln(n), \tag{5.60}$$

ΔV_{BE} is PTAT. To find out the temperature coefficient of ΔV_{BE}, we differentiate (5.60) with respect to T

$$\frac{\partial \Delta V_{BE}}{\partial T} = \frac{k}{q}\ln(n) = 0.086\ln(n) \quad (\text{mV/°K}). \tag{5.61}$$

Eq.(5.61) shows that the temperature coefficient of ΔV_{BE} is positive and is much smaller as compared with that of V_{BE}, which is approximately -1.45 mV/°K derived earlierA total compensation of the temperature effect on V_{BE} using $\frac{\partial \Delta V_{BE}}{\partial T}$ requires that

$$\frac{\partial \Delta V_{be}}{\partial T} + \frac{\partial V_{be}}{\partial T} = 0, \tag{5.62}$$

from which we obtain

$$0.086 \ln(n) - 1.45 = 0.$$

Solve for n

$$n = e^{16.86} = 2.1 \times 10^7.$$

The ratio of the emitter area of Q1 to that of Q2 must be prohibitively large in order to achieve the total compensation. Fortunately, this difficulty can be avoided by using circuit configurations, as demonstrated by Widlar [153, 154] and Brokaw [155]. We will study these circuits shortly in Section 5.3.

The circuit of Fig.5.5 possesses a unique characteristic that the value of ΔV_{BE} is independent of the collector currents of Q1 and Q2. This is important as it allows us to minimize the power consumption of the circuit by reducing J without sacrificing the PTAT of ΔV_{BE}.

Because the bipolar junction transistors in the preceding PTAT circuit are used for temperature sensing only, they can be implemented in standard CMOS technology. A vertical PNP transistor can be implemented using a n-well, as shown in Fig.5.6 [156]. Note the collector of the transistor is the substrate, which is always connected to the ground. It was shown in [157] the absolute temperature accuracy of substrate PNP transistors can be $\pm 0.1°$C in the temperature range $-50{\sim}130°$C. Similarly, a vertical NPN transistor can be implemented in a deep n-well when triple-well CMOS technologies are used, as shown in Fig.5.7 [158, 159].

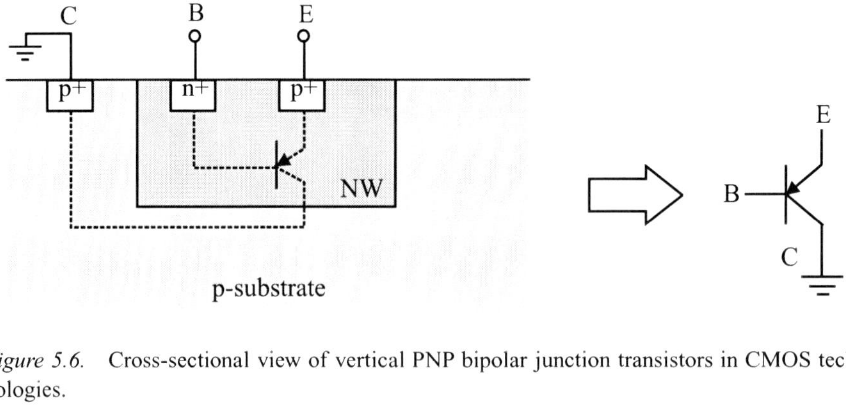

Figure 5.6. Cross-sectional view of vertical PNP bipolar junction transistors in CMOS technologies.

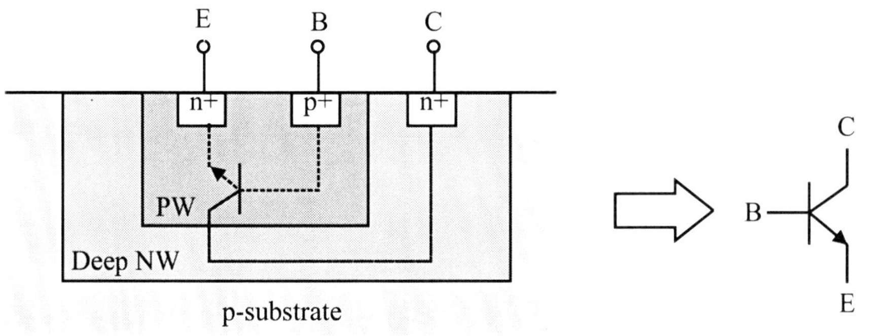

Figure 5.7. Cross-sectional view of vertical NPN bipolar junction transistors in CMOS technologies.

5.2.6 Zero-Temperature-Coefficient Bias Point

The channel current of nMOS transistors biased in saturation is given by

$$I_D \approx \frac{1}{2}\mu_n C_{ox} S (V_{GS} - V_T)^2. \tag{5.63}$$

Note that we have neglected the effect of channel length modulation to simplify analysis. Since both the mobility of electrons given by (5.9) and the threshold voltage of MOSFETs given by (5.42) are the functions of temperature and both have negative temperature coefficients, it is possible to make I_D independent of temperature by properly choosing V_{GS}. To illustrate this point, we differentiate (5.63) with respect to temperature T and impose

$$\begin{aligned} \frac{\partial I_D}{\partial T} &= 0, \\ \frac{\partial V_{GS}}{\partial T} &= 0. \end{aligned} \tag{5.64}$$

The result is given by

$$V_{GS} = V_T + 2\mu_n \left(\frac{\partial V_T / \partial T}{\partial \mu_n / \partial T} \right). \tag{5.65}$$

Substitute (5.9) and (5.42) into (5.65)

$$V_{GS} = V_T(T_o) - \alpha_{V_T} T_o + \alpha_{V_T} \left(1 + \frac{2}{m} \right) T. \tag{5.66}$$

The first two terms on the right hand side of (5.66) are functions of the reference temperature while only the last term varies with temperature. To ensure that V_{GS} is independent of temperature, $m = -2$ is required. It was shown in [139, 140, 160] that the value of m is indeed in the vicinity of -2. If we assume $m = -2$, (5.66) becomes

$$V_{GS}(T) = V_T(T_o) - \alpha_{V_T} T_o. \tag{5.67}$$

Eq.(5.67) shows that V_{GS} in this case is independent of temperature. The corresponding current is obtained from (5.63)

$$I_D = \frac{1}{2}\mu_n(T_o)C_{ox}S\alpha_{V_T}^2 T_o^2. \tag{5.68}$$

The operating point of the device in this case is independent of temperature, it is called the zero-temperature-coefficient (ZTC) bias point, as illustrated in Fig.5.8. Clearly if the device is biased at its ZTC bias point, the effect of temperature on V_{GS} and I_D is removed completely.

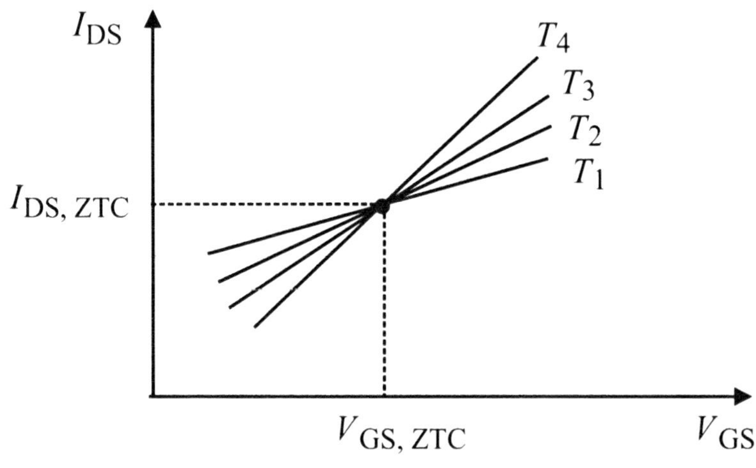

Figure 5.8. Zero-temperature-coefficient (ZTC) point. $T_k > T_{k+1}$.

5.3 First-Order Voltage References

It was shown in the preceding sections that V_{BE} of BJTs is a nonlinear function of temperature. V_{BE} can be linearized at a reference temperature T_o, as shown graphically in Fig.5.9. In order to generate a temperature-independent reference voltage V_{ref}, a correction voltage V_c that compensates the effect of the temperature on V_{BE} is required. The reference voltage in this case is given by

$$V_{ref} = aV_{BE} + bV_c, \tag{5.69}$$

where a and b are constants, with the constraint

$$\left.\frac{\partial V_{ref}}{\partial T}\right|_{T=T_o} = 0. \tag{5.70}$$

Since V_{BE} varies with temperature in a nonlinear fashion, the correction voltage V_c must also vary with temperature in a nonlinear fashion accordingly. Further, V_{BE} can be broken down into a linear portion and a nonlinear portion. The former varies with temperature linearly while the latter changes with temperature in a nonlinear fashion. V_c needs also to be broken down into a linear component denoted by V_{c1} and a nonlinear component denoted by V_{c2}, as shown in Fig.5.9. V_{c1} is used to cancel out the linear dependence of V_{BE} on temperature while V_{c2} offsets the nonlinear dependence of V_{BE} on temperature [134]. When only the linear dependence of V_{BE} is compensated, these voltage references are termed the first-order voltage references. Since we are only interested in the first-order bandgap references in this section, our attention will be on how to create V_{c1}. The presentation of how to generate V_{c2} will be given in Section 5.4.

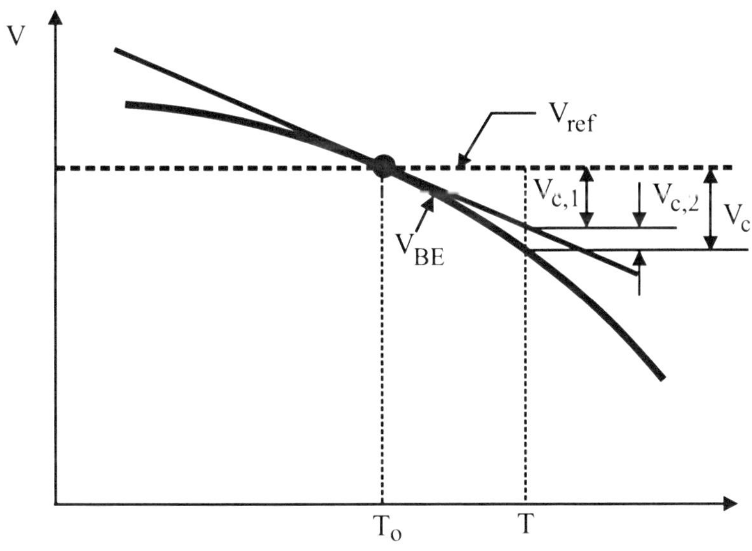

Figure 5.9. Compensation of base-emitter voltage of BJTs.

5.3.1 Widlar Voltage Reference

Consider the circuit shown in Fig.5.10. As compared with Fig.5.5, two resistors R_1 and R_2, and an auxiliary amplifier are added. We continue to assume that $A_{e2} = nA_{e1}$. In addition, we assume that the auxiliary amplifier is ideal, and M1 and M2 are perfectly matched.

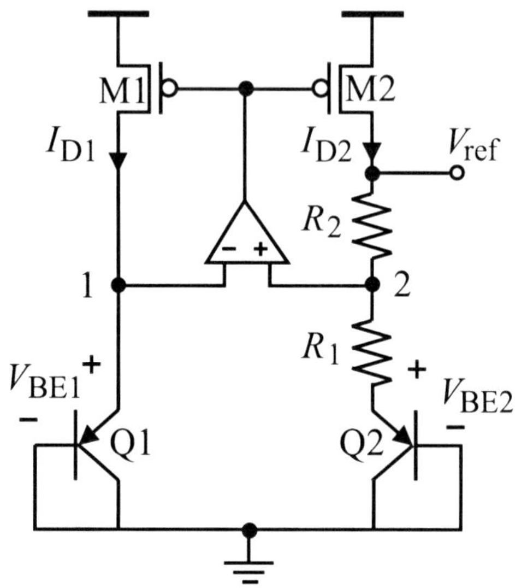

Figure 5.10. PTAT-compensated voltage reference.

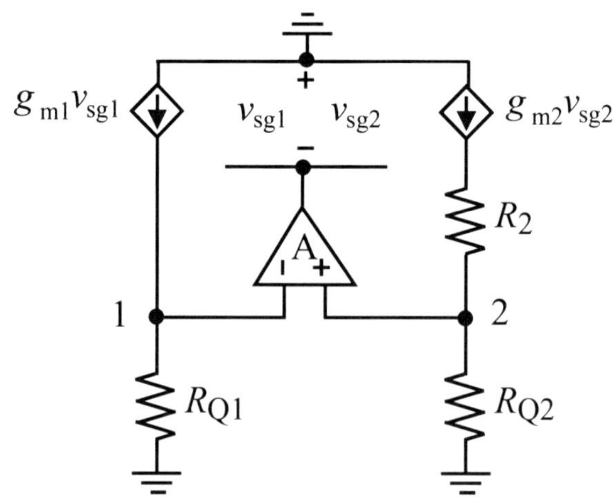

Figure 5.11. Simplified small-signal equivalent circuit of PTAT-compensated voltage reference.

To show that the negative feedback formed by M1, M2, and the auxiliary amplifier ensures $V_1 = V_2$, we replace Fig.5.10 with its simplified small-signal equivalent circuit shown in Fig.5.11. Here we use two resistors R_{Q1} and R_{Q2} to replace Q1, and the combination of Q2 and R_1, respectively to simplify analysis. Let the voltage gain of the amplifier be A. Write KCL at nodes 1 and 2

$$\begin{aligned} &\frac{V_1}{R_{Q1}} - g_{m1}\left[0 - A(V_2 - V_1)\right] = 0, \\ &\frac{V_2}{R_{Q2}} - g_{m2}\left[0 - A(V_2 - V_1)\right] = 0. \end{aligned} \quad (5.71)$$

Solving for V_1 and V_2

$$\begin{aligned} V_1 &= \left(1 + \frac{1}{R_{Q2}g_{m2}A}\right)V_2, \\ V_2 &= \left(1 - \frac{1}{R_{Q1}g_{m1}A}\right)V_1. \end{aligned} \quad (5.72)$$

In the limit $A \rightarrow \infty$, $V_1 = V_2$ is established. Having derived $V_1 = V_2$, let us now find the output voltage V_{ref}. Making use of $V_1 = V_2$, we have

$$V_{BE2} + R_1 I_{D1} = V_{BE1}, \quad (5.73)$$

where I_{D1} is the current of M1. Solving (5.73) for I_{D1}

$$I_{D1} = \frac{V_{BE1} - V_{BE2}}{R_1} = \frac{V_t}{R_1}\ln(n). \quad (5.74)$$

The current mirror M1-M2 ensure that $I_{D1} = I_{D2}$. It is seen from (5.74) that I_{D1} and I_{D2} are PTAT. The output voltage V_{ref} is obtained from

$$V_{ref} = V_{BE1} + \frac{R_2}{R_1}V_t \ln(n). \quad (5.75)$$

It is important to note that since R_1 and R_2 are realized using the same type of resistors, the temperature-dependence of $\frac{R_2}{R1}$ is removed. This is demonstrated in (5.76)

$$\frac{R_2(T)}{R_1(T)} = \frac{R_2(T_o)\,[1+\alpha(T-T_o)]}{R_1(T_o)\,[1+\alpha(T-T_o)]} = \frac{R_2(T_o)}{R_1(T_o)}. \tag{5.76}$$

The derivative of (5.75) with respect to T is given by

$$\frac{\partial V_{ref}}{\partial T} = \frac{\partial V_{BE1}}{\partial T} + \frac{R_2}{R_1}\frac{k}{q}\ln(n). \tag{5.77}$$

It was shown earlier that the first term on the right hand side of (5.77) is approximately -1.45 mV/°K at 300 °K while the second term equals to $+0.086\frac{R_2}{R_1}\ln(n)$ mV/°K. By properly choosing the resistance ratio and the emitter area ratio, $\frac{\partial V_{ref}}{\partial T} = 0$ can be achieved for a given temperature, typically the room temperature. Since V_{BE} is compensated by a PTAT correction voltage, only the first-order compensation of the dependence of V_{BE} on temperature can be achieved. The output voltage of Fig.5.10 provides the first-order correction to the temperature effect of V_{BE}.

The minimum supply voltage of the reference circuit is given by

$$V_{DD,min} = V_{BE,min} + V_{R_1} + V_{R_2} + V_{SD2}. \tag{5.78}$$

Since M1 and M2 are in saturation, $V_{SD2,min} = V_{sat}$, where V_{sat} is the pinch-off voltage of transistors. Also, because the minimum V_{BE} of BJTs, especially the vertical BJTs implemented in standard CMOS technologies, is not small with its typical value around 0.7 V at room temperature, $V_{DD,min}$ of the reference circuit is in the vicinity of 1 V. Further lowering the supply voltage of the circuit becomes difficult.

Since $V_{BE}\approx 0.7$ V at room temperature and it varies with temperature at a rate of approximately -2 mV/°C over the temperature range of -40°C $\sim$ 125°C, V_{BE} will vary from 830 mV to 500 mV. It becomes difficult to design amplifiers with such a large common-mode input voltage range while V_{DD} is scaled down to the neighborhood of 1 V [161, 162].

The currents of the transistors are computed from

$$I_{D1,2} = \frac{\Delta V_{BE}}{R_1} = \frac{V_t}{R_1}\ln(n). \tag{5.79}$$

With $V_t\approx 25.6$ mV at 300°K, $I_{D1,2}$ will be quite large unless a large R_1 is used. For example, with $n = 10$ and if we want $I_{D1,2} = 10$ nA, $R_1\approx 6.1$ MΩ is required. Clearly, Fig.5.10 is not suitable for applications such as passive wireless microsystems where a stringent constraint on power consumption exists.

5.3.2 Banba Voltage Reference

Banba *et al.* proposed a CMOS voltage reference that can operate with a sub-1V supply voltage [163, 161, 104]. Fig.5.12 is the simplified schematic of Banba voltage reference. As compared with Fig.5.10, two resistors R_1 and R_2 are added. In addition, the output voltage is sampled using an additional current branch formed by M3 and R_3 with $S_3 = kS_1 = kS_2$. The feedback formed by M1, M2, and the auxiliary amplifier ensures $V_1 = V_2$. This leads to $I_{R_1} = I_{R_2}$. The current mirror formed by M1 and M2 also ensures that $I_{D1} = I_{D2}$. Since $I_{R_1} = I_{R_2} = \frac{V_{BE1}}{R_1}$ and $I_{D1} = I_{D2}$, $I_{C1} = I_{C2}$ follows. Further because $A_{e2} = nA_{e1}$, we have

$$\Delta V_{BE} = V_{BE1} - V_{BE2} = V_t \ln(n), \tag{5.80}$$

and

$$I_{C2} = \frac{\Delta V_{BE}}{R_0} = \frac{V_t \ln(n)}{R_0}. \tag{5.81}$$

Thus I_{C1} and I_{C2} are PTAT. The channel current of M2 is determined from

$$I_{D2} = I_{C2} + I_{R_2} = \frac{V_{BE1}}{R_1} + \frac{V_t \ln(n)}{R_0}. \tag{5.82}$$

Once I_{D2} determined, it is mirrored to the output branch with $I_{D3} = kI_{D2}$. The reference voltage V_{ref} is obtained from

$$V_{ref} = R_3 I_{D3} = k\frac{R_3}{R_1}\left[V_{BE1} + V_t \ln(n)\frac{R_1}{R_0}\right]. \tag{5.83}$$

Eq.(5.83) shows that V_{ref} can be made independent of temperature at a given temperature by adjusting R_0, R_1, and n. Also, the reference voltage can be adjusted conveniently by varying R_3 and k.

The minimum supply voltage of Banba voltage reference is given by

$$V_{DD,min} = V_{BE,min} + V_{R0} + V_{sat}, \tag{5.84}$$

where V_{R0} is the voltage drop across R_0. As compared with (5.78), the minimum supply voltage is reduced. The additional current paths provided by R_1 and R_2 also reduce the currents flowing through Q1 and Q2, lowering V_{BE}. As demonstrated by Lin *et al.*, the bandgap reference implemented in a 0.35μm CMOS technology can operate reliably with $V_{DD} = 0.88$ V [104].

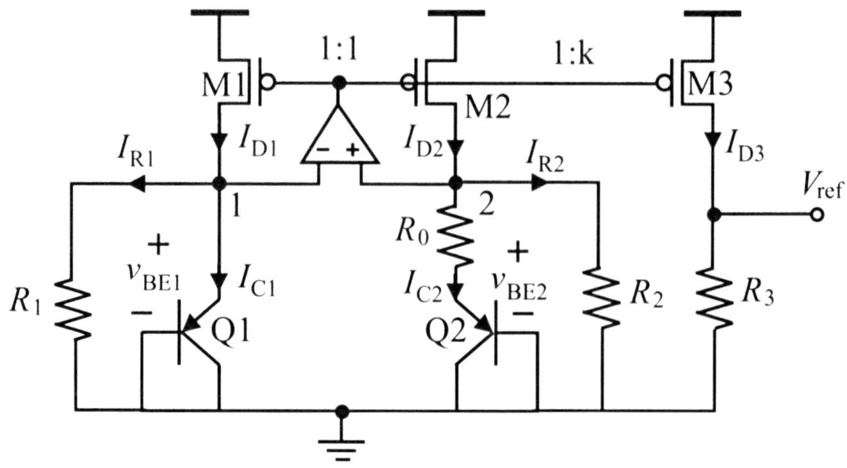

Figure 5.12. Voltage reference proposed by Banba *et al.* [163].

5.3.3 Waltari-Halonen Voltage Reference

Difficulties are encountered when lowering the supply voltage of the auxiliary amplifier of Banba voltage reference to sub-1 V and at the same time accommodating the high common-mode input voltage of the auxiliary amplifier lower-bounded by V_{BE}. As shown in (5.25), V_{BE} of vertical BJTs is mainly set by the bandgap voltage of silicon, which is approximately 0.7 V at room temperature. Waltari and Halonen used resistor voltage dividers to effectively lower the common-mode input voltage of the amplifier, as shown in Fig.5.13 [161]. Since

$$V_1 = \frac{R_{1B}}{R_{1A} + R_{1B}} V_{BE1}, \tag{5.85}$$

and

$$V_2 = \frac{R_{2B}}{R_{2A} + R_{2B}} (V_{BE2} + R_0 I_{C2}), \tag{5.86}$$

the common-mode input voltage of the amplifier is reduced. Because $V_1 = V_2$, $I_{R_{1B}} = I_{R_{2B}}$ follows. As a result, $V_3 = V_4$. A PTAT current I_{R_0} is generated by Q1, Q2, and R_0. The operation principle of Waltari-Halonen voltage reference is the same as that of Banba voltage reference. This simple design technique has gained its broad popularity since its inception [164, 165, 151].

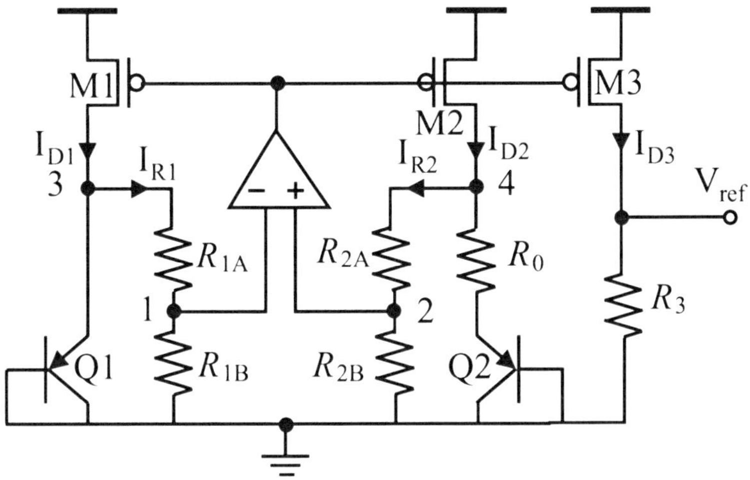

Figure 5.13. Voltage reference proposed by Waltari and Halonen [161].

5.3.4 Jiang-Lee Voltage Reference

Jiang and Lee showed that the difficulties associated with the high common-mode input voltage of the auxiliary amplifier in Banba voltage reference can also be avoided by replacing the auxiliary amplifier with a transimpedance amplifier, as shown in Fig.5.14 [166, 160]. The transimpedance amplifier is designed in such a way that it has a low input impedance and a large transimpedance gain, ensuring $V^- = V^+$ and $I_{R1} = I_{R2}$. With $R_1 = R_2$, we have $V_1 = V_2$. As a result,

$$I_{R_0} = \frac{V_{BE1} - V_{BE2}}{R_0} = \frac{V_t \ln(n)}{R_0}. \tag{5.87}$$

Since

$$I_{R_1} = \frac{V_{BE1} - V^-}{R_1}, \tag{5.88}$$

we have

$$\begin{aligned} I_{D4} &= I_{R_0} + I_{R_2} \\ &= \frac{V_t \ln(n)}{R_0} + \frac{V_{BE1} - V^-}{R_1}. \end{aligned} \tag{5.89}$$

The output voltage is obtained from

$$V_{ref} = \frac{S_5}{S_4} R_3 \left[\frac{V_t \ln(n)}{R_0} + \frac{V_{BE1} - V^-}{R_1} \right]. \tag{5.90}$$

It is seen that the only difference between (5.90) and the output voltage of Widlar voltage reference given in (5.75) is the term associated with V^-. To remove this term from (5.90), an auxiliary current source whose current is $I = \frac{V^-}{R_1}$ is injected into the output node. As a result, the output voltage becomes

$$V_{ref} = \frac{S_5}{S_4} R_3 \left[\frac{V_t \ln(n)}{R_0} + \frac{V_{BE1}}{R_1} \right]. \tag{5.91}$$

The need for a transimpedance amplifier with a large gain and the need for an auxiliary current source increase the power consumption. In order to lower the supply voltage, the transimpedance amplifier must be configured in such a way that it can operate at a low supply voltage. The minimum supply voltage of the voltage reference is therefore lower bounded by $V_{BE} + V_{sat}$.

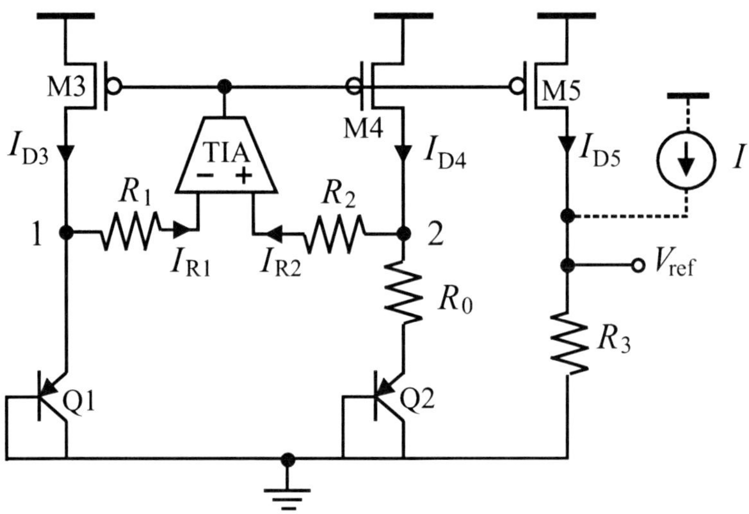

Figure 5.14. Voltage reference proposed by Jiang and Lee [166].

5.3.5 Threshold Voltage Based Voltage References

A critical issue associated with the preceding BJT-based references is that the characteristics of vertical BJTs in CMOS technologies are usually poorly modeled in most CMOS technologies. This creates a great difficulty in the accurate prediction of the performance of reference circuits in the design stage

[167]. It was shown earlier that the two temperature-dependent parameters of MOS transistors are the threshold voltage and the mobility of minority charge carriers, Dai *et al.* showed that if two separate circuits, one nMOS-based and one pMOS-based, are constructed in such a way that their output voltage varies with temperature in an identical linear fashion, then the difference of the output voltage of these two voltages can be made independent of temperature [168]. Fig.5.15 shows the simplified schematic of such a design. Let us focus on the pMOS sub-circuit first. Because M1 is diode-connected while M2 is in triode, we have

$$I_{D1} = \frac{1}{2}\mu_p C_{ox} S_1 \left(V_a - |V_{tp}|\right)^2, \tag{5.92}$$

and

$$I_{D2} = \mu_p C_{ox} S_2 \left[(V_P - |V_{tp}|)(V_P - V_a) - \frac{1}{2}(V_P - V_a)^2 \right], \tag{5.93}$$

where

$$V_a = \left(\frac{R_2}{R_1 + R_2} \right) V_P = a_p V_p.$$

Making use of $I_{D1} = I_{D2}$, we obtain

$$\left[1 - a_p^2(1 + b_p)\right] \left(\frac{V_P}{|V_{tp}|} \right)^2 - 2\left[1 - a_p(1 + b_p)\right] \left(\frac{V_P}{|V_{tp}|} \right) - b = 0, \tag{5.94}$$

where $b_p = S_1/S_2$, and V_{tp} is the threshold voltage of pMOS transistors. Solve (5.94) for V_P

$$V_P = \left[\frac{1 - a_p(1 + b_p) + (1 - a_p)\sqrt{1 + b_p}}{1 - a_p^2(1 + b_p)} \right] |V_{tp}|. \tag{5.95}$$

It is seen from (5.95) that the dependence of V_P on mobility μ_p is completely removed. Also observed is that V_P is a linear function of V_{tp}. Similarly, one can show that the output voltage of the nMOS sub-circuit is given by

$$V_N = \frac{V_{tn} - V_{tno}\sqrt{b_n}}{1 - a_n(1 + \sqrt{b_n})}, \tag{5.96}$$

where

$$a_n = \frac{R_4}{R_3 + R_4}, \tag{5.97}$$

$b_n = S_3/S_4$, V_{tn} is the threshold voltage of nMOS transistors and V_{tno} is the threshold voltage of nMOS transistors without body effect. It is evident from (5.96) that V_N is a linear function of V_{tn}.

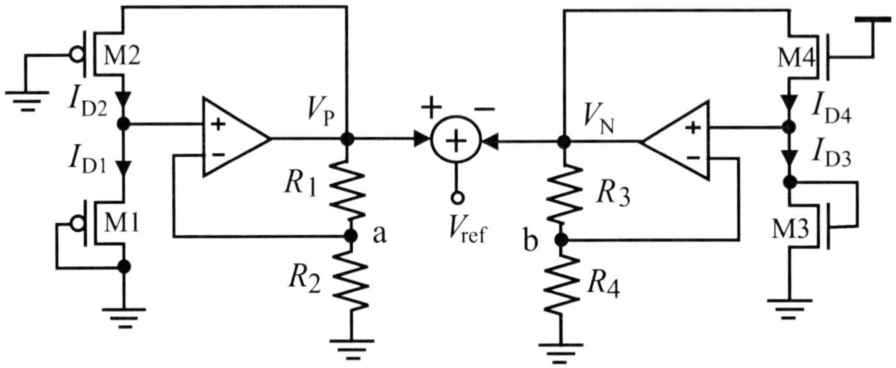

Figure 5.15. Threshold voltage based voltage reference proposed by Dai *et al.* [168].

As detailed in Section 5.2 that the threshold voltage of MOS transistors varies with temperature nearly linearly. V_P and V_N are therefore linear functions of temperature with their rate of dependence on temperature adjusted by varying a_n, b_n, a_p, and b_p. Finally, the reference voltage V_{ref} is obtained from

$$V_{ref} = k_1 V_P - k_2 V_N, \tag{5.98}$$

where k_1 and k_2 are the coefficients of the subtracter. By adjusting a_n, b_n, a_p,b_p, k_1, and k_2, V_{ref} can be made independent of temperature.

The threshold-voltage based voltage reference is not particularly attractive for low-power applications. This is due to the fact that two auxiliary amplifiers and a subtracter, often implemented using an operational amplifier, are needed. Also, since $V_{a,min} = |V_{tp}|$ and $V_{b,min} = V_{tn}$, the current drawn by the resistor networks will be large, further increasing the power consumption.

5.3.6 Buck Voltage Reference

A common drawback of the preceding voltage references is that their accuracy is severely affected by the accuracy of the resistance of resistors. It is well known that the accuracy of silicide resistors is poor. Although non-silicide resistors offer an improved accuracy, their accuracy is still in the neighborhood of $\pm 15\%$. As a result, a precision reference voltage can only be obtained using resistance trimming, which is both area and power hungry. Also, non-silicide resistors are in general not available in digitally-oriented CMOS technology while diffusion resistors, which are available in digitally-oriented CMOS technology, suffer from the drawbacks of a poor accuracy, a poor linearity, and a strong interaction with the substrate, further complicating design.

Buck *et al.* proposed a resistor-less CMOS bandgap reference by employing an inverse function technique [169, 167]. Fig.5.16 shows the simplified schematic of Buck bandgap reference. All MOS transistors are in saturation. Bipolar transistors Q1 and Q2 are used to sense temperature with $A_{e2} = nA_{e1}$. The following relations

$$\begin{aligned} I_{D1} &= GI_{D4}, \\ I_{D2} &= GI_{D3} \end{aligned} \tag{5.99}$$

are established from $S_6 = GS_5$, $I_{D1} = GI_{D4} = GBI_{D3}$, and $I_{D2} = I_{D1}/B = GI_{D3}$. Substituting the expressions of I_{D1}, I_{D2}, I_{D3}, and I_{D4} into (5.99) yields

$$V_{SG1} - |V_{tp}| = \sqrt{AG}\left(V_{SG4} - |V_{tp}|\right), \tag{5.100}$$

and

$$V_{SG2} - |V_{tp}| = \sqrt{AG}\left(V_{SG3} - |V_{tp}|\right). \tag{5.101}$$

Further subtracting (5.101) from (5.100) and noting that

$$V_{SG4} - V_{SG3} = V_{BE2} - V_{BE1} = -\Delta V_{BE} = -V_t \ln(n), \tag{5.102}$$

we arrive at

$$V_{ref} = V_{BE2} + \frac{kT}{q}\sqrt{AG}\ln(n). \tag{5.103}$$

Eq.(5.103) shows that by varying $\sqrt{AG}$, V_{ref} can be made independent of temperature.

It is important to note that the current relations given in (5.99) form the foundation for the elimination of the mobility of the minority charge carriers, which is temperature-dependent, in (5.100) and (5.101). The terms associated with the threshold voltage of pMOS transistors in (5.100) and (5.101) are also eliminated from the subtraction of (5.101) and (5.100). As a result, the reference voltage is only determined by the aspect ratio of the transistors and the emitter area ratio of Q1 and Q2. The weighting factor $\sqrt{AG}$ is a function of the aspect ratio of the transistors and can be adjusted using digital trimming. The cost of digital trimming both in terms of power and silicon resources, however, must be tightly controlled if one wants to minimize the overall power consumption of the voltage reference.

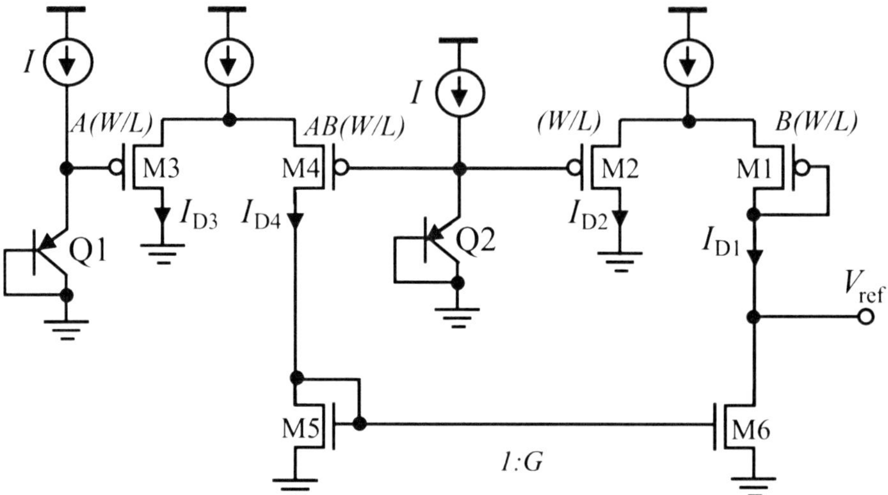

Figure 5.16. Resistor-less voltage reference proposed by Buck *et al.* [167].

The preceding Buck bandgap reference was modified by Gu *at el.* in [170] where two voltage references, one with a positive temperature coefficient and the other with a negative temperature coefficient, are generated in a way similar to that of Buck bandgap reference. The output of the two voltage references are fed to two temperature coefficient tuning circuits where two distinct reference voltages are generated. As compared with Buck bandgap reference, the output voltages of Gu voltage reference are lower and tunable. The need for two voltage reference circuits and their corresponding temperature coefficient tuning circuits, however, makes it difficult to lower its power consumption.

New topologies and design techniques for resistor-less bandgap references continue to emerge, aiming at either improving the accuracy [171, 172] or lowering power consumption [173]. We will examine these designs in detail in Section 5.4 and Section 5.5 where curvature-compensated references and ultra-low power references are investigated.

5.3.7 Comparison of First-Order Voltage References

Table 5.3 compares the performance of the first-order references reported recently. It is seen that the temperature coefficient of these first-order references is in general high with a lower bound around 15 ppm/°C [174]. A further reduction of the temperature coefficient requires the compensation of the high-order terms of V_{BE}.

Table 5.3. Performance comparison of first-order voltage references.

Reference	Tech.	TC [ppm/°C]	Temp. [°C]	PSRR [dB]	Min. V_{DD} [V]
Banba *et al.* (99)[163]	0.4μm	118	27~125	–	0.84
Waltari-Halonen (00) [161]	0.5μm	16.2	-20~100	–	0.95
Jiang-Lee (00) [166]	1.2μm	60	0~100	20	1.2
Hoon *et al.* (02) [175]	0.8μm	17.7	-40~120	110	–
Leung-Mok(02) [164]	0.6μm	15	0~100	44	–
Buck *et al.* (02) [167]	0.5μm	122	0~70	45	3.7
Chen-Shi(03) [165]	0.18μm	50	0~110	–	1
Dai *et al.* (04) [168]	0.5μm	32	-10~80	40	–
Lin *et al.* (05) [104]	0.35μm	13.6	-20~120	75	0.88
Crovetti-Fiori (07) [176]	0.35μm	30	-40~110	65	0.85
Falconi *et al.* (07)[177]	0.13μm	191	-10~80	52	1
Bendali-Audet (07) [160]	0.18μm	185	0~100	27.5	1
Gu *et al.* (08) [170]	0.35μm	100	0~80	41	2

5.4 High-Order Voltage References

It was shown in Section 5.2 that the base-emitter voltage of bipolar junction transistors contains both a linear portion whose value is proportional to temperature and a nonlinear portion whose value is a nonlinear function of temperature. The linear portion dominates. We further showed in Section 5.3 that first-order voltage references only compensate the linear portion of V_{BE} using a PTAT correction voltage. Because the nonlinear temperature dependence of V_{BE} is not considered in these references, the temperature coefficient of first-order voltage references is rather high and is lower-bounded by 15 ppm/°C approximately, as evident in Table 5.3. To improve the accuracy of voltage references, the nonlinear dependence of V_{BE} must be compensated. These voltage references are termed the high-order voltage references, also known as curvature-compensated voltage references.

A number of design techniques have been developed to provide curvature compensation for V_{BE}. The early investigation on curvature-compensated voltage references by Song and Gray utilizes switched-capacitor techniques to provide the 2nd-order compensation for V_{BE} [156]. The voltage reference proposed by Lin and Salama and implemented in a 5μm CMOS technology achieves a temperature coefficient of 15.1 ppm/°C over $0 \sim 70°C$ by using three differential amplifiers and dual power supplies [137]. The exponentially curvature-compensated voltage reference proposed by Lee *et al.* utilizes the temperature characteristics of the current amplification coefficient of BJTs to construct a correction voltage [174]. This method provides better compensation as compared with the 2nd-order compensation approach by Song and Gray. Implemented in a 1.5μm BiCMOS technology, the temperature coefficient of Lee voltage reference is reduced to below 10 ppm/°C. These early implementations of curvature-compensated references, however, either require Bipolar or BiCMOS technologies, or employ complex configurations such as switched-capacitor networks, making them less attractive for ultra-low power applications, especially for passive wireless microsystems where a stringent constraint on power consumption exists. In what follow, we investigate high-order voltage references that are specifically tailored for low-voltage low-power applications. Ultra-low power voltage references where devices operate in week inversion will be dealt with separately in Section 5.5.

5.4.1 Piecewise-Linear Voltage Reference

Rincon-Mora and Allen proposed a current-mode piecewise-linear curvature-compensated voltage reference [178]. Huang *et al.* improved the design by lowering the power consumption and improving the temperature coefficient [179]. The simplified schematic of Huang voltage reference is shown in Fig.5.17. All transistors are in strong inversion except M1, which operates in the deep triode and behaves as a resistor. The PTAT circuit consisting of Q1, Q2, M1-M3, and the auxiliary amplifier ensures that

$$I_{D4} = \frac{V_t \ln(n)}{R_{ds1}}, \tag{5.104}$$

where n is the ratio of the emitter area of Q2 to that of Q1, and R_{ds1} is the channel resistance of M1. When temperature rises, I_{D4} increases accordingly. $I_{D5} = \left(\frac{S_5}{S_4}\right) I_{D4}$ also increases. As a result, $V_{GS2} = V_{DS2}$ rises. Since M6 is diode-connected, I_{D6} increases. This is echoed with a drop of I_{D2} as $I_{D2} = I_{D5} - I_{D6}$. As a result, $V_{GS2} = V_{ref}$ also drops. The preceding analysis shows that there is no open-loop compensation for the linear or nonlinear portions of V_{BE}. This differs from the preceding voltage references. Instead, Huang voltage reference uses negative feedback to stabilize V_{ref}.

5.4.2 Malcovati Voltage Reference

The simplified schematic of the voltage reference proposed by Malcovati *et al.* is shown Fig.5.18 [180–182]. M1 and M2 are identical, so are Q1 and Q3. The emitter area of Q2 is n times that of Q1, i.e. $A_{e2} = nA_{e1}$. Further

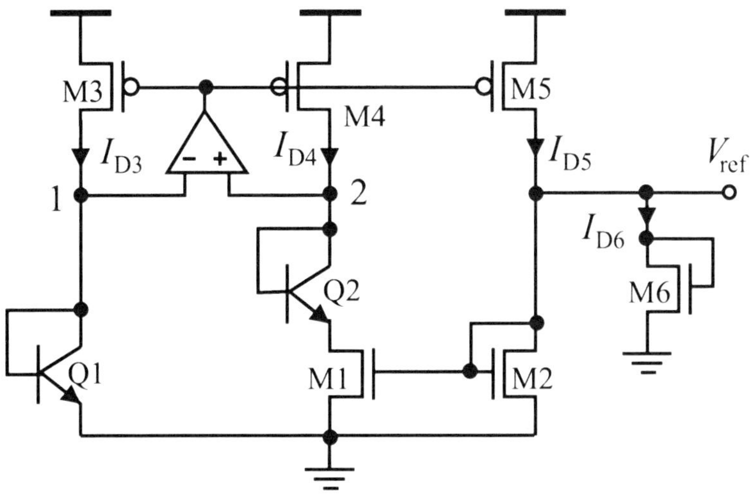

Figure 5.17. Piecewise linear curvature-compensated voltage reference proposed by Huang *et al.* [179].

$R_1 = R_2$ and $R_3 = R_4$ are set. Since $V_1 = V_2$, we have $I_{R_1} = I_{R_2}$. The current mirror M1-M2 ensures $I_{D1} = I_{D2}$. The collector current of Q2 is obtained from

$$I_{C2} = \frac{V_{BE1} - V_{BE2}}{R_0} = \frac{1}{R_0}\frac{kT}{q}\ln(n). \tag{5.105}$$

Eq.(5.105) confirms that I_{C2} is PTAT. Further because

$$I_{R_4} = \frac{V_1 - V_3}{R_4}, \tag{5.106}$$

and

$$I_{R_5} = \frac{V_2 - V_3}{R_5}, \tag{5.107}$$

$I_{R_4} = I_{R_5}$ and $I_{C1} = I_{C2}$ follow. The preceding analysis reveals that both I_{C1} and I_{C2} are PTAT. Writing (5.27) for Q1 yields

$$\begin{aligned} V_{BE1}(T) = V_G(T) &+ \frac{T}{T_o}\left[V_{BE1}(T_o) - V_G(T_o)\right] \\ &- (m+3)\frac{kT}{q}\ln\left(\frac{T}{T_o}\right). \end{aligned} \tag{5.108}$$

Note that since I_{C1} is PTAT, $\alpha = 1$ was used in simplifying (5.108). Because the current of Q3 is independent of temperature, i.e. $\alpha = 0$, we have

$$\begin{aligned} V_{BE3}(T) = V_G(T) \quad &+ \quad \frac{T}{T_o}\left[V_{BE3}(T_o) - V_G(T_o)\right] \\ &- \quad (m+4)\frac{kT}{q}\ln\left(\frac{T}{T_o}\right). \end{aligned} \tag{5.109}$$

Subtracting (5.109) from (5.108) and noting that Q1 and Q3 are identical, we arrive at

$$V_{BE1} - V_{BE3} = \frac{kT}{q}\ln\left(\frac{T}{T_o}\right). \tag{5.110}$$

I_{R_4} is obtained from

$$I_{R_4} = \frac{V_{BE1} - V_{BE3}}{R_4} = \frac{1}{R_4}\frac{kT}{q}\ln\left(\frac{T}{T_o}\right). \tag{5.111}$$

Eq.(5.111) shows that I_{R_4} contains the nonlinear term of (5.27). The current of M1 is obtained from

$$\begin{aligned} I_{D1} &= \frac{V_{BE1}}{R_1} + I_{C1} + I_{R_4} \\ &= \frac{1}{R_1}\left[V_G(T) + \frac{T}{T_o}\left[V_{BE1}(T_o) - V_G(T_o)\right] - (m+3)\frac{kT}{q}\ln\left(\frac{T}{T_o}\right)\right] \\ &+ \frac{kT}{qR_0}\ln(n) + \frac{kT}{qR_4}\ln\left(\frac{T}{T_o}\right). \end{aligned} \tag{5.112}$$

Re-group the terms in (5.112)

$$\begin{aligned} I_{D1} \quad &= \quad \frac{V_G(T)}{R_1} + \left[\frac{V_{BE1}(T_o) - V_G(T_o)}{R_1 T_o} + \frac{k}{qR_0}\ln(n)\right]T \\ &+ \quad \left[\frac{1}{R_4} - \frac{(m+3)}{R_1}\right]\frac{kT}{q}\ln\left(\frac{T}{T_o}\right). \end{aligned} \tag{5.113}$$

Eq.(5.113) reveals that by varying the coefficient of T and that of $\ln\left(\frac{T}{T_o}\right)$, I_{D1} can be made independent of temperature.

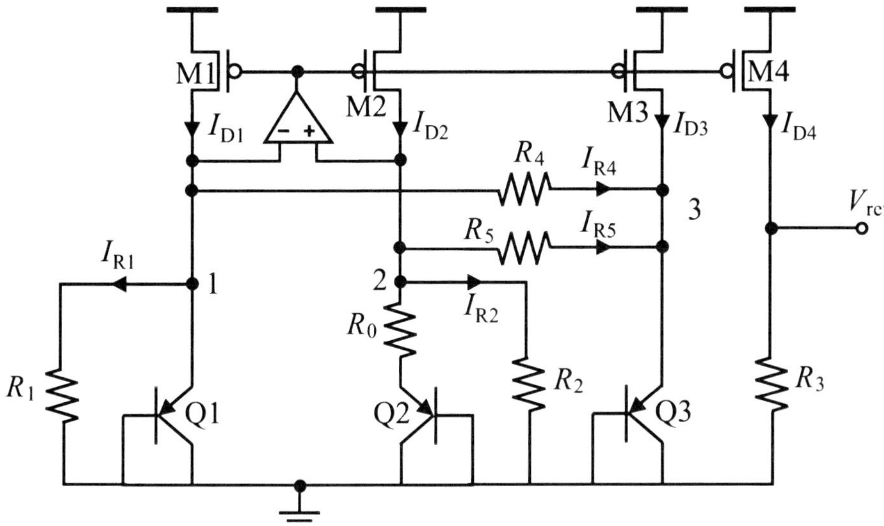

Figure 5.18. Curvature-compensated voltage reference proposed by Malcovati *et al.* [180].

5.4.3 Resistor Curvature-Compensated Voltage Reference

It was shown in Section 5.2 that poly and diffusion resistors exhibit distinct temperature characteristics. The resistance of diffusion resistors typically increases with temperature while that of poly resistors decreases with temperature. Also, the temperature coefficient of diffusion resistors is several times that of poly resistors. The distinct temperature characteristics of diffusion and poly resistors can be utilized to provide an economical and yet effective way to compensate the nonlinear dependence of V_{BE} on temperature.

Leung *et. al* proposed a curvature-compensated voltage reference by utilizing the distinct temperature characteristics of diffusion and high-resistive poly resistors [152]. Fig.5.19 shows the simplified schematic of Leung voltage reference. The types of the resistors are explicitly indicated in the schematic. The PTAT current generator consists of Q1, Q2, and R_1 with $A_{e1} = nA_{e2}$. M1 and M2 are sized to be identical to yield $I_{D1} = I_{D2}$. Since

$$I_{D1} = I_{D2} = \frac{V_t \ln(n)}{R_1}, \tag{5.114}$$

we have

$$V_{ref} = V_{BE2} + \frac{R_2}{R_1} V_t \ln(n) + \frac{R_3}{R_1} V_t \ln(n). \tag{5.115}$$

The resistance ratio $\frac{R_2}{R_1}$ is independent of temperature because they are of the same type of resistors while $\frac{R_3}{R_1}$ is temperature-dependent because they are made of different types of resistors. The second term on the right hand side of (5.115) is PTAT. It compensates the linear dependence of V_{BE2} on temperature. The third term on the right hand side of (5.115) can be adjusted by varying R_1 and R_3 such that it can compensate the nonlinear dependence of V_{BE2} on temperature. Clearly, the degree of curvature compensation depends upon the temperature characteristics of R_1 and R_3. To illustrate this, let us assume

$$R_1(T) = R_1(T_o)\left[1 + \alpha_1(T - T_o)\right], \tag{5.116}$$

and

$$R_3(T) = R_3(T_o)\left[1 + \alpha_3(T - T_o)\right], \tag{5.117}$$

where α_1 and α_3 are the temperature coefficient of R_1 and R_3, respectively.

$$\frac{R_3(T)}{R_1(T)} = \frac{R_3(T_o)[1 + \alpha_3(T - T_o)]}{R_1(T_o)[1 + \alpha_1(T - T_o)]}. \tag{5.118}$$

Since

$$\alpha_1(T - T_o) \ll 1, \tag{5.119}$$

we have

$$\frac{1}{1 + \alpha_1(T - T_o)} \approx 1 - \alpha_1(T - T_o). \tag{5.120}$$

Eq.(5.118) becomes

$$\frac{R_3(T)}{R_1(T)} \approx \frac{R_3(T_o)}{R_1(T_o)}\left[1 + \alpha_3(T - T_o) - \alpha_1(T - T_o) - \alpha_1\alpha_3(T - T_o)^2\right]. \tag{5.121}$$

The reference voltage is obtained from

$$\begin{aligned} V_{ref} &= V_{BE2} + (R_2 + R_3)\, I_{C2} \\ &= V_{BE2} + \left(\frac{R_2}{R_1} + \frac{R_3}{R_1}\right) V_t \ln(n) \end{aligned}$$

$$
\begin{aligned}
&= V_{BE2} + \frac{R_2}{R_1} V_t \ln(n) \\
&+ \frac{R_3(T_o)}{R_1(T_o)} \left[1 + \alpha_3(T - T_o) - \alpha_1(T - T_o) - \alpha_1 \alpha 3(T - T_o)^2\right] V_t \ln(n).
\end{aligned}
\tag{5.122}
$$

It is apparent that the nonlinear term in (5.122) can be used to compensate the nonlinear dependence of V_{BE2} on temperature. It should be noted that since the resistance of both poly and diffusion resistors has a large variation, digital trimming is often required to tune R_1 and R_3 such that the effect of temperature on V_{ref} is minimized [183].

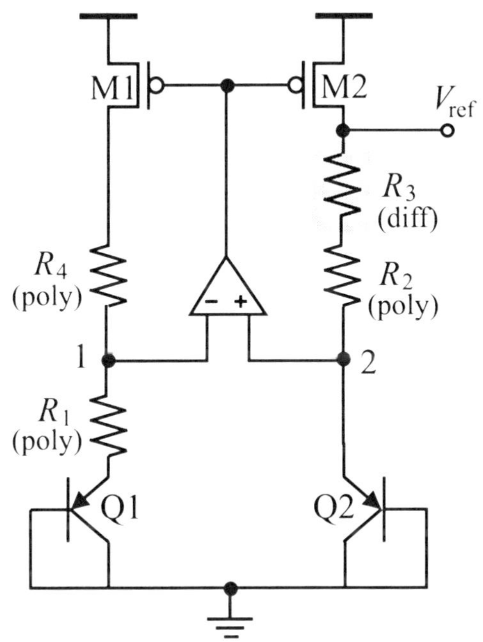

Figure 5.19. Curvature-compensated voltage reference proposed by Leung *et al.* [152].

5.4.4 Ker-Chen Voltage Reference

Ker and Chen showed that a curvature-compensated reference voltage can be constructed by subtracting the PTAT-compensated currents of two first-order voltage references [158]. Fig.5.20 is the simplified schematic of Ker-Chen voltage reference. It consists of two first-order voltage references, a current subtracter, and a current-to-voltage converter (R_7). let us consider the first-order voltage reference on the left. $R_{1a} = R_{1b}$, $R_{2a} = R_{2b}$, M1, M2 and M5 are identical. Because $V_1 = V_2$, $I_{R_{1b}} = I_{R_{2b}}$ follows. Since

$$V_3 = V_{DD} - (R_{1a} + R_{1b}) I_{R_{2b}}, \tag{5.123}$$

and

$$V_4 = V_{DD} - (R_{2a} + R_{2b}) I_{R_{1b}}, \tag{5.124}$$

we have $V_3 = V_4$. Further from

$$I_{C1} = \frac{V_{BE2} - V_{BE1}}{R_3}, \tag{5.125}$$

we obtain the current of M1

$$\begin{aligned} I_{D1} &= I_{R_{1b}} + I_{C1} \\ &= \frac{V_{BE2}}{R_1} + \frac{V_t \ln(n_{12})}{R_3}, \end{aligned} \tag{5.126}$$

where $R_1 = R_{1a} + R_{1b}$ and n_{12} is the ratio of the cross-sectional area of Q1 to that of Q2. Similarly, one can show that

$$I_{D3} = \frac{V_{BE4}}{R_5} + \frac{V_t \ln(n_{34})}{R_6}, \tag{5.127}$$

where $R_3 = R_{3a} + R_{3b}$ and n_{34} is the ratio of the cross-sectional area of Q3 to that of Q4. It is seen from (5.126) and (5.127) that I_{D1} and I_{D3} are PTAT-compensated only. Their nonlinear dependence on temperature is left uncompensated. The effect of the nonlinear dependence of the base-emitter voltage can be reduced if we subtract (5.126) from (5.127), provided that the effect of temperature on V_{BE2} and that on V_{BE4} have a similar profile. The output voltage of the reference is given by

$$\begin{aligned} V_{ref} &= \left[\frac{S_7}{S_6} I_{D6} - \frac{S_{10}}{S_9} I_{D8} \right] R_7 \\ &= R_7 \left[\frac{S_7}{S_6} \frac{V_{BE2}}{R_1} - \frac{S_{10}}{S_9} \frac{V_{BE4}}{R_5} \right] + R_7 \left[\frac{S_7}{S_6} \frac{\ln(n_{12})}{R_3} - \frac{S_{10}}{S_9} \frac{\ln(n_{34})}{R_6} \right] V_t. \end{aligned} \tag{5.128}$$

Eq.(5.128) shows that the effect of temperature on V_{ref} can be minimized by adjusting both the resistances and the aspect ratios of the current mirrors.

One disadvantage of this reference is that the resistance values and aspect ratios affect both the linear and nonlinear compensation, making the optimal compensation of both difficult. Also, since two first-order references are used, both the power consumption and silicon cost of Ker-Chen voltage reference are high.

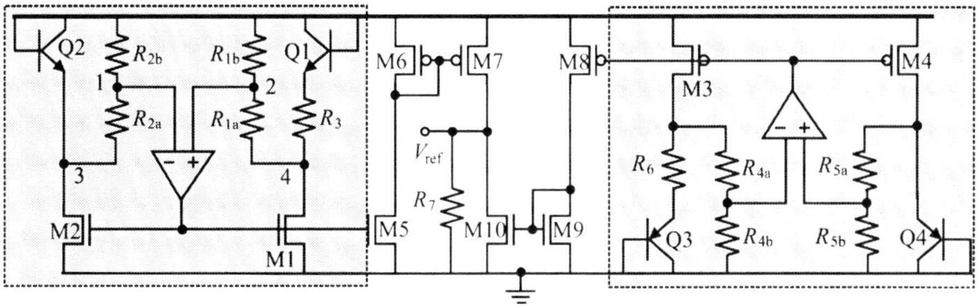

Figure 5.20. Curvature-compensated voltage reference proposed by Ker and Chen [158].

5.4.5 Comparison of High-Order Voltage References

Table 5.4 compares the performance of high-order references reported recently. It is seen that the temperature coefficient of high-order references is generally smaller as compared with that of the first-order voltage references given in Table 5.3. The power consumption of high-order voltage references is generally higher as compared with that of the first-order voltage references due to the need for additional circuitry to generate a correction voltage to compensate the nonlinear portion of V_{BE} of temperature-sensing BJTs.

Table 5.4. Performance comparison of high-order voltage references.

Reference	Tech.	TC [ppm/°C]	Temp. [°C]	Power [μW]	PSRR [dB]	Min. V_{DD} [V]
Stanescu *et al.* (02)[184]	0.8μm	10	−40∼125	12.8	66	1.6
Leung *et al.* (03)[152]	0.6μm	5.3	0∼100	46	47	2
Tom-Alvandpour (05)[181]	0.18μm	7	0∼60	5	–	0.9
Hsiao *et al.* (06)[183]	0.35μm	10	25∼100	82.5	–	1.5
Guan *et al.* (06)[182]	0.35μm	3.1	−20∼100	111	80	1.5
Ker-Chen (06)[158]	0.25μm	19.5	0∼100	45	25.5	0.9
Huang *et al.* (08)[179]	0.18μm	1.7	−40∼125	4.4	75	1.8

5.5 Sub-threshold Voltage References

Biasing transistors in weak inversion is perhaps the most effective means to lower the power consumption of analog circuits where continuous biasing currents are typically present [131]. As pointed out in Section 5.2 that the gate-source voltage of MOSFETs in weak inversion exhibits a similar temperature characteristic as that of the base-emitter voltage of bipolar junction transistors. In this section, we show that the temperature-dependence of V_{GS} of MOSFETs in weak inversion can be utilized to construct ultra-low power voltage references. The behavior of MOSFETs in weak inversion is well documented in most CMOS technologies. This enables an accurate analysis of the behavior of these devices when temperature varies. Vertical BJTs in CMOS technologies, on the contrary, are typically poorly characterized. The operation of MOSFETs in weak inversion can be preciously controlled by varying their biasing conditions. The characteristics of vertical BJTs, on the other hand, are fixed once the geometrical dimensions of the devices are set. Since V_{GS} must be smaller than the threshold voltage in order to force MOSFETs to operate in the weak inversion region, the minimum supply voltage of circuits, which is often lower-bounded by V_{GS}, can also be reduced effectively. This added low-voltage characteristics, in addition to low-power operation, make them particularly attractive for passive wireless microsystems. In this section, we investigate voltage references with devices in weak inversion.

It was shown in Section 5.2 that the gate-source voltage of MOSFETs in weak inversion contains both a linear and a nonlinear portions. The former decreases with temperature linearly while the latter decreases with temperature in a nonlinear fashion. A first-order sub-threshold voltage reference can therefore be construct by creating a PTAT correction voltage with a positive temperature coefficient to compensate the linear portion of gate-source voltage of MOSFETs.

5.5.1 Ytterdal Voltage Reference

Fig.5.21 shows the simplified schematic of the voltage reference proposed by Ytterdal [185]. A similar topology was used by Yu and Zou in the development of current references [149]. This reference is very similar to Banba voltage reference in Fig.5.12 studied earlier except that BJTs in Banba voltage reference are now replaced with MOSFETs in weak inversion. The feedback formed by M3, M4, and the auxiliary amplifier yields $V_1 = V_2$. This leads to $I_{R_1} = I_{R_2}$. The current mirror formed by M3 and M4 ensures that $I_{D3} = I_{D4}$. Since $I_{R_1} = I_{R_2} = \frac{V_{GS1}}{R_1}$, $I_{D1} = I_{D2}$ follows. Transistors M1 and M2 both operate in weak inversion with their aspect ratio $S_2/S_1 = n$. Re-write (5.43) here for convenience

$$I_D \approx 2n\mu_n C_{ox} S V_t^2 e^{\frac{V_{GS}-V_T}{nV_t}}. \tag{5.129}$$

Solve for V_{GS}

$$V_{GS} = V_T + nV_t \ln\left(\frac{I_D}{2n\mu_n C_{ox} S V_t^2}\right). \tag{5.130}$$

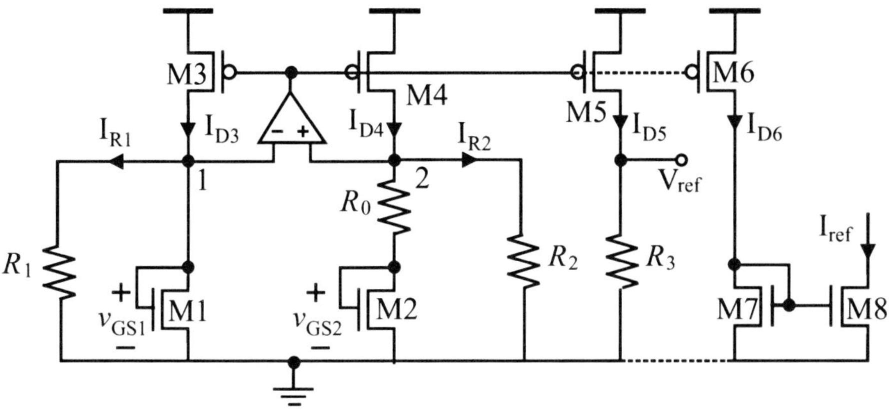

Figure 5.21. Current reference proposed by Yu and Zou [149].

The current flowing though R_0 is obtained from

$$I_{D2} = \frac{V_{GS1} - V_{GS2}}{R_0} = \frac{nV_t}{R_0}\ln(n). \tag{5.131}$$

It is evident from (5.131) that I_{D2} is PTAT. Writing KCL at node 2 yields

$$I_{D4} = \frac{nV_t}{R_0}\ln(n) + \frac{V_{GS1}}{R_1}. \tag{5.132}$$

Since

$$I_{D5} = \frac{S_5}{S_4} I_{D4}, \tag{5.133}$$

the output voltage is given by

$$V_{ref} = \frac{S_5}{S_4}\left[\frac{R_3}{R_0} nV_t \ln(n) + \frac{R_3}{R_1} V_{GS1}\right]. \tag{5.134}$$

Since V_{GS1} decreases with temperature while the first term on the right hand side of (5.132) is PTAT with a positive temperature coefficient, V_{ref} can be made independent of temperature to the first order. Also note that since only the ratios of the resistances appear in (5.134), the effect of temperature on the resistance of the resistors vanishes desirably. Evaluate the derivative of V_{ref} with respect to temperature

$$\frac{\partial V_{ref}}{\partial T} = \frac{S_5}{S_4}\left[\frac{R_3}{R_0}n\frac{k}{q}\ln(n) + \frac{R_3}{R_1}\frac{\partial V_{GS1}}{\partial T}\right]. \tag{5.135}$$

Since the first term on the right hand side of (5.135) is constant, only the constant of $\frac{\partial V_{GS}}{\partial T}$ needs to be considered in the first-order compensation. Set (5.135) to zero at room temperature (300°K)

$$\frac{R_0}{R_1} = -\frac{nk}{a_1 q}\ln(n), \tag{5.136}$$

where a_1 is defined in (5.57). Eq.(5.136) is the design equation that guides the choice of the value of R_0, R_1, and n. Once R_0, R_1, and n are available, the value of R_3 can be set by the desired output voltage V_{ref} using (5.134).

5.5.2 Cheng-Wu Voltage Reference

Voltage references that operate on the principle of peaking current mirrors developed in [131] were proposed by Cheng and Wu to achieve low power consumption [186]. Fig.5.22 is the simplified schematic of Cheng-Wu voltage reference. M3 and M4 are in weak inversion. Writing (5.47) for M3 and M4 yields

$$I_{D4} = I_{D3}\frac{S_4}{S_3}e^{-\frac{R_1 I_{D3}}{nV_t}}, \tag{5.137}$$

As I_{D3} rises, V_{DS3} will increase, so will I_{D4}. Since the voltage drop across R_1 also increases in this case, it can be shown that I_{D4} will drop once I_{D3} becomes large [131]. There hence exists a current peak in I_{D4}. The current I_{D3} at which I_{D4} reaches its maximum is obtained from

$$\frac{\partial I_{D4}}{\partial I_{D3}} = 0. \tag{5.138}$$

The result is given by

$$R_1 I_{D3} = nV_t. \tag{5.139}$$

The output voltage of the reference is obtained from

$$V_{ref} = R_2 I_{D3} + V_{GS3}. \tag{5.140}$$

Making use of (5.139), we arrive at

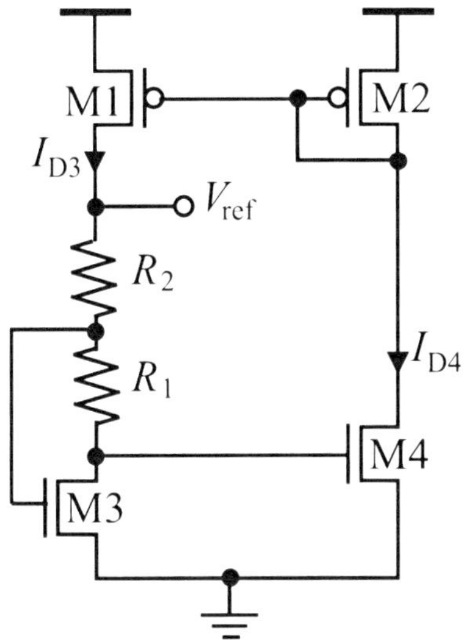

Figure 5.22. Voltage reference proposed by Cheng and Wu [186].

$$V_{ref} = \frac{R_2}{R_1} nV_t + V_{GS3}. \tag{5.141}$$

We observe that V_{ref} can be made independent of temperature if the first term on the right hand side of (5.141), which is PTAT, compensates the effect of temperature on V_{GS3}. Also observed is that only the first-order compensation is achieved. An attractive advantage of Cheng-Wu voltage reference is its low power consumption, attributive to the sub-threshold operation of the transistors and the absence of the auxiliary amplifier used in voltage references such as Widlar voltage reference.

5.5.3 Huang Voltage Reference

Giustolisi *et al.* proposed a low-power voltage reference where MOSFETs in weak inversion are employed [187]. Giustolisi voltage reference was simplified by Huang *et al.* [188]. The simplified schematic of Huang voltage reference is shown in Fig.5.23. M3, M8, and M9 operate in weak inversion, and are sized to be identical. M1, M6, and M7 also have the same dimensions. This ensures that $I_{D1} = I_{D6} = I_{D7}$. Let us focus on the PTAT circuit consisting of M8, M9, and R_2 first. Since

$$V_{GS8} = V_{GS9} + R_2 I_{D9}, \tag{5.142}$$

we have

$$I_{D9} = \frac{1}{R_2} (V_{GS8} - V_{GS9}). \quad (5.143)$$

Since M8 and M9 operate in weak inversion, we have from (5.47) that

$$V_{GS} = V_T + nV_t \ln \left(\frac{I_D}{2n\mu_n C_{ox} S V_t^2} \right). \quad (5.144)$$

Utilizing (5.144) in (5.143) yields

$$I_{D9} = \frac{nV_t}{R_2} \ln \left(\frac{S_9}{S_8} \right). \quad (5.145)$$

It becomes evident that M8, M9, and R_2 create a PTAT current I_{D9}. Having obtained the wanted PTAT variable, we now need to have a variable whose temperature coefficient is negative in order to create a temperature-independent reference voltage. Let us shift our focus to M3 and R_1. The current of M4 is determined from

$$I_{D4} = \frac{V_{GS3}}{R_1} - NI_{D9}, \quad (5.146)$$

where $N = S_5/S_7$. Eq.(5.146) reveals that I_{D4} contains the PTAT variable I_{D9} and V_{GS3}. I_{D4} can be made independent of temperature by properly choosing the value of R_2 and R_1, and the transistor aspect ratios. To improve the design flexibility, I_{D4} is mirrored to M11 and the output voltage is taken from R_3

$$V_{ref} = R_3 \left[\frac{S_{10}}{S_7} I_{D9} + \frac{S_{11}}{S_2} I_{D4} \right]. \quad (5.147)$$

Making use of (5.145) and (5.146), we arrive at

$$V_{ref} = \frac{R_3}{R_2} \left(\frac{S_{10}}{S_7} - N \frac{S_{11}}{S_2} \right) \ln \left(\frac{S_9}{S_8} \right) nV_t + \frac{R_3}{R_1} \frac{S_{11}}{S_2} V_{GS3}. \quad (5.148)$$

It is evident from (5.148) that the first term on the right hand side is PTAT while the second term drops with temperature. V_{ref} can therefore be made independent of temperature (to the first-order). It should be noted that sizing M3, M8, and M9 to be identical is important. The same holds for M1, M6, and M7 as well. This is because M3, M8, and M9 are temperature sensors, specifically, M3 and R_1 create I_{R1} that decreases with temperature while M8, M9 and R_2 create I_{R2} that is PTAT with a positive temperature coefficient.

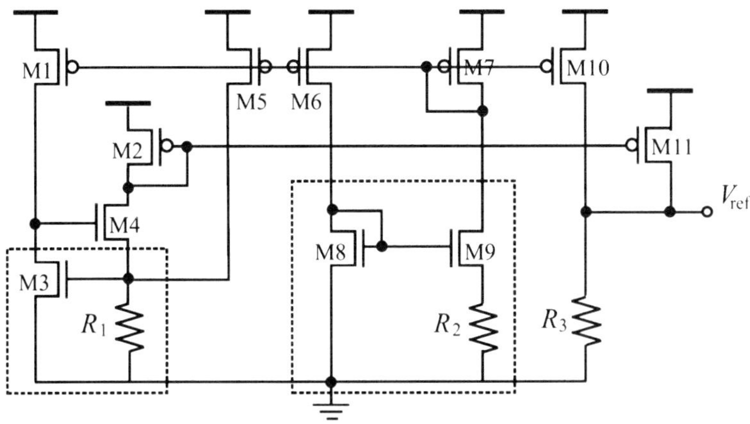

Figure 5.23. Voltage reference proposed by Huang *et al.* [188].

5.5.4 Ueno Voltage Reference

Since the variation of the resistance of both diffusion and poly resistors is large, it is desirable to avoid using passive resistors in design of voltage references. The voltage reference proposed by Ueno *et al.* and shown in Fig.5.24 removes the need for passive resistors [144]. In addition, all transistors except M8, which operates in the deep triode and behaves as a resistor, are in weak inversion. Transistors M9-M13 are sized to have the same dimension such that all the branches of these transistors carry the same current. M1, M2, and M8 generate I that is PTAT, similar to that shown in Fig.5.23. The key of Ueno voltage reference is the use of the cascaded stage consisting of M3-M7. Because

$$V_{ref} = V_{GS7} + V_{DS6} + V_{DS4}, \tag{5.149}$$

and

$$\begin{aligned} V_{DS6} &= V_{GS6} - V_{GS5}, \\ V_{DS4} &= V_{GS4} - V_{GS3}, \end{aligned} \tag{5.150}$$

(5.149) becomes

$$V_{ref} = V_{GS7} + (V_{GS6} - V_{GS5}) + (V_{GS4} - V_{GS3}). \tag{5.151}$$

Making use of (5.47) and noting that $I_{D6} = 2I_{D5}$, we obtain

$$V_{ref} = V_{GS4} + nV_t \ln\left(\frac{S_5 S_3}{S_6 S_7}\right). \tag{5.152}$$

It is evident that the first term on the right hand side of (5.152) drops with temperature while the second term is PTAT with a positive temperature coefficient. V_{ref} can therefore be made independent of temperature (to the first-order).

A notable advantage of Ueno voltage reference is its low power consumption, owing to the sub-threshold operation of all the transistors except M8. This differs from other sub-threshold references studied earlier where only temperature sensing transistors operate in the sub-threshold region. The low V_{GS} of sub-threshold MOSFETs also lowers the minimum supply voltage, allowing Ueno voltage reference implemented in a 0.35μm CMOS technology to operate with a sub-1V supply voltage.

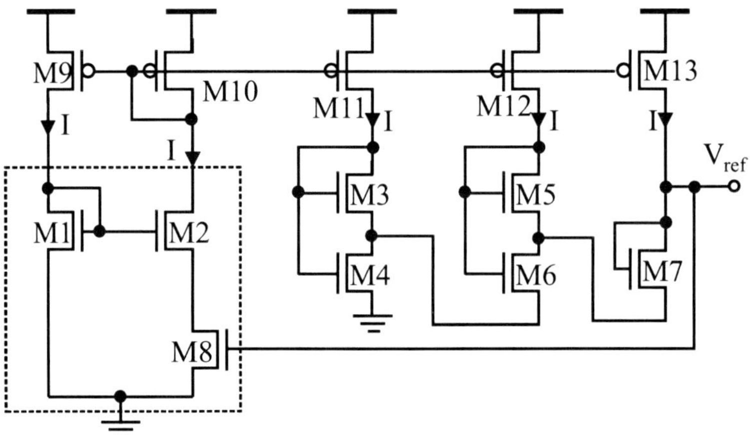

Figure 5.24. Voltage reference proposed by Ueno *et al.* [144].

5.5.5 De Vita - Iannaccone Voltage Reference

The voltage reference proposed by De Vita and Iannaccone employs MOS devices of different threshold voltages, which are typically available in standard CMOS technologies, to achieve a low-voltage operation [32]. In addition, the power and silicon consumption are minimized by eliminating passive resistors. The simplified schematic of De Vita - Iannaccone voltage reference is shown in Fig.5.25. M1 and M3 are implemented using nMOS devices with a high threshold voltage while M2 and M4 are implemented using nMOS devices with the nominal threshold voltage. The purpose of doing so is to allow M1 and M3 to operate in weak inversion, and M2 and M4 to operate in strong inversion at the same time. M5-M6 is a unity-gain current mirror. The same is for M7-M9. Making use of (5.47) for M1 and M3 yields

$$V_{GS1,3} = V_{T1,3} + nV_t \ln\left(\frac{I_{D1,3}}{2n\mu_n C_{ox} S_{1,3} V_t^2}\right). \quad (5.153)$$

For M2 and M4, we have

$$V_{GS2,4} = V_{T2,4} + \sqrt{\frac{2I_{D2,4}}{\mu_n C_{ox} S_{2,4}}}. \quad (5.154)$$

Because $V_{GS1} = V_{GS2}$ and $V_{GS3} = V_{GS4}$, we have

$$V_{T1} + nV_t \ln\left(\frac{I_{D1}}{2n\mu_n C_{ox} S_1 V_t^2}\right) = V_{T2} + \sqrt{\frac{2I_{D2}}{\mu_n C_{ox} S_2}}, \quad (5.155)$$

and

$$V_{T3} + nV_t \ln\left(\frac{I_{D3}}{2n\mu_n C_{ox} S_3 V_t^2}\right) = V_{T4} + \sqrt{\frac{2I_{D4}}{\mu_n C_{ox} S_4}}. \quad (5.156)$$

Subtract (5.155) from (5.156) and noting that $V_{T1} = V_{T3}$ and $V_{T2} = V_{T4}$, we arrive at

$$I_{D4} = \frac{1}{2(N-1)^2} \mu_n C_{ox} n^2 V_t^2 S_4 \ln^2\left(\frac{S_3}{S_1}\right), \quad (5.157)$$

where $N = \sqrt{\frac{S_4}{S_2}}$. Since

$$I_{D10} = \frac{1}{2} \mu_n C_{ox} S_{10} \left(V_{ref} - V_T\right)^2, \quad (5.158)$$

we have

$$V_{ref} = V_T + \sqrt{\frac{2I_{D10}}{\mu_n C_{ox} S_{10}}}. \quad (5.159)$$

Substituting (5.157), we obtain the output voltage $V_{ref} = V_{GS10}$

$$V_{ref} = V_T + \frac{nV_t}{N-1} \sqrt{\frac{S_4}{S_{10}}} \ln\left(\frac{S_3}{S_1}\right). \quad (5.160)$$

A notable difference between (5.160) and the expression of the reference voltages of other voltage references in sub-threshold obtained earlier is that the term with a negative temperature coefficient is the threshold voltage rather than the gate-source voltage. Since the former exhibits a fairly linear dependence on temperature while the latter contains a non-negligible nonlinear term, the temperature coefficient of the output voltage of De Vita - Iannaccone voltage reference will be smaller.

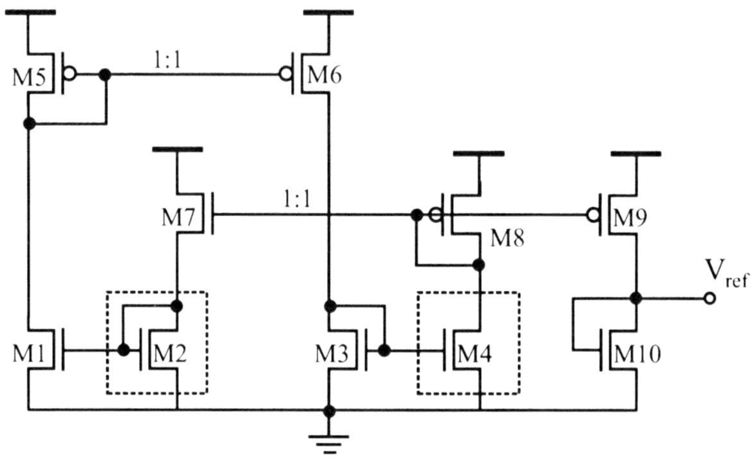

Figure 5.25. Voltage reference proposed by De Vita and Iannaccone [32].

Qu *et al.* modified De Vita-Iannaccone voltage reference by employing a bulk-driven technique such that a sub-threshold operation can be obtained without using devices of different threshold voltages [189]. The simplified schematic of Qu voltage reference is shown in Fig.5.26. M1 and M3 are bulk-driven by V_b such that their threshold voltage is reduced. The bulk of M2 and M4, on the other hand, is connected to V_{DD} and their threshold voltages remain unchanged. The gate voltages of M1-M4 are controlled in such a way that M1 and M3 operate in strong inversion while M2 and M4 are in weak inversion. The operation of this voltage reference is similar to that of De Vita-Iannaccone voltage reference. Readers can derive the output voltage of this reference by following the steps detailed in depicting De Vita-Iannaccone voltage reference.

5.5.6 Sub-threshold Voltage References Without Amplifiers

A common drawback of many of the preceding voltage references in weak inversion is that an auxiliary amplifier is needed to generate the required PTAT current. Although exhibiting superior performance, these references generally consume more power. It is interesting to note that the original voltage reference proposed by Widler does not use auxiliary amplifiers [153, 154]. Chang *et al.*

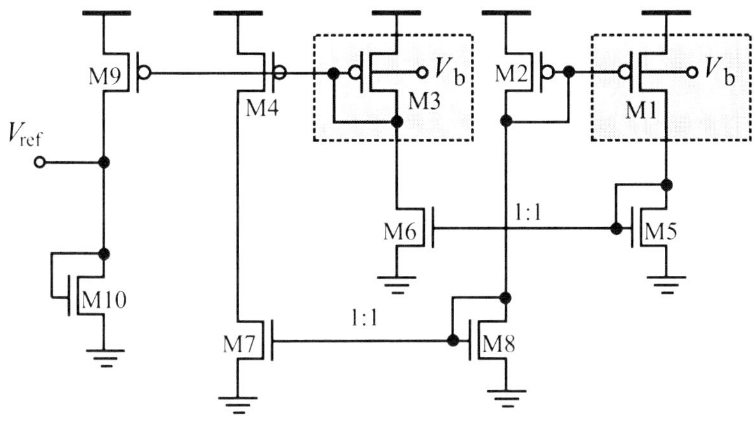

Figure 5.26. Voltage reference proposed by Qu *et al.* [189].

showed that sub-threshold voltage references can be constructed without using an auxiliary amplifier [190]. Fig.5.27 shows the simplified schematic of the voltage reference proposed by Chang *et al.*. M1 and M2 are identical. M3 and M4 are in weak inversion. Together with R_1, they create a PTAT current I_{D3}

$$I_{D3} = \frac{V_{GS4} - V_{GS3}}{R_1} = \frac{nV_t \ln(n)}{R_1}, \tag{5.161}$$

where $n = S_3/S_4$. M4 and R_2 ensure that

$$V_A = V_{GS4} + R_2 I_{R2}. \tag{5.162}$$

Because

$$I_{R2} = I_{D3} + I_{D4} = 2I_{D3}, \tag{5.163}$$

we have

$$V_A = V_{GS4} + 2nV_t \frac{R_2}{R_1} \ln(n). \tag{5.164}$$

As a result, V_A can be made independent of temperature at reference temperature T_o. The output voltage is given by

$$V_{ref} = \left(1 + \frac{R_3}{R_4}\right) V_A. \tag{5.165}$$

It is important to note that V_A and V_{ref} in (5.164) and (5.165) are only related to the ratio of the resistance of the resistors rather than the absolute value of the resistance of the resistors. This ensures that temperature-induced variation of the resistance of the resistors has a little impact on the reference voltage. Also noted is that the use of passive resistors in this voltage reference makes it difficult to lower the power consumption unless large resistors are used. This will be at the cost of a large silicon consumption and a strong interaction with the substrate if diffusion resistances are employed.

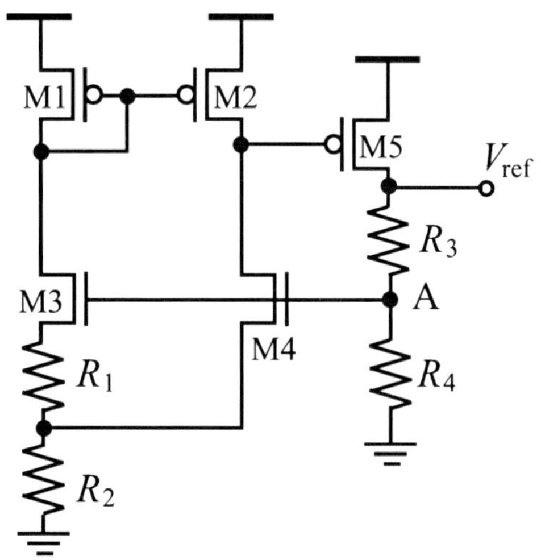

Figure 5.27. Voltage reference proposed by Chang *et el.* [190].

Another example of voltage references without an auxiliary amplifier is the voltage reference proposed by Lin and Liang with its simplified schematic shown in Fig.5.28 [159]. The ratio of the emitter area of Q1 and Q2 is $A_{e1} = nA_{e2}$. Since

$$I_{R2} = \frac{V_{BE2}}{R_2}, \tag{5.166}$$

by neglecting the base current of Q2, we have

$$V_A \approx V_{BE2} + \frac{R_1}{R_2} V_{BE2}. \tag{5.167}$$

The output voltage is obtained from

$$V_{ref} \quad = \quad V_A - V_{BE1}$$

$$= (V_{BE2} - V_{BE1}) + \frac{R_1}{R_2} V_{BE2}. \quad (5.168)$$

Making use of (5.60), we can write (5.168) as

$$V_{ref} = V_t \ln(n) + \frac{R_1}{R_2} V_{BE2}. \quad (5.169)$$

The compensation of V_{BE2} using a PTAT quantity is evident in (5.169). Similar to Chang voltage reference studied earlier in Fig.5.27, the use of passive resistors in Lin-Liang voltage reference makes it difficult to lower its power consumption unless large resistors are employed. For example, if $V_{BE2} = 0.7$ V, to have $I_{R2} = 10$ nA, $R_2 = 70$ MΩ is required.

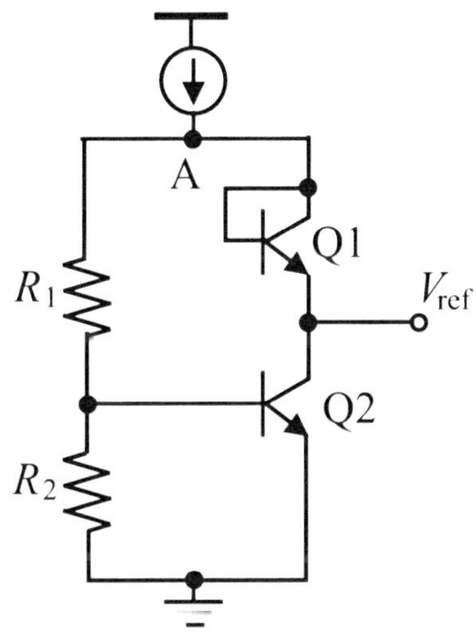

Figure 5.28. Voltage reference proposed by Lin and Liang [159].

5.5.7 Comparison of Sub-Threshold Voltage References

Table 5.5 compares the performance of sub-threshold voltage references reported recently. As compared with those in strong inversion, the power consumption of voltage references in the sub-threshold is much lower. The temperature coefficient of these voltage reference, however, is in general larger as compared with that of voltage references in strong inversion. Some elegant designs, such as Yu-Zhou voltage reference [149], Qu voltage reference [189], and Ueno voltage reference [144] offer both low temperature coefficients and low power consumption.

Table 5.5. Performance comparison of sub-threshold voltage references.

Reference	Tech.	TC [ppm/°C]	Temp. [°C]	Power [μW]	PSRR [dB]	Min. V_{DD} [V]
Ytterdal (03)[185]	0.13μm	93	-40∼100	–	–	0.6
Giustolisi *et al.* (03)[187]	1.2μm	119	-25∼125	4.32	40	–
Cheng-Wu (05)[186]	0.35μm	62	0∼70	4.6	84	1.4
Lin and Liang (05)[159]	0.18μm	63	-20∼100	2.4	–	1
Huang *et al.* (06)[188]	0.18μm	194	-20∼120	3.3	–	0.85
De Vita-Iannaccone (07)[32]	0.35μm	10	0∼80	0.28	53	0.9
Chang *et al.* (07)[190]	0.18μm	11.5	-20∼120	17.25	–	1.5
Ferreira *et al.* (08)[191]	0.35μm	39	-20∼80	0.39	25	0.95
Yu-Zou(08)[149]	0.18μm	7.7	-40∼150	2.55	126	1.2
Qu *et al.* (09)[189]	0.18μm	4.6	-40∼85	0.35	40.45	0.58
Ueno *et al.* (09)[144]	0.35μm	15	-20∼80	0.3	45	<1

5.6 Chapter Summary

In this chapter, we have examined the widely used figure-or-merits that quantify the performance of voltage references. Among them, the normalized average temperature coefficient or the effective temperature coefficient is perhaps the most important parameter depicting the effect of temperature on the performance of voltage references. The temperature-dependent characteristics of MOS devices have been investigated in detail. We have shown that the base-emitter voltage of bipolar junction transistors decreases with temperature in a nonlinear fashion, however, with its linear temperature-dependence dominates. Both NPN and PNP bipolar junction transistors can be constructed using vertical configurations in CMOS technologies. Since these BJTs are used for temperature sensing only, there is no constraint on their current gain. The only concern is the lack of precision mathematical models for these devices in standard CMOS as they are seldom used in other applications. We have also shown that the gate-source voltage of MOSFETs that operate in weak inversion decreases with temperature. The temperature-dependence of the gate-source voltage of MOSFETs in weak inversion bears a strong resemblance to that of the base-emitter voltage of BJTs. They can also be used to sense temperature. The low-current characteristics of MOSFETs in weak inversion also make them particularly attractive for passive wireless microsystems. Further, the low gate-source voltage requirement of MOSFETs in weak inversion also reduces the minimum supply voltage, making it attractive for applications where only a low supply voltage is available. We have further shown that the threshold voltage of MOSFETs also decreases with temperature. As compared with the temperature characteristics of other parameters, such as mobility, the dependence of the threshold voltage on temperature is fairly linear. As a result, high-precision voltage references can be constructed by utilizing this unique property.

The dependence of the mobility of minority charge carriers on temperature has been studied. We have shown that the mobility drops with temperature in a nonlinear fashion. The combined effect of mobility reduction and threshold voltage reduction yields the zero-temperature-coefficient bias point of a transistor at which the channel current of the transistor becomes independent of temperature. The resistance of poly resistors and that of diffusion resistors are also functions of temperature. Not only the temperature coefficient of diffusion resistors is typically positive while that of poly resistors is negative, the dependence of the resistance of diffusion resistors is typically several times that of poly resistors.

A temperature-independent reference voltage can be constructed by utilizing the negative temperature dependence of the base-emitter voltage of BJTs, the negative temperature dependence of the gate-source voltage of MOSFETs in weak inversion, and proportional-to-absolute-temperature circuits that generate a PTAT voltage or current with a positive temperature coefficient. Since the output of PTAT circuits is directly proportional to temperature, only first-order voltage references can be constructed in this way. A notable characteristic of first-order voltage references is their large temperature coefficient. To reduce the effect of temperature on the output voltage of voltage references, the nonlinear dependence of V_{BE} of BJTs or V_{GS} of MOSFETs in weak inversion must also be compensated.

Widlar voltage reference requires a large supply voltage, mainly due to the large bandgap voltage of silicon. Banba voltage reference lowers the supply voltage to below 1 V. The design of Banba voltage reference becomes difficult when the supply voltage is low. This is due to the large variation of the common-mode input voltage of its auxiliary amplifier over the desired temperature range. Waltari and Halonen showed that this deficiency can be removed effectively by employing simple resistor dividers. A common issue associated with Widlar, Banba, and Waltari-Halonen references is their need for vertical BJTs to sense temperature. Often either there is no models for these devices or they are poorly modeled in standard CMOS technologies. As a result, it becomes difficult if not possible to accurately predict the performance of voltage references in the design stage. Threshold voltage-based voltage references cleverly avoid this obstacle. Since the dependence of the threshold voltage of MOSFETs on temperature is fairly linear, these voltage references can provide a better accuracy. The use of these voltage references for low-power applications, especially passive wireless microsystems, however, is hampered by their high power consumption. Another common deficiency of Widlar, Banba, Waltari-

Halonen, and threshold voltage-based voltage references is their dependence of passive resistors. Since both diffusion and poly resistors have a poor resistance accuracy in most CMOS technologies especially low-cost CMOS technologies, excessive resistance trimming becomes mandatory in these voltage references in order to achieve a high accuracy. On top of that, it is extremely costly to use these references for ultra-low power applications as the resistance of these resistors must be prohibitively large. This is echoed with a large silicon consumption and a strong interaction with the substrate if diffusion resistors are used. Buck voltage reference removes the need for passive resistors, making it an ideal candidate for low-power applications.

To improve the accuracy of reference voltages, the nonlinear dependence of V_{BE} of BJTs or V_{GS} of MOSFETs in weak inversion on temperature must be accounted for. These voltage references are curvature-compensated. Two key approaches, namely closed-loop and open-loop approaches to achieve curvature-compensation have been studied. The former makes use of negative feedback to stabilize the output voltage while the latter utilizes another circuit whose output voltage or current is a nonlinear function of temperature to offset the effect of temperature on V_{BE} or V_{GS}. The piecewise linear curvature-compensated voltage reference proposed by Huang *et al.* is a closed-loop approach while Malcovati voltage reference, Ker-Chen voltage reference, and resistor-based curvature-compensated voltage references are the members of the family of open-loop approaches. Because high-order voltage references are often more complex in configuration as compared with their first-order counterparts, these voltage references in general consume more power. Innovations are needed in design of low-voltage low-power curvature-compensated voltage references.

To lower the power consumption and at the same time to achieve a high accuracy, sub-threshold voltage references have attracted a lot of attention recently. We have shown that these references operate on the principle that V_{GS} of MOSFETs in weak inversion drops with temperature nonlinearly in a very similar way as that of V_{BE} of BJTs. The small V_{GS} of MOSFETs in weak inversion also opens a door for the low-voltage design of these references. With more research is being conducted on this subject worldwide, it is certain that more innovative and elegant sub-threshold voltage references will emerge.

Chapter 6

CLOCK GENERATION AND CALIBRATION

The operation of the baseband units of a passive wireless microsystem and the up-link from the microsystem to its base station are controlled by its system clock. To ensure a reliable communications between the microsystem and its base station, a stringent requirement on the frequency of the system clock exists. For example, EPC radio-frequency identity protocols class-1 generation-2 UHF RFID protocols require that the accuracy of the frequency of the backscattered data be upper-bounded by $\pm 4\%$ [53]. Often, the system clock of a passive wireless microsystem is generated by a local oscillator residing in the microsystem. Arising from the effect of process, supply voltage, and temperature change (PVT) variations, the frequency of the local oscillator exhibits a high degree of uncertainty [33]. For example, the variation of the frequency of a free-running oscillator can be as large as $\pm 20\%$ [192]. The uncertainty of the frequency of the system clock of passive wireless microsystems is further escalated by the fact that these systems are usually fabricated using low cost CMOS technologies, which typically have a high degree of process variation. Although the effect of temperature variation on the oscillation frequency of the local oscillator can be minimized by utilizing high-precision reference circuits such as those presented in Chapter 5, their effectiveness is largely hindered by the limited power resource of passive wireless microsystems, which prevents the deployment of complex reference circuits.

This chapter deals with generation and calibration of the system clock of passive wireless microsystems. Section 6.1 focuses on the generation of the system clock of passive wireless microsystems directly from the carrier of the received RF signal. The system clock of the baseband units of the microsystems can also be generated from the envelope of the received RF signal, as detailed in Section 6.2. The system clock of passive wireless microsystems can be generated using a local oscillator directly. Since the frequency of the oscillator

F. Yuan, *CMOS Circuits for Passive Wireless Microsystems*,
DOI 10.1007/978-1-4419-7680-2_6,

is subject to the effect of PVT, the calibration of the frequency of the oscillator prior to data communications becomes indispensable. The calibration of the frequency of the system clock of the passive wireless microsystems using injection-locking with the carrier of the received RF signal as the injection signal is investigated in Section 6.3. Frequency calibration of the local oscillator using digital trimming techniques is studied in Section 6.4. Frequency calibration using phase-locked loops is examined in Section 6.5. Frequency calibration using digital frequency-locked loops is investigated in Section 6.6. Section 6.7 explores the calibration of the frequency of the local oscillator of passive wireless microsystems using injection-locking with the envelope of the received RF signal as the injection-locking signal. Integrating feedback that computes the accumulated difference between the frequency of the injection-locking reference and that of the local oscillator is used to pull the frequency of the local oscillator towards that of the injection-locking reference such that the effective lock range can be dramatically increased. The chapter is concluded in Section 6.9.

6.1 Clock Generation From Carrier

As pointed out in earlier chapters that amplitude-shift-key is widely used in passive wireless microsystems due to its ease of implementation subsequently low power consumption. The ASK-modulated signal received by a passive wireless microsystem from its base station typically consists of a high-frequency carrier with its frequency located in ISM bands and a low-frequency envelope representing the baseband data or timing information. Due to the unavailability of crystal oscillators and other precision reference components in a passive wireless microsystem, the timing reference required for calibrating the frequency of the local oscillator of the passive wireless microsystem can only be provided by the base station. Both the carrier and envelope of the received RF signal can be used as the timing reference to calibrate the local oscillator. The former contains the needed timing information while the latter might not have the timing information due to the randomness of the baseband data. Timing information must therefore be embedded in the envelope of the carrier by encoding the baseband data if the envelope is to be used as the timing reference.

A system clock can be generated directly from the carrier using a comparator-based clock recovery system, as illustrated graphically in Fig.6.1 [34, 193]. When the voltage of the received RF signal exceeds the reference voltage, a logic-1 state is set at the output of the comparator. Otherwise, a logic-0 state is set. Since the frequency of the recovered clock is the same as that of the carrier, a chain of frequency dividers are often required to lower the frequency of the recovered clock to the desired base band frequency at which the microsystems

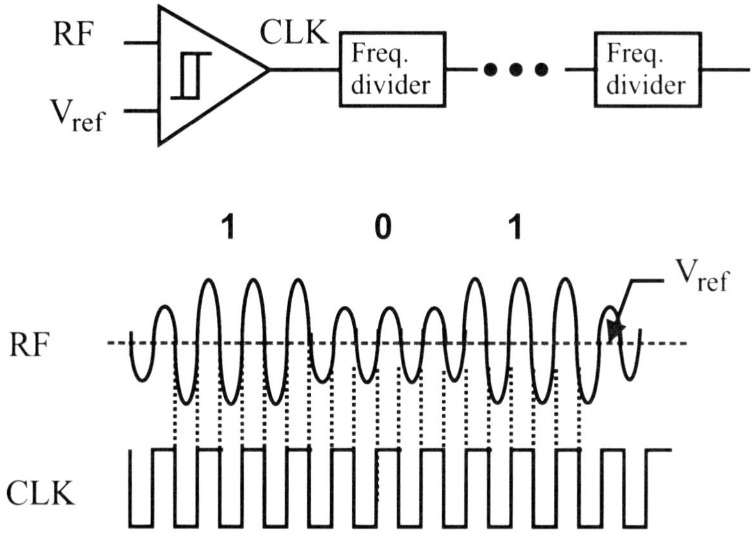

Figure 6.1. Clock generation from the carrier of an ASK-modulated RF signal.

operate. Although this approach is straightforward, a number of issues must be considered :

- Because the comparator must operate at the carrier frequency, its power consumption will be high, especially when the carrier frequency is high. Also, when the carrier frequency is exceedingly high, it becomes difficult to design comparators that can operate at the carrier frequency.

- Since the frequency of the recovered clock is high, a large number of frequency dividers are needed to lower the frequency of the recovered clock to the baseband frequency, typically in kHz or low MHz ranges. The power consumption of these frequency dividers, especially the first frequency divider in the frequency division chain that operates at the carrier frequency, will be significant.

- Because the amplitude of the received RF signal is usually small, especially in logic-0 states when the modulation index of the incoming RF signal is large or when the distance between the microsystem and its base station is large, a large timing error will exist in the recovered clock. In the extreme case where the modulation index of the received RF signal is 100%, no clock can be extracted when the received RF signal is at the logic-0 state, as shown in Fig.6.2 unless the baseband data are properly encoded at the base station prior to their transmission.

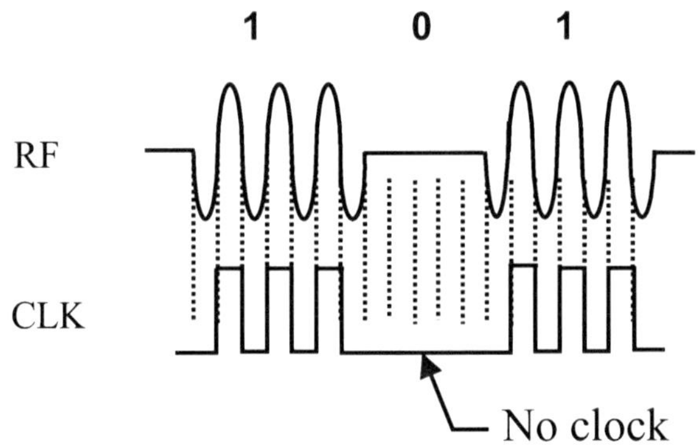

Figure 6.2. Clock generation from the carrier of an OOK-modulated RF signal.

It is evident from the preceding analysis that the generation of the baseband clock directly from the carrier should only be used if the received RF signal is strong and the frequency of the RF signal is not exceedingly high.

To improve the frequency accuracy and at the same time to reduce the power consumption of the preceding clock generation directly from the carrier, clock generation from the carrier with injection-locked frequency division was proposed by Leung and Luong [194], as shown in Fig.6.3. The principle of this approach is that when the incoming RF signal is strong, the system clock is extracted directly from the carrier using the comparator-based clock recovery approach discussed earlier to take the advantage of its simple configuration. In this case, an injection-locked divided-by-2 block is used to lower the frequency. The main purpose of using the injection-locked frequency divider is to take the advantage of the low power consumption and high frequency accuracy of injection-locked frequency dividers. This is because an injection-locked frequency divider is a closed-loop system while a conventional frequency divider is an open-loop system. Once frequency is divided by 2, it it further lowered using a set of TSPC (true-single-phase-clocked) conventional frequency dividers to take the advantages of their simple configuration subsequently low power consumption. Since these conventional frequency dividers operate at lower frequencies, the overall power consumption of the frequency division chain is reduced.

When the incoming RF signal is weak, a local free-running relaxation oscillator is used to generate the system clock. The threshold of the RF signal at which the input of the multiplexer is routed to the local oscillator is determined by an ASK demodulator and a baseband processor, as shown in Fig.6.3.

A number of issues must be addressed when using this approach for clock generation. First, when the incoming RF signal is weak, the clock is generated using a local free-running oscillator instead of generating from the carrier

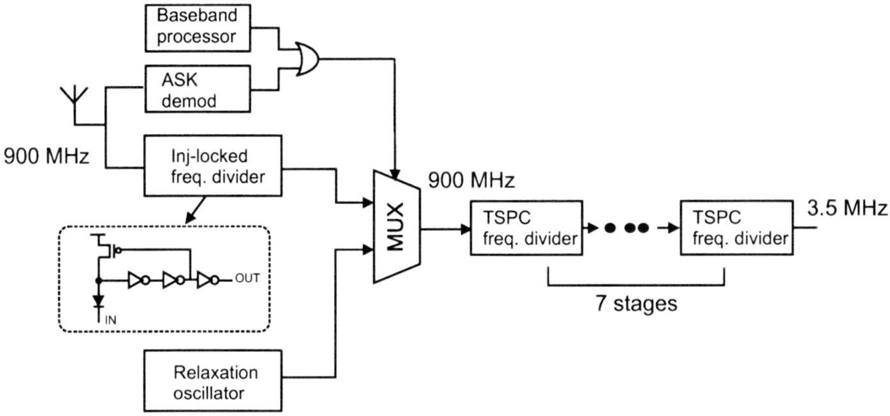

Figure 6.3. Frequency generation using injection-locked frequency division and local oscillator.

directly. No frequency calibration is performed in this case. The frequency of the local oscillator is thus subject to the effect of PVT and its accuracy can not be ensured. Secondly, because the local oscillator must oscillate at the carrier frequency, reducing its power consumption will be challenging. Further, a large number of frequency dividers are still needed in order to lower the frequency of the recovered clock to the base band. The power and silicon consumption of these frequency dividers can not be neglected. Finally, the lock range of the injection-locked frequency divider must be sufficiently large in order to ensure that the frequency divider will lock to the carrier in the presence of a strong PVT effect. The lock range of the injection-locked frequency divider in [194] is 180 MHz at the minimum input power of -22 dBm at 950 MHz. The lock range is increased to 595-1120 MHz when the input power is -12 dBm. The power consumption of the frequency divider in this case is 5.5 μW. The average power consumption of the system is found to be 7 μW.

If the injection-locked frequency divider is LC oscillator-based, the frequency lock range of these oscillators is typically small. Although the lock range can be increased by boosting the amplitude of the locking signal, given the fact that the amplitude of the received RF carrier signal is typically small, it will be difficult to increase the lock range of the injection-locked frequency divider. The injection-locked frequency divider might not be able to lock to the carrier if the effect of PVT is strong.

6.2 Clock Generation From Envelope

As pointed out earlier that the baseband frequency of passive wireless microsystems is typically in kHz or low MHz ranges, and the modulation scheme used for data communications between a passive wireless microsystem and its base station is often ASK to take the advantage of its ease of implementation subsequently low power consumption. A baseband timing reference can therefore be embedded in the envelope of a RF signal through encoding and transmitted from a base station to a microsystem. The envelope of the received RF signal can be extracted using an envelope detector at the receiving end and used to recover the clock [195, 196]. Note that the carrier is periodic naturally and contains the required timing information for clock generation. The envelope, on the other hand, might not contain a timing reference unless it is encoded with timing information.

EPC radio-frequency identity protocols class-1 generation-2 UHF RFID protocols use pulse-interval-encoding (PIE) for interrogator-to-tag communications [53]. As shown in Fig.6.4, the logic state of data is represented by the length of the interval. A short interval represents binary 0 while a long interval represents binary 1. The duration of a binary 0 is one Tari (*T*ype *A* *R*eference *I*nterval, ISO 18000-6 type A) while that of a binary 1 is between 1.5 Tari to 2 Tari where Tari is the unit of time duration with its value varying from 6.35μs to 25μs. The pulses of binary 0 and binary 1 are terminated with a logic-0 pulse of width between 0.265 Tari and 0.525 Tari. PIE coding guarantees that there always exists a state transition in each symbol.

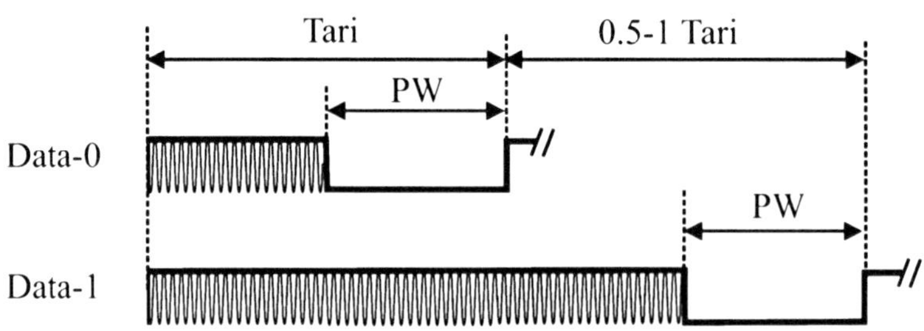

Figure 6.4. Pulse-interval-encoding (PIE) for interrogator-to-tag communications EPC radio-frequency identity protocols class-1 generation-2 UHF RFID protocol for communications at 860-960 MHz. 0.365 Tari $\leq$ PW $\leq$ 0.525 Tari [53].

Fig.6.5 shows the configuration of clock recovery directly from the envelope of the received RF signal [193]. The received RF signal is first fed to an envelope detector where the envelope of the modulated RF signal is extracted. The extracted envelope of the received RF signal is then fed to a comparator to restore the voltage swing of the envelope. Comparators with hysteresis are preferred as the output of the envelope detector contains a large number of ripples, as detailed in Chapter 4. Since there always exists a transition in each of the symbol, the clock is recovered by the comparator. The data is obtained by employing an integrator whose main function is two-fold : (i) To minimize the effect of high-frequency disturbances coupled to the output of the trigger, similar to the approach used for Gb/s serial links [197]. The integrator functions as a low-pass filter. (ii) To recover the data through integration. Since the voltage of the integrating capacitor at the end of the pulse corresponding to binary 1 is larger as compared with that at the end of the pulse corresponding to binary 0, the value of the baseband data can be determined by measuring the voltage of the integrating capacitor with another comparator.

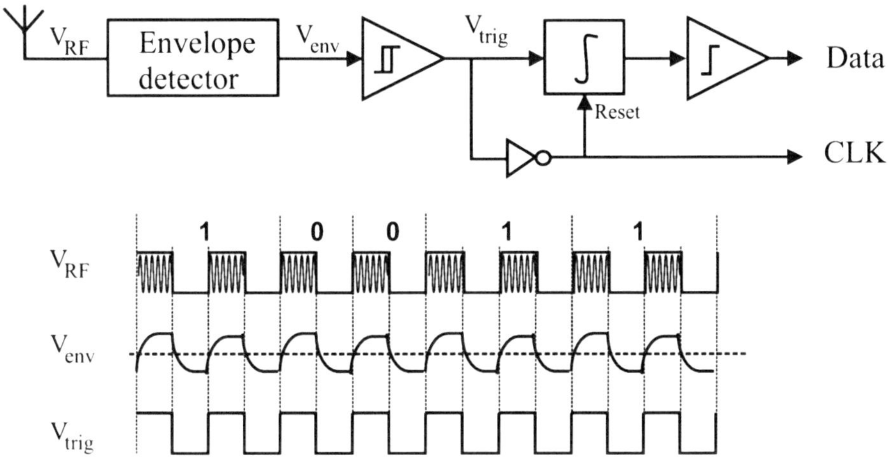

Figure 6.5. Clock generation from the envelope of a RF signal.

As compared with clock generation from the carrier, the frequency of the clock generated from the envelope of the received RF signal is much lower. It can be used directly for baseband operation. Frequency dividers are no longer needed. As a result, the power and silicon consumption is reduced. A drawback of this approach is that the RF signal must be properly encoded such that a timing reference is continuously available during the operation of the microsystem. Since data encoding is done at base stations, no computational overhead is imposed on microsystems.

6.3 Clock Generation Using Carrier Injection-Locking

Unlike the methods where the system clock is extracted from either the carrier or the envelope of the received RF signal, the clock of the passive wireless microsystem in [11, 12] is generated using a local LC tank oscillator that oscillates at 900 MHz, as shown in Fig.6.6. To ensure that the local oscillator will oscillate at the desired frequency preciously, an injection-locking operation with the carrier of the received RF signal as the injection signal is performed on the local oscillator. Because harmonic oscillators, such as LC tank oscillators, can be injection-locked to an injection signal at either a super-harmonic or a sub-harmonic frequency, an injection signal at the second sub-harmonic frequency (450 MHz) is used in [11, 12] to lock the frequency of the local oscillator at 900 MHz such that the up-link (microsystem to base station) and down-link (base station to microsystem) can take place at different frequencies. An advantage of using different frequencies for up-links and down-links is to avoid frequency collision when a power amplifier is employed at the microsystem end for up-links over a long distance. Once the local LC oscillator is locked to the incoming RF carrier, its output is fed to a chain of frequency dividers to generate the desired baseband frequency.

The key advantage of carrier-based injection-locking for frequency calibration is the high frequency accuracy of the local oscillator, which is ensured by the intrinsic property of injection-locking [198]. Another notable advantage of this method is that the amplitude of the injection-locking signal needs not to be large. As a result, wireless microsystems can reside at a large distance away from the base station. This also allows the use of a RF signal with a small amplitude variation from the base station in the calibration, which in turn maximizes the amount of power transferred from the base station to the passive wireless microsystem during the calibration phase.

A number of factors need to be considered when using carrier-based injection-locking for frequency calibration of the system clock of passive wireless microsystems. The frequency lock range of LC tank oscillators is typically small. This will prevent the local LC oscillator from locking to the RF carrier once the effect of PVT on the frequency of the local oscillator is strong. For example, the lock range of the LC-VCO reported in [11, 12] is 12 MHz when locking the 900 MHz oscillator to a 450 MHz 140 mV injection signal. The lock range is only 1.33% of the oscillation frequency. In addition to the small frequency lock range, the local LC oscillator must oscillate at the carrier frequency or the super-harmonics of the carrier frequency. As a result, the oscillator will consume a large amount of power. For example, the power consumption of the LC-VCO in [11, 12] is 6 mW. This is clearly undesirable. Moreover, the high frequency of the local oscillator requires a large number of frequency dividers to generate the base band frequency at which baseband units such as ADCs and DSPs operate. This further increases the demand for power and silicon area.

Since the frequency of the local oscillator is calibrated using the timing reference from the base station, we term this approach remote frequency calibration to distinguish it from the methods that generate the local clock from the received RF signal directly without using a local oscillator. Remote frequency calibration is of a critical importance to embedded or implanted microsystems. Since the system clock of the these microsystems is generated locally, there is no need for the base station to continuously supply a timing reference to the microsystems during their operation. As a result, the constraint on encoding data to embed a timing reference is relaxed. Further, the calibration of the

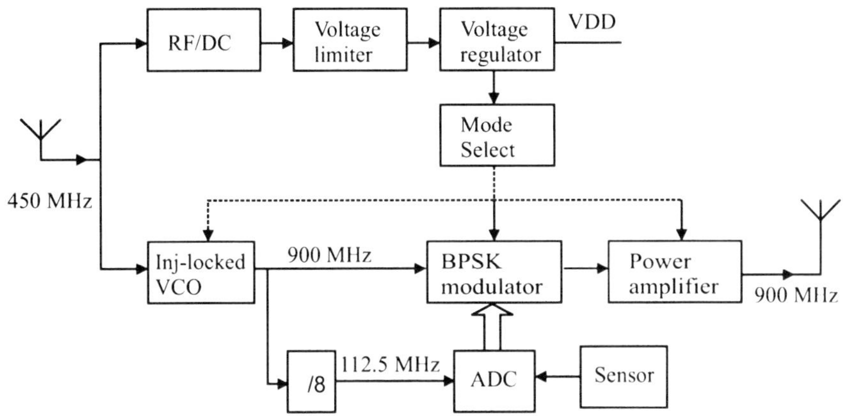

Figure 6.6. Remote frequency calibration of local oscillator of wireless microsystems using carrier-based injection-locking [11, 12].

local oscillator can be performed prior to data transmission between the base station and the microsystems. As an example, the signal from a reader to a tag in EPC class-1 generation-2 UHF RFID communication protocols starts with a preamble of 12.5 μs, as shown in Fig.6.7. The 12.5 μs delimiter can be used to perform frequency calibration [199].

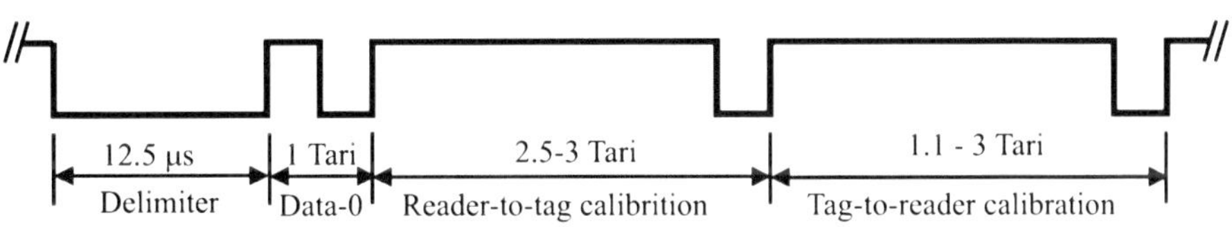

Figure 6.7. Reader-to-tag preamble of EPC UHF RFID communication protocol for communications at 860-950 MHz [53].

The power consumption of the clock generation and calibration of the local oscillator of a microsystem using carrier-based injection-locking can be reduced by conducting remote frequency calibration of the local oscillator using envelope-based injection-locking. To do so, the envelope of the RF signal transmitted in the frequency calibration phase must contain the required timing information. We will detail this in Section 6.7.

6.4 Clock Generation Using Digital Trimming

The drawback of the small locking range of frequency calibration using carrier-based injection-locking can be eliminated by using digital trimming techniques [199, 28, 30, 200, 106]. Fig.6.8 shows the configuration of this approach. The system clock of a wireless microsystem is generated using a local oscillator, often a current-starving voltage-controlled oscillator to take the advantage of its simple configuration, low power consumption, and a low sensitivity to supply voltage fluctuation. The frequency of the oscillator is calibrated in a calibration phase prior to the start of data transmission. The time duration of the calibration phase is set to be much larger than the period of the oscillator such that the number of the oscillation cycles of the oscillator in the calibration phase can be used to measure the frequency of the oscillator. As an example, if the 12.5μs delimiter of EPC UHF class-1 generation-2 RFID communication protocols is used as the calibration phase and the desired frequency of the local oscillator is 10 MHz, the number of the cycles of the oscillator in the delimiter will be 125.

The control current of the oscillator is supplied by a digital-to-analog converter whose input is fed by a successive approximation register (SAR). The control current of the local oscillator is preset to a certain value by the SAR and the oscillator oscillates at the corresponding frequency. A counter is used to record the number of the oscillation cycles of the oscillator in the delimiter. The counter starts counting at the onset of the delimiter and continues its counting until the end of the delimiter. If the content of the counter at the end of the delimiter is 125, the frequency of the oscillator is 10 MHz with an error $\Delta f/2^N$ where Δf is the frequency tuning range and N is the number of the bits of the SAR. If the content of the counter is less than 125, the most significant bit of the SAR is set to 1. Otherwise, it is set to 0. The output of the SAR is fed to the digital-to-analog converter whose analog output adjusts the control current of the oscillator subsequently the frequency of the oscillator. The second round of frequency calibration test is then performed and the content of the counter at the end of the delimiter is compared with 125 again. Depending upon the result of the comparison, the second most significant bit of the SAR is set in a similar way as that of the most significant bit. This process continues until all bits of the SAR are set.

It becomes apparent that the accuracy of frequency calibration using digital trimming is set by the number of the bits of the SAR that controls the resolution of the control voltage of the oscillator and the frequency tuning range of the oscillator. The larger the number of the bits of the SAR, the higher the accuracy of frequency calibration. This, however, is at the cost of a longer calibration time.

A key advantage of frequency calibration using digital trimming is the large frequency adjustment range, which is upper bounded by the frequency tuning range of the oscillator. As long as the frequency tuning range of the oscillator is sufficiently large, frequency calibration using digital trimming will guarantee that the frequency of the oscillator be tuned to the desired frequency. Frequency calibration using injection-locking, however, is upper-bounded by the lock range of the local oscillator, which is typically small.

The use of frequency calibration using digital trimming is affected by a number of factors. First, this approach requires a successive approximation register, a digital-to-analog converter, and other logic for pulse counting, comparison, and control signal generation. The power consumption of these blocks will add to the total power consumption. For example, the power consumption of the current-starving VCO in [200] is 9.44 μW while the power consumption of the digital frequency calibration block is 31 μW. Other blocks such as the bias generator, address decoder and voltage regulator consume additional 4 μW. Second, the accuracy of frequency calibration is set by the number of the bits of the SAR and the frequency tuning range of the oscillator. The larger the number of the bits of the SAR, the higher the calibration accuracy. Since each bit of the SAR is set in one cycle of counting over the delimiter and comparison operation, the larger the number of the bits of the SAR, the longer the calibration time. For example, the digital frequency calibration in [200] uses an 8-bit SAR and a 9-bit counter. The value of each bit of the SAR requires approximately 116 μs to set. A total of approximately 928 μs is required to set all bits of the SAR. Finally, the length of the calibration phase and the frequency of the local oscillator must be known a prior such that the preset value with which the content of the counter is compared can be determined.

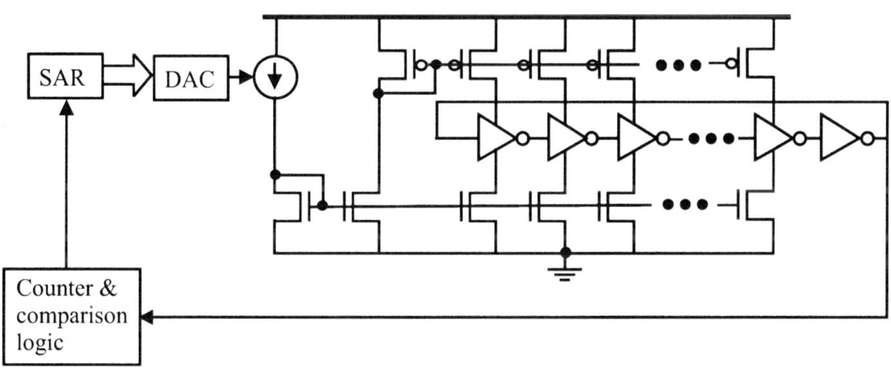

Figure 6.8. Frequency calibration using digital trimming.

6.5 Clock Generation Using Phase-Locked Loops

It is well understood that the output of a phase-locked loop has the ability to lock to a reference in both frequency and phase when a phase/frequency detector is employed. As compared with injection-locking, phase-locked loops provide a much larger frequency lock range. This, however, is at the cost of high power consumption and a long lock time. If the power consumption can be tightly controlled and the frequency that PLLs lock to is not overly high, PLLs can be an excellent candidate for generating the system clock of passive wireless microsystems. In [201], a PLL-based clock generation and calibration system was proposed for RFID tags operating at 13.56 MHz. As shown in Fig.6.9, the system harvests its operating power from an interrogator via a coupling coil. The clock generation and calibration system consists of an envelope detector that extracts the envelope of the received RF signal, a clock extractor that generates a full-swing square wave whose frequency is the same as that of the carrier, and a phase-locked loop that enables the local oscillator to lock to the output of the clock extractor.

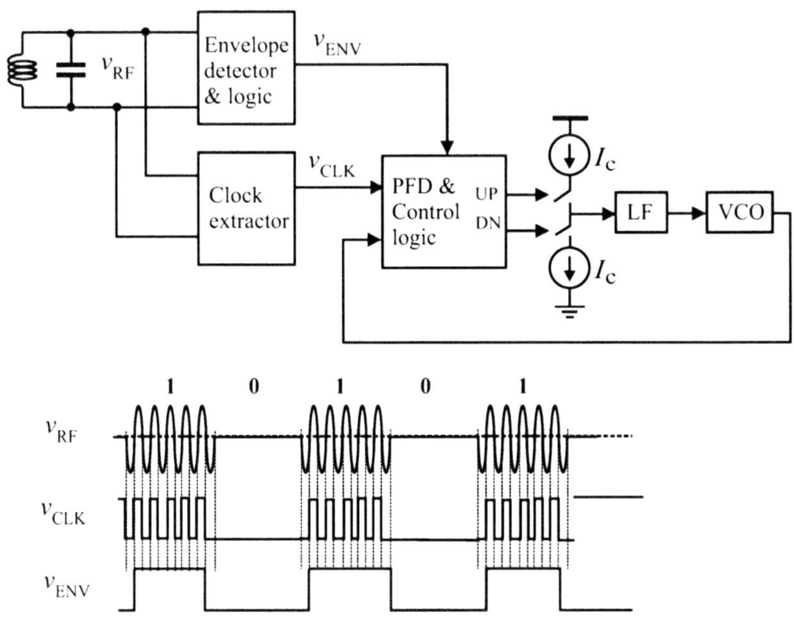

Figure 6.9. Frequency calibration using a phase-locked loop [201].

The clock extractor consisting of a voltage comparator and a voltage buffer generates a full-swing clock whose frequency is the same as the carrier, as shown in the figure. The generated square wave v_{CLK} is used as the reference of the downstream PLL. Once the PLL is locked to the reference, a local system clock whose frequency is the same as that of the carrier is generated by the PLL. Since ASK is typically used for passive wireless microsystems, a large modulation index is preferred to minimize the bit-error-rate. When an 100%

modulation index is used, no carrier is available when data transmitted are 0. The frequency of the local oscillator will drift due to the absence of the reference. To eliminate this drawback, the conventional DFF phase/frequency detector is modified to allow the recovered envelope signal v_{ENV} to disable the phase/frequency detector, as shown in Fig.6.10, such that both UP and DN of the phase/frequency detector will be held unchanged during binary 0 intervals subsequently the frequency of the oscillator of the PLL.

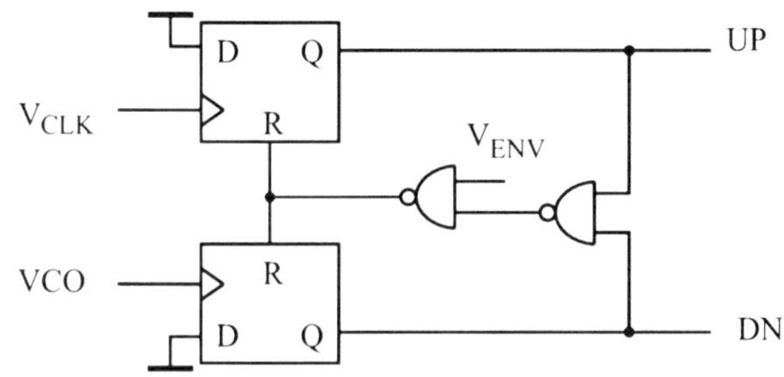

Figure 6.10. Modified phase/frequency detector.

The oscillator is implemented using a single-ended current-steering configuration to take the advantage of its simple configuration subsequently a large frequency tuning range and a low level of power consumption. Implemented in a 0.35μm CMOS technology, the power consumption of the phase/frequency detector and charge pump is 10.4 μW during the locking state and only 0.4 μW during the locked state. The power consumption of the oscillator is 2.6 μW. The total power consumption of the clock generation and calibration system is 17 μW during the locking state and 3 μW during the locked state. It should be noted that the PLL operates at the frequency of the extracted clock, which is the frequency of the carrier, the power consumption of the clock recovery circuit will become exceedingly large once the carrier frequency is high. The low power consumption of the design in [201] is simply due to the fact that the carrier frequency is 13.56 MHz. Should the carrier frequency be in UHF or even microwave ranges, the power consumption would be significant.

6.6 Clock Generation Using Frequency-Locked Loop

The preceding PLL-based clock generation and calibration method utilizes the extracted clock of the received RF signal as the reference of the PLL. The frequency of the recovered clock is the same as that of the carrier. When the envelope of the carrier becomes zero, it utilizes the logic state of the envelope of the received RF signal to sustain the oscillation of the oscillator of the PLL so as to sustain the frequency of the system clock. As pointed out earlier that this approach will become power hungry once the frequency of the carrier is high. For example, the carrier frequency of UHF RFID EPC is in the range of 860-960 MHz, the preceding PLL-based clock generation method will be power-greedy. In [192], a digital frequency-locked loop based method was proposed to generate the system clock of UHF RFID EPC systems. Since PIE-coding is used for reader-to-tag communications with binary 0 and binary 1 having different pulse width as shown in Fig.6.4, a counter that records the number of the oscillation cycle of a local oscillator during a binary 0 or a binary 1 interval is used. The content of the counter is then compared with the number of the oscillation cycle of the oscillator during a binary 0 and a binary 1 intervals at the desired frequency. The selection of the reference value to which the content of the counter is compared at the end of a binary 0 or binary 1 interval is determined by the baseband processor that measures RTcal duration. If symbol duration > RTcal/2, a binary 1 is received. Otherwise, a binary 0 is received. A digital frequency-locked loop is then used to adjust the frequency of the local oscillator in accordance with the difference between the counter and the preset value until the frequency difference becomes zero.

A key advantage of this approach is that the frequency of the local oscillator is continuously calibrated as long as a data flow from the base station to the RFID tag exists. As a result, a high frequency accuracy can be obtained. Also, it can distinguish the different pulse durations of binary 0 and binary 1 PIE symbols. The frequency accuracy of the system in [192] is 1.2/-3.2% at 2.56 MHz with power consumption of only 1.8 μW.

6.7 Clock Generation Using Envelope Injection-Locking

As pointed out earlier that frequency calibration using carrier-based injection-locking suffers from the drawback of high power consumption and a small frequency lock range. In [202], a remote frequency calibration technique using envelope-based injection-locking was proposed to utilize the fact that the base-band frequency of passive wireless microsystems is typically in kHz and low MHz ranges, and the modulation schemes used for passive wireless microsystems are often ASK. A baseband timing reference can therefore be embedded in the envelope of a RF signal and transmitted from the base station to the microsystem. The envelope of the received RF signal can be extracted using an envelope detector and used as the injection-locking signal to lock the local oscillator. Since the local oscillator only needs to oscillate at the baseband frequency, its power consumption is significantly lower as compared with the power consumption of oscillators that oscillate at the carrier frequency. The low frequency of the local oscillator also removes the need for frequency dividers, further lowering power and silicon consumption. The power consumption can be further reduced by operating the devices of the local oscillators in the sub-threshold. It should be emphasized that if the frequency of the local oscillator were at the carrier frequency, it would be difficult to operate the devices of the

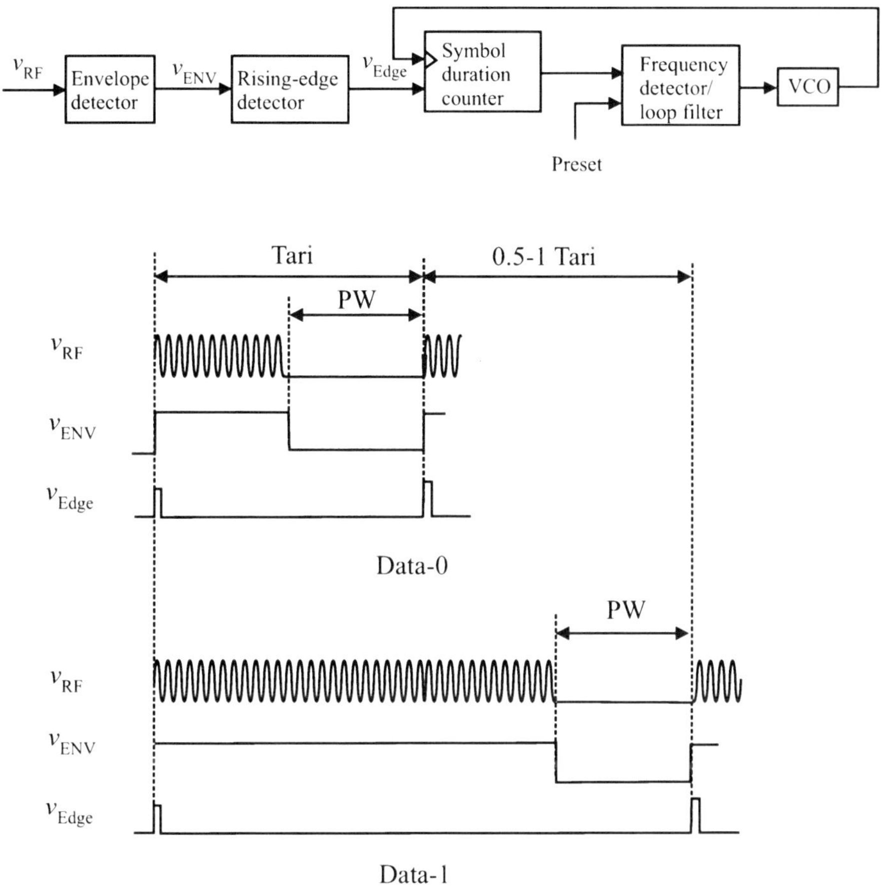

Figure 6.11. Frequency calibration using a digital frequency-locked loop [192].

oscillator in the sub-threshold while meeting the frequency requirement at the same time.

Frequency calibration using envelope-based injection-locking has the intrinsic advantages of lower power consumption and high frequency accuracy. Since the local oscillator oscillates at the baseband frequency, it can not be realized using an LC tank oscillator due to the need for a prohibitively large inductor. Instead, current-starving ring oscillators and relaxation oscillators are typically used. Relaxation oscillators are in general preferred over ring oscillators. This is because ring oscillators are prone to PVT effect as the frequency of these oscillators is determined by active components. The oscillation frequency of current-starving oscillators exhibit a larger sensitivity to PVT effect as compared with conventional ring oscillator. This is because the devices of current starving oscillators dwell longer in the sub-threshold region where the channel current of transistors is more sensitive to PVT effect.

When the effect of PVT is strong, the difference between the free-running frequency of the local oscillator and the reference frequency from the base station will exceed the lock range of the oscillator and injection-locking will fail. Clearly, a mechanism that pulls the frequency of the local oscillator towards the desired frequency when the frequency difference is large is critically needed. If such a mechanism can narrow the frequency difference down to the lock range of the local oscillator, a lock state can be established.

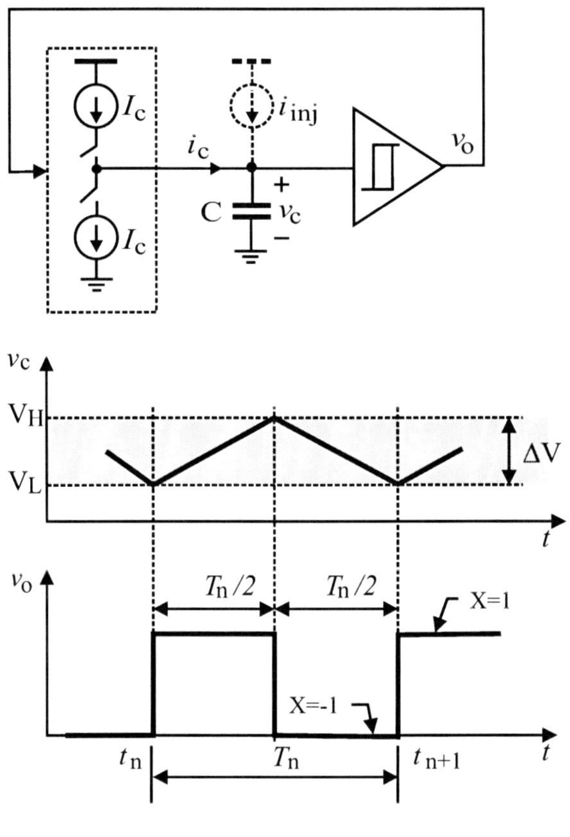

Figure 6.12. Configuration of relaxation oscillators.

Soltani and Yuan showed in [203] that integrating feedback that integrates the difference between the reference frequency and the frequency of the local oscillator can be used to adjust the control voltage of the local oscillator so as to pull the frequency of the oscillator towards the reference frequency. Integrating feedback will be in action as long as a frequency difference exists and the strength of integration feedback increases with time. To provide a mathematical treatment of integrating feedback so as to gain a much needed insight of its effectiveness in frequency pulling, let the free-running period of the relaxation oscillator be T_n. The operation of the relaxation oscillator is dominated by the charging and discharging processes of the timing capacitor C, as shown in Fig.6.12. The oscillation of the relaxation oscillator requires that the net variation of the charge of the timing capacitor C over one oscillation period be zero, i.e.

$$\int_t^{t+T_n} X i_c(t)dt = 0, \tag{6.1}$$

where X is a Boolean variable depicting the logic state of the output of the comparator. $X = 1$ when the output voltage of the comparator is high and $X = -1$ otherwise.

When an injection current $i_{inj}(t)$ is injected into the timing capacitor, as shown in Fig.6.12, (6.1) becomes

$$\int_t^{t+T} X\left[i_c(t) + i_{inj}(t)\right] dt = 0. \tag{6.2}$$

Note that the oscillation period is changed from T_n without injection current $i_{inj}(t)$ to T when $i_{inj}(t)$ is applied. If we let $T = T_n + \Delta T$ where ΔT quantifies the change of the period of the oscillator caused by $i_{inj}(t)$, as shown in Fig.6.12, (6.2) can be written as

$$\int_t^{t+T_n} X i_c(t)dt + \int_{t+T_n}^{t+T_n+\Delta T} X i_c(t)dt + \int_t^{t+T} X i_{inj}(t)dt = 0. \tag{6.3}$$

The first integral in (6.3) vanishes due to (6.1). Since $X = -1$ when $t+T_n \leq t < t + T_n + \Delta T$ and $i_c(t) = I_c$, (6.3) becomes

$$\Delta T = \frac{1}{I_c}\int_t^{t+T} X i_{inj}(t)dt. \tag{6.4}$$

It is evident from (6.4) that the variation of the oscillation period of the oscillator is solely due to the injection of $i_{inj}(t)$ and is determined by the amount of net charge ΔQ that $i_{inj}(t)$ provides over one oscillation period $[0, T]$.

$$\Delta Q = \int_t^{t+T} X i_{inj}(t)dt. \tag{6.5}$$

Eq.(6.4) can be written as

$$\Delta T = \frac{\Delta Q}{I_c}. \tag{6.6}$$

The total variation of the oscillation period of the oscillator due to the injection current $i_{inj}(t)$ over an interval $[0, t_k]$ is obtained from

$$\sum_{j=1}^{k} \Delta T_j = \frac{1}{I_c}\sum_{j=1}^{k} \Delta Q_j = \sum_{j=0}^{k-1}\int_{t_j}^{t_{j+1}} X i_{inj}(t)dt. \tag{6.7}$$

Eq.(6.7) reveals that the total variation of the period of the oscillator over the interval $[0, t_k]$ can be obtained by integrating the injection current modulated by the output of the comparator. The product $X i_{inj}(t)$ can be obtained using a transmission-gate passive mixer where X is the modulating signal and $i_{inj}(t)$ is the modulated signal. The integration can be carried out using a capacitor connected to the output of the transmission-gate passive mixer, as shown in Fig.6.13(a). To ensure that there is not an abrupt voltage drop across the switch upon closing, the circuit shown in Fig.6.13(b) is employed. When $X = 1$, the circuit behaves in the same way as that in Fig.6.13(a). When $X = 0$, the unity-gain voltage buffer ensures that the voltage of node A remains unchanged until the arrival of the next sampling clock.

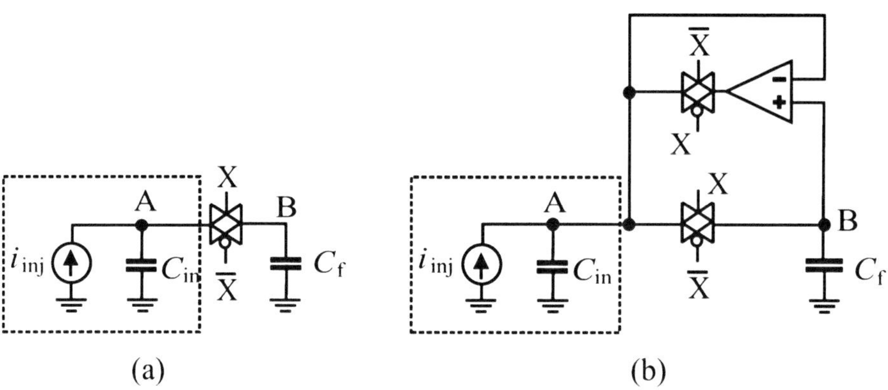

Figure 6.13. Simplified circuit of integrating feedback.

Integrating feedback becomes less effective when the frequency difference becomes small, i.e. when the frequency of the local oscillator is pulled close to the reference frequency. In this case, injection-locking should be activated to further drive the frequency difference to zero. In fact, both injection-locking and integrating feedback can be employed simultaneously. When the frequency difference is large and outside the lock range of the oscillator, only integrating feedback is in action and the frequency of the local oscillator is pulled towards the reference frequency by integrating feedback. Once the frequency of the local oscillator is reduced by a sufficiently large amount and falls into the lock range of the oscillator, injection-locking will be activated and dominate. Clearly, the use of integrating feedback in conjunction with injection-locking will ensure that the frequency of the local oscillator be locked to the reference frequency. Because integrating feedback will be in action as long as the accumulated difference between the frequency of the extracted injection-locking signal and that of the local oscillator is not zero, a large effective frequency lock range can be obtained.

The strength of integrating feedback is bounded by the stability of the system. The stronger the integrating feedback, the faster the frequency pulling and the less stable the system. A more detailed mathematical treatment of the stability of the system with injection-locking and integrating feedback is available in [203].

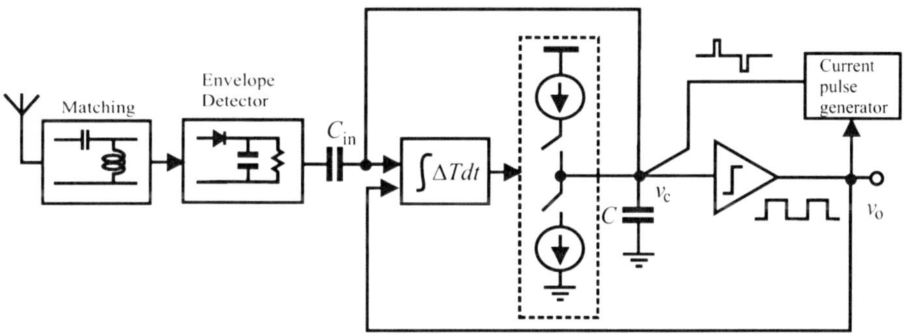

Figure 6.14. Frequency calibration using envelope-based injection-locking with integrating feedback.

Fig.6.14 shows the configuration of the remote frequency calibration using envelope-based injection-locking with integrating feedback. C_{in} is an isolation capacitor whose function is to prevent any dc signal from loading the antenna. The envelope of the received RF signal is extracted using an envelope detector and injected into the timing capacitor as the injection-locking signal. A current pulse generator is employed to generate large current pulses to charge and discharge the timing capacitor in a very short period of time. The purpose of using the current pulse generator will be detailed shortly. An integrating block that integrates the difference between the frequency of the local oscillator and that of the envelope of the received RF signal drives the charge pump. The charge pump provides a constant current to charge or discharge the timing capacitor.

The local oscillator, which consists of a charge pump, a comparator, and a current pulse generator, is a relaxation oscillator whose frequency is determined by the charging and discharging processes of the timing capacitor implemented as a MIM (metal-insulator-metal) capacitor. A key advantage of relaxation oscillators is their low sensitivity to PVT effect. This is because the frequency of relaxation oscillators is determined by the current of the constant current sources and the capacitance of MIM capacitors. The frequency of ring oscillators, on the other hand, is sensitive to PVT effect due to the fact that their frequency is determined by active devices only [204].

Schmitt triggers are often used in constructing relaxation oscillators. As pointed out earlier that the oscillation of a relaxation oscillator is established by charging and discharging the timing oscillator, as shown in Fig.6.12. When the voltage of the timing oscillator exceeds V_H, the output of Schmitt trigger of the relaxation oscillator changes its logic state. The charging process of the timing capacitor is replaced with a discharging process and the voltage of the timing capacitor drops. When the voltage of the timing capacitor drops below V_L, the output of the Schmitt trigger switches from low to high and the timing capacitor is charged. This process repeats and oscillation is sustained. Clearly, the hysteresis width $\Delta V = V_H - V_L$ of Schmitt trigger directly affects the frequency of these oscillators. The voltage of the timing capacitor C is given by

$$\begin{aligned} v_c(t) &= V_L + \frac{I_c}{C}t \quad \text{(charging)}, \\ v_c(t) &= V_H - \frac{I_c}{C}t \quad \text{(discharging)}. \end{aligned} \tag{6.8}$$

The output voltage of the oscillator has a 50% duty cycle with oscillation period T_n given by

$$T_n = 2\left(\frac{C\Delta V}{I_c}\right). \tag{6.9}$$

The sensitivity of the frequency of the oscillator to V_{DD} is obtained from

$$\frac{\partial f}{\partial V_{DD}} = -\frac{I_c}{2C\left(\Delta V\right)^2}\frac{\partial \Delta V}{\partial V_{DD}}. \tag{6.10}$$

It is evident that the smaller the hysteresis width ΔV, the more sensitive the frequency of the oscillator to V_{DD} fluctuation. Since the supply voltage of passive wireless microsystems is typically low and fluctuates and the hysteresis of a Schmitt trigger is upper-bounded by

$$\Delta V_{max} = V_{DD} - V_{tn} - |V_{tp}|, \tag{6.11}$$

the frequency of Schmitt trigger-based relaxation oscillators will exhibit a high degree of uncertainty when V_{DD} is low and fluctuates. Schmitt trigger-based relaxation oscillators are therefore not particularly attractive for passive wireless microsystems.

One way to generate a large hysteresis without using a Schmitt trigger is to use latches in conjunction with two timing capacitors and two voltage comparators, as shown in Fig.6.15 [205]. It is, however, difficult to find an appropriate injection node in this oscillator so that the charging and discharging of both timing capacitors can be controlled by a single injection signal. Relaxation oscillators with one timing capacitor and two voltage comparators are also proposed [206, 207]. Although an appropriate injection node exists in this oscillator, the need for two comparators makes it a rather difficult task to lower the power consumption.

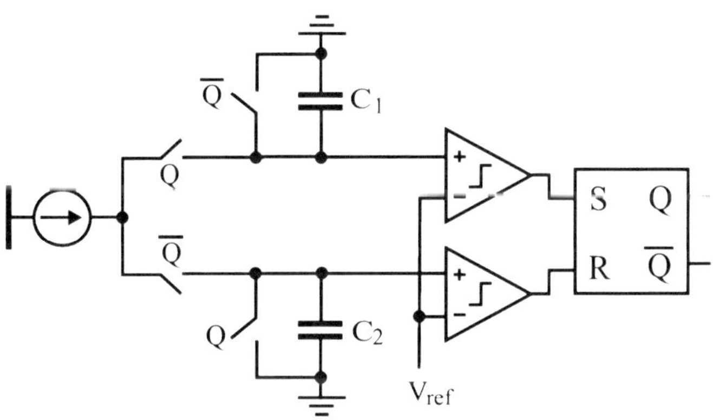

Figure 6.15. Relaxation oscillator proposed by Barnett [205].

Yuan and Soltani showed that a relaxation oscillator can be constructed using one timing capacitor and one voltage comparator with a low supply voltage, as shown in Fig.6.16 [204]. The operation of the current pulse generator is depicted as follows : Assume X=0, M1 switches off and M2 turns on. M5 turns on and M6 is off. C_1 is discharged via the path provided by M5. M3 is off and M4 is on. M4 sources the current $i_p = i_{ds4}$ to the output. In the mean time, C_2 is charged by V_{DD} via the M2-R_2 path. When v_A reaches $V_{DD} - |V_{tp}|$, M4 turns off and i_p vanishes. A current pulse is generated. The width of the current pulse can be determined as the follows : Assume $v_X = 0$ when X is at logic-0 and the voltage drop cross the source and drain of M2 (in triode) is constant and is denoted by V_{sd2}. From

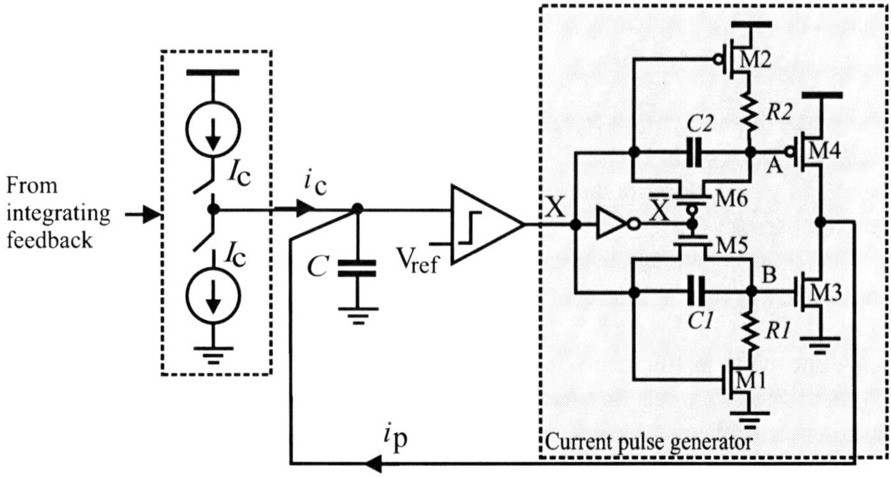

Figure 6.16. Relaxation oscillator proposed by Yuan and Soltani [204]. C is the timing capacitor. Time constants R_1C_1 and R_2C_2 set the width of current pulses.

$$V_{sd2} + R_2C_2\frac{dv_A}{dt} + v_A = V_{DD}, \qquad v_A(0^-) = 0, \tag{6.12}$$

we have

$$v_A(t) = (V_{DD} - V_{sd2})\left[1 - e^{-\frac{t}{R_2C_2}}\right]. \tag{6.13}$$

The width of the positive current pulse, denoted by τ_p, can be obtained by letting $v_A(t) = V_{DD} - |V_{tp}|$

$$\tau_p = R_2C_2\ln\left(\frac{V_{DD} - V_{sd2}}{|V_{tp}| - V_{sd2}}\right) \approx R_2C_2\ln\left(\frac{V_{DD}}{|V_{tp}|}\right). \tag{6.14}$$

In a similar way, one can show that when X=1, M3 will sink a current pulse from the output node with the pulse width

$$\tau_n \approx R_1C_1\ln\left(\frac{V_{DD}}{V_{tn}}\right). \tag{6.15}$$

The current pulse generator will generate a current pulse each time the output of the comparator changes its logic state. The amplitude of the current pulses must be sufficiently large such that it will charge or discharge the timing capacitor completely. The width of the current pulses must be much smaller as compared with the relaxation time of the timing capacitor such that it will have a little impact on the duty cycle and the period of the output voltage of the oscillator. Fig.6.17 shows the waveforms of the relaxation oscillator.

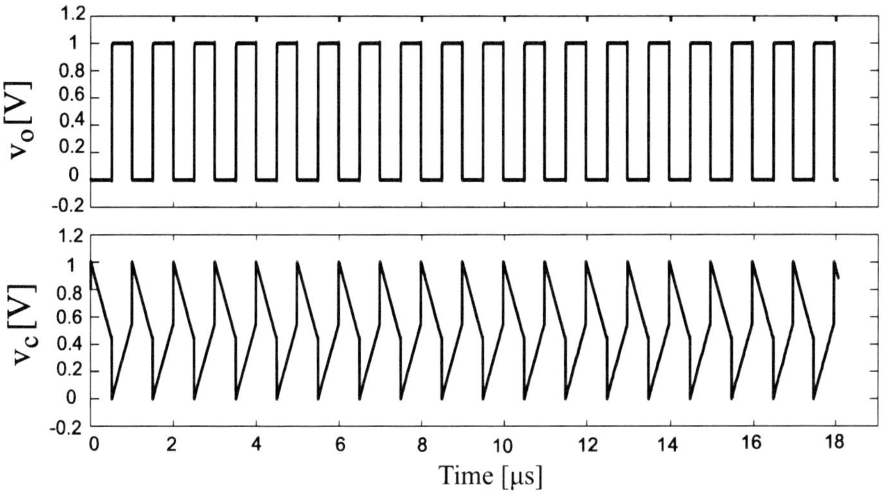

Figure 6.17. Top - Output voltage of comparator. Bottom - Voltage of timing capacitor (Copyright © IEEE).

To examine the injection locking of the relaxation oscillator with and without integrating feedback, the relaxation oscillator is designed in TSMC 0.18μm 6-metal 1.8V CMOS technology and analyzed using Spectre from Cadence Design Systems with BSIM3.3 device models. The free-running frequency of the relaxation oscillator is varied from 975.9 kHz to 902.9 kHz. A 10-mV 1-MHz sinusoidal signal is injected. In the first sub-plot of Fig.6.18, the free-running frequency of the oscillator is 975.9 kHz. The oscillator is able to lock to injection signal. In the second sub-plot, the free-running frequency is set to 969.1 kHz. The oscillator nearly locks to the injected signal but its phase slips eventually. In the the third sub-plot, the free-running frequency is set to 962.4 kHz. The injection signal is outside the lock range. The oscillation frequency of the oscillator approaches the injection frequency but falls back after a short time. In the last sub-plot, the free-running frequency is set to 902.9 kHz. The locking signal in this case is outside the lock range of the oscillator. No lock state can be established.

Fig.6.19 compares the locking process of the relaxation oscillator with and without integrating feedback when the free-running period of the oscillator is set to 902.9 kHz and with the same injection signal. It is seen that when integrating feedback is absent, the oscillator fails to lock to the injection signal.

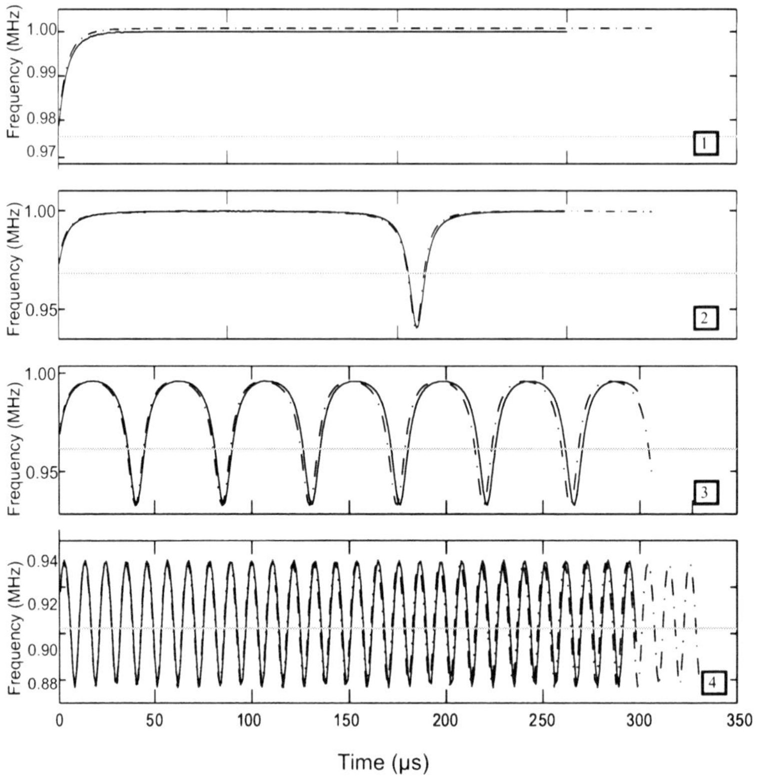

Figure 6.18. Simulated effect of the injection of a sinusoidal current on the frequency of the relaxation oscillator without integrating feedback at various frequency offsets from the injection frequency. Sub-plot 1 : free-running frequency of the oscillator is 975.9 kHz. Sub-plot 2 : free-running frequency of the oscillator is 969.1 kHz. Sub-plot 3 : free-running frequency of the oscillator is 962.4 kHz. Sub-plot 4 : free-running frequency of the oscillator is 902.9 kHz. Legends : Solid line - simulation results. Dotted line - Analytical results (Copyright © IEEE).

However, when integrating feedback is present, the oscillator successfully locks to the injection signal in less than 100 μs.

Fig.6.20 shows the measured lock range of the oscillator with and without the integrating feedback. It is seen that the lock range of the oscillator with integrating feedback increases with the amplitude of the injection signal initially and the rate of increase gradually levels off. The lock range of the oscillator without integrating feedback, on the other hand, increases with the amplitude of the injection signal. Also observed is that the lock range of the oscillator with integrating feedback is approximately one order of magnitude of that without integrating feedback. The increased lock range is critical to ensure that the local oscillator will lock to the injection signal.

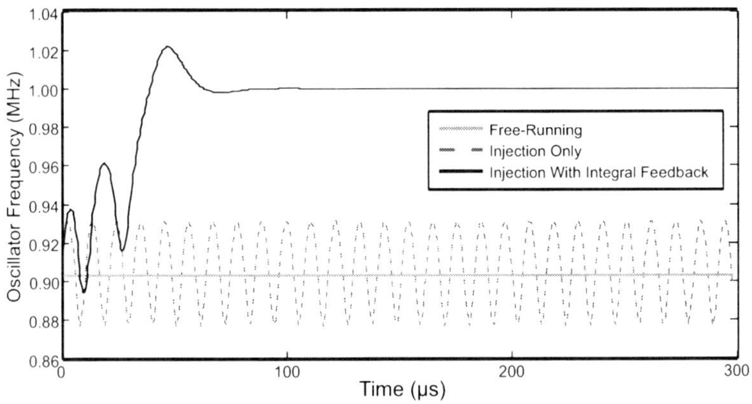

Figure 6.19. Comparison of locking process of relaxation oscillator with and without integrating feedback in the nominal process conditions. Legends : Light-colored solid line - Without both injection and integrating feedback. Dashed line - With injection only. Dark-colored solid line - With both injection and integrating feedback (Copyright © IEEE).

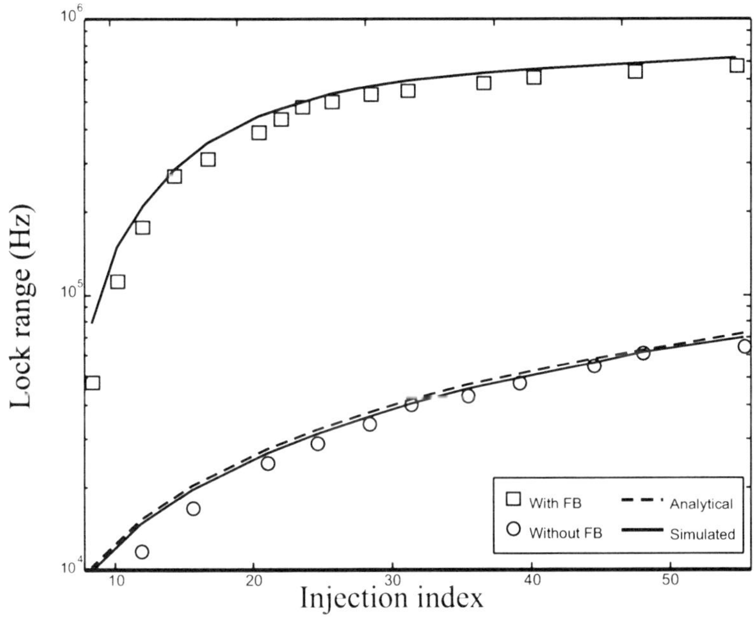

Figure 6.20. Measured locking range. Injection index $m = I_{inj}/I_c$ (Copyright © IEEE).

When the injection signal is removed, the frequency of the oscillator with integrating feedback drifts at a slow rate of 5 Hz/ms. This drift is negligible as backscattering from a passive wireless microsystem to its base station typically takes only a few milliseconds to complete. For the same argument, the effect of temperature variation on the frequency of the oscillator during backscattering can also be neglected safely.

6.8 Performance Comparison

Table 6.8 compares the performance of recently reported clock generation and calibration methods in terms of carrier frequency, system clock frequency, power consumption, and calibration accuracy.

Table 6.1. Performance comparison of clock generation and calibration methods.

Ref.	Tech.	Method	Carrier freq. [MHz]	System clock [MHz]	Power	Calib. accuracy
Kocer-Flynn[11](06)	0.25μm	Car-Inj.	900	112.5	6 mW	–
Leung-Luong[194](07)	0.18μm	Inj-Div	900	3.5	7 μW	–
Pillai *et al.*[30](07)	1μm	DT	915/2.4 GHz	0.032	10.5μW	–
Ma *et al.*[196](08)	0.18μm	Env.	902-928	902-928	21μW	–
Najafi *et al.*[199](08)	0.18μm	DT	860-960	2.56	8 μW	–
Lee-Lee[200](09)	0.35μm	DT	900	2.2	31 μW	0.91%
Fan *et al.*[201](09)	0.35μm	PLL	13.56	6.78	17 μW	7.5%
Chan *et al.*[192](10)	0.18μm	FLL	860-960	2.56	2 μW	$-3.2\sim1.2\%$
Soltani-Yuan[203](10)	0.18μm	Inj-Int	900	1	0.96 μW	–

Legends :
Car-Inj. - Clock generation/calibration using carrier-based injection-locking.
DT - Clock generation/calibration using digital trimming.
Env. - Clock generation from the envelope.
FLL - Clock generation/calibration using a digital frequency-locked loop.
Inj-Int - Clock generation/calibration using an envelope-based injection-locking and integrating feedback.
PLL - Clock generation/calibration using a phase-locked loop.

6.9 Chapter Summary

In this chapter, we have presented the methods for the generation and calibration of the system clock of passive wireless microsystems. We have shown that clock generation directly from the carrier of the received RF signal offers the advantages of a simple design but suffers from high power consumption and large timing jitter. Also, no clock can be received during binary 0 intervals if modulation index is 100%. This approach should only be used if the received RF signal is strong and the frequency of the RF signal is not exceedingly high.

As compared with clock generation from the carrier, the frequency of the clock generated from the envelope of the received RF signal is much lower. It can therefore be used directly for baseband operation without the need for frequency dividers. As a result, the power and silicon consumption of clock recovery are reduced. A drawback of this approach is that RF signals must be encoded with timing information such that a timing reference is continuously available during the operation of the microsystem.

Clock generation and calibration using injection-locked frequency division offer the same advantages as clock generation from the carrier but suffer from the drawback of no frequency calibration when the received RF signal is weak. The frequency of the local oscillator is thus subject to the effect of PVT. As a result, the frequency accuracy of the system clock can not be guaranteed. The power consumption of this approach is high due to the high frequency of the local oscillator and the need for frequency dividers. In addition, the lock range of the injection-locked frequency divider must be sufficiently large in order to ensure that the frequency divider will lock to the received RF signal in the presence of a strong PVT effect.

Frequency calibration using carrier-based injection-locking offers the advantage of a high frequency accuracy. Since the clock in this case is generated locally, there is no need for the base station to continuously supply a timing reference to the microsystem during its operation. As a result, the constraint on encoding data to embed timing information is relaxed. Further, the calibration of the local oscillator can be performed prior to the commence of the data communications between the base station and the microsystem as long as the frequency drift of the local oscillator during the operation of the microsystem is small. This approach, however, suffers from a small frequency lock range and high power consumption due to the high frequency of the local oscillator and the need for frequency dividers. Also, no injection signal is available during a binary 0 interval if the modulation index is 100%.

Frequency calibration using digital trimming offers a large frequency tuning range. As long as the frequency tuning range of the oscillator is sufficiently large, frequency calibration using digital trimming will guarantee that the frequency of the oscillator be tuned to the desired frequency. This approach requires a SAR, a digital-to-analog converter, and other logic for pulse counting, comparison, and control signal generation. As a result, the power consumption is high. In addition, the accuracy of frequency calibration is set by the number of the bits of the SAR and the frequency tuning range of the local oscillator. The larger the number of the bits of the SAR, the higher the calibration accuracy. This is at the cost of a longer calibration time.

Clock generation and calibration using phase-locked loops offer the advantage of a large lock range and a high frequency accuracy. The lock time can be tuned by varying the loop dynamics of the PLL. This approach is power-hungry as the frequency of the reference of the PLL is the carrier frequency. The use of this approach must therefore be cautious when a stringent constraint on power consumption exists.

Clock generation and calibration using digital frequency-locked loops offer the advantage of a large lock range, a high frequency accuracy, and continuous frequency calibration during data transmission. In addition, it can handle pulse-width modulated data. Since the counter used for frequency calibration is driven by the baseband clock, the power consumption is low.

Frequency calibration using injection-locking and integrating feedback offers the advantage of a large frequency lock range, a high frequency accuracy, a fast lock time, and low power consumption. The use of a current pulse generator in the relaxation oscillator eliminates the drawbacks of Schmitt trigger-based relaxation oscillators when the supply voltage is low and time-varying. The integrating nature of this approach also minimizes the drift of the frequency of the local oscillator when the injection-locking signal is absent.

Chapter 7

LOW-POWER ANALOG-TO-DIGITAL CONVERTERS

Analog-to-digital converters (ADCS) are an essential block of passive wireless microsystems. Although there are a number of ways to conduct analog-to-digital conversion such as flash ADCs, pipelined ADCs, over-sampled sigma-delta ADCs, etc., the low voltage and low power constraints of passive wireless microsystems only warrant a few architectures of ADCs that can be used. Integrating ADCs, charge-redistribution successive approximation ADCs, oscillation-based ADCs, time-to-digital ADCs, and frequency-to-digital ADCs are possible candidates for these applications. This chapter focuses on both the architecture and design of low-power ADCs. The chapter is organized as follows : Section 7.1 briefly examines the fundamentals of ADCs. The figure-of-merits that are widely used to quantify the performance of ADCs are studied. In Section 7.2, integrating ADCs are investigated. Both single-slope and dual-slope integrating ADCs, their advantages and disadvantages are examined. Section 7.3 focuses on the design of oscillation-based ADCs for temperature measurement. Attention is paid to both relaxation oscillator and ring oscillator based temperature ADCs. Time-to-digital converter based ADCs for temperature measurement is explored in Section 7.4. As compared with oscillator-based temperature ADCs, these ADCs have the advantage of low power consumption due to the elimination of oscillators. Section 7.5 concentrates on frequency-to-digital based ADCS for temperature measurement. Charge redistribution successive approximation ADCs are investigated in a great detail in Section 7.6. A focus is given to the design of charge-scaling digital-to-analog converters (DACs). Three configurations of charging-scaling capacitor arrays, namely single-stage binary-weighted capacitor arrays, two-stage binary-weighted capacitor arrays, and C-2C capacitor arrays are studied in detail and their pros and cons are examined. Section 7.7 compares the per-

F. Yuan, *CMOS Circuits for Passive Wireless Microsystems*,
DOI 10.1007/978-1-4419-7680-2_7,

formance of some recently reported ultra-low power ADCs. The chapter is summarized in Section 7.8.

7.1 Fundamentals of Analog-to-Digital Converters

An analog-to-digital converter (ADC) accepts an analog signal v_{in} and produces an N-bit binary output code $D_N ... D_1$ such that

$$v_{in} = \left(\frac{D_N}{2^1} + \frac{D_{N-1}}{2^2} + ... + \frac{D_2}{2^{N-1}} + \frac{D_1}{2^N} \right) V_{ref}, \tag{7.1}$$

where V_{REF} is the reference voltage, and $D_k, k = 1, 2, ..., N$ are binary numbers. D_1 is the least significant bit (LSB) and D_N is the most significant bit (MSB). FSR $= V_{ref}$ is termed the full-scale range of the input. It is the input voltage range over which the ADC will digitize the input. The contribution of the MSB is $\frac{\text{FSR}}{2}$ while that of the LSB is $\frac{\text{FSR}}{2^N}$. LSB is the resolution of the ADC.

There are a number of metrics that quantify the performance of ADCs. Among them, quantization noise, offset errors, gain errors, differential nonlinearity (DNL), integral nonlinearity (INL), dynamic range (DR), signal-to-noise ratio (SNR), signal-to-noise plus distortion ratio (SNDR), and effective number of bits (ENOB), are the most widely used. We examine each of them in detail in the following sections.

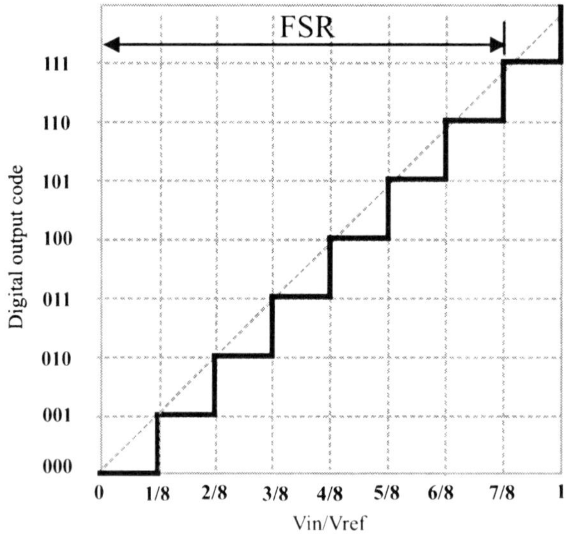

Figure 7.1. Full-scale range of ADCs.

7.1.1 Quantization Error

The quantization error of an ADC is defined as the difference between the analog input of the ADC and the value of its corresponding digital code, as shown graphically in Fig.7.2 where an 8-bit ADC is used to illustrate an uncompensated quantization process. It is seen from the figure that the maximum quantization error is 1 LSB. Also observed is that the quantization error of the uncompensated quantization is not centered at zero volt. The quantization error can be reduced to 0.5 LSB if an offset of $\frac{V_{ref}}{16}$ is introduced to the input. This is termed 1/2 LSB compensated quantization [208]. In this case, the offset moves the center of the quantization error from $\frac{\text{LSB}}{2}$ without the offset to zero volt, as shown in Fig.7.3.

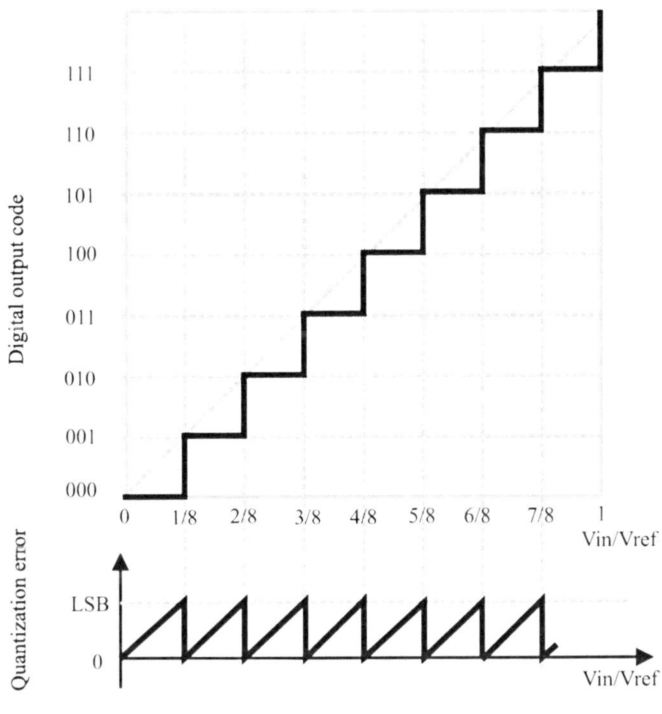

Figure 7.2. Quantization and quantization errors of ADCs - uncompensated quantization.

Since the quantization error is random in nature, its rms (root-mean-square) value is used to quantify its power. The rms value of the quantization error can be obtained with the aid of Fig7.4 [209]

$$\text{rms} = \sqrt{\frac{1}{T}\int_0^T \text{LSB}^2 \left(\frac{x}{T} - 0.5\right)^2 dx}$$

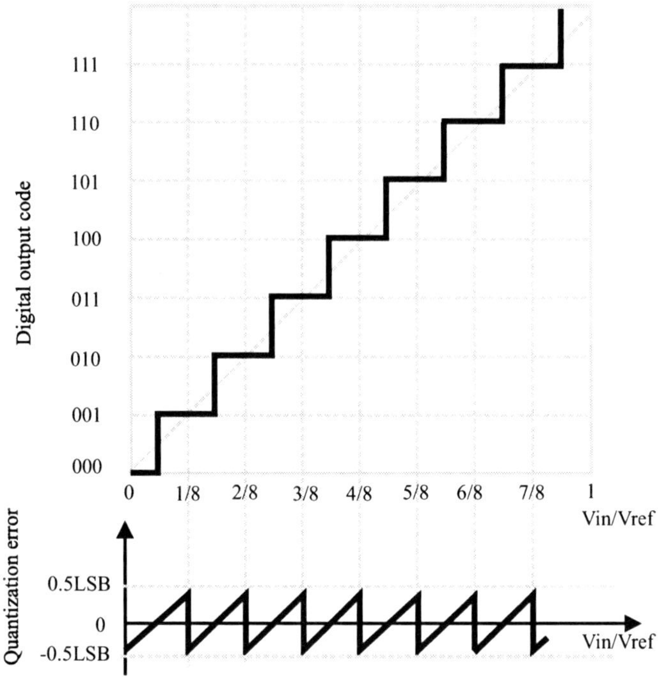

Figure 7.3. Quantization and quantization errors of ADCs - 1/2 LSB compensated quantization.

$$= \frac{\text{LSB}}{\sqrt{12}}$$
$$= \frac{\text{FSR}}{2^N\sqrt{12}}. \tag{7.2}$$

Clearly, for a given FSR, the larger the number of the bits of ADCs, the smaller the quantization error.

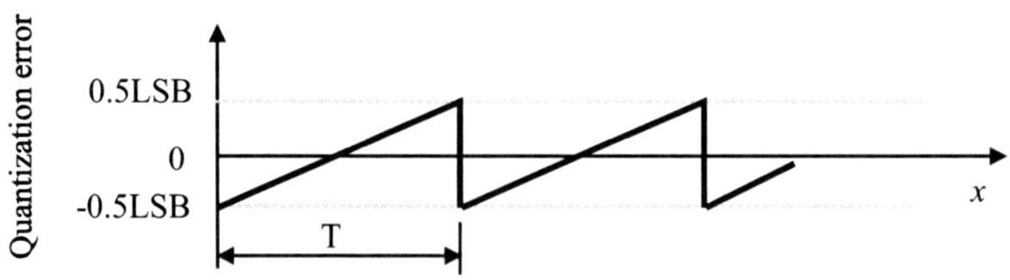

Figure 7.4. Quantization error of ADCs.

7.1.2 Offset Error

The offset error of an ADC is due to the mismatch of the devices of the ADC, as shown in Fig.7.5. The offset error of the ADC does not vary with the input signal of the ADC. It simply shifts the quantization error. Since the offset error of the ADC is a systematic error, it can be eliminated by employing mismatch compensation techniques.

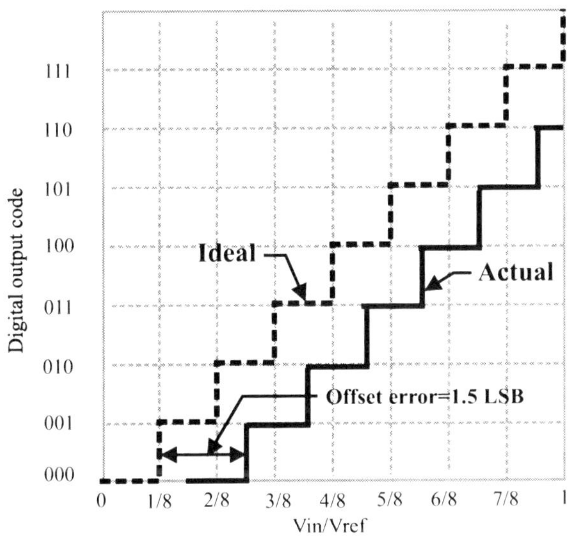

Figure 7.5. Offset error of ADCs.

7.1.3 Gain Error

The gain error of an ADC is due to the variation of the conversion gain of the ADC when the amplitude of the input changes, as shown in Fig.7.6. It is defined as the difference between the nominal gain and the actual gain of the ADC after the offset error of the ADC is removed. This error represents the difference in the slope of the actual and ideal transfer functions of the ADC. The gain error increases the quantization error of the ADC,

7.1.4 Differential Nonlinearity

The differential nonlinearity (DNL) of an ADC is the maximum difference between the step width of an actual ADC and that of an ideal ADC, as illustrated in Fig.7.7. The nonuniform step size of the ADC is caused by the nonlinearity of the ADC. DNL of the kth-step is obtained from

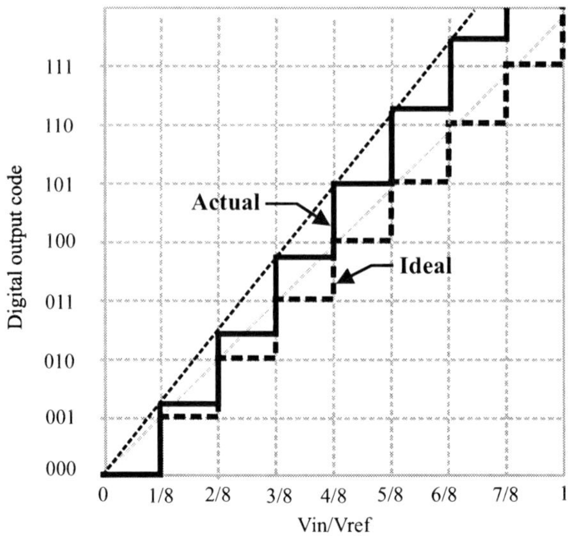

Figure 7.6. Gain error of ADCs.

$$\text{DNL} = \left| \frac{(V_{in,k} - V_{in,k-1}) - LSB}{LSB} \right|, \tag{7.3}$$

where $V_{in,k}$ and $V_{in,k-1}$ are the input voltages of the actual ADC at steps k and $k-1$, respectively. It is seen from (7.3) that DNL is normalized to LSB. DNL is a measure of the maximum deviation of the actual step size from the ideal step size. It provides an effective means to quantify the effect of the nonlinearity on the step width of ADCs. The larger the differential nonlinearity, the larger the quantization error. If the DNL is less than 1 LSB, the ADC will have a monotonic transfer function with no missing code. If the DNL exceeds 1 LSB, the converter will become non-monotonic, meaning the output of the ADC will drop even though its input actually increases. In this case, a missing code will occur.

7.1.5 Integral Nonlinearity

The integral nonlinearity (INL) of an ADC is the accumulated difference between the actual step size and the ideal step size, i.e. it is the integral of DNL. Fig.7.8 depicts INL graphically. INL is measured after both the static gain error and offset error are removed.

7.1.6 Dynamic Range

The dynamic range (DR) of an ADC is defined as the ratio of the full-scale-range value of the input to the smallest input that the ADC can detect. Since

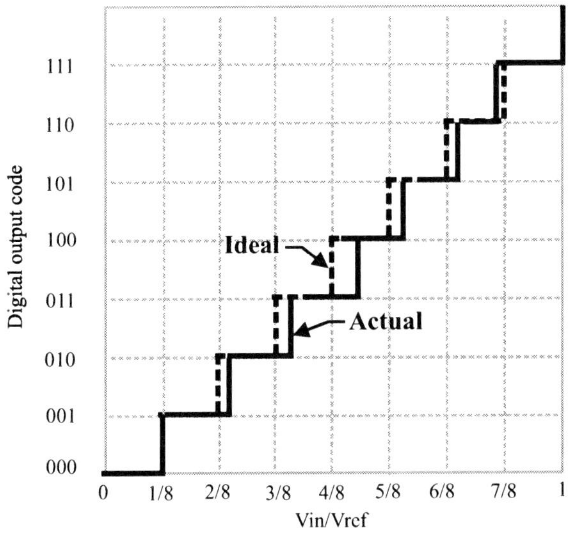

Figure 7.7. Differential nonlinearity of ADCs.

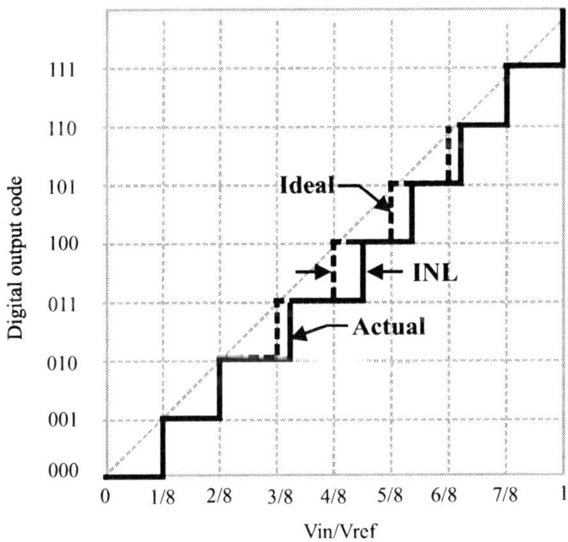

Figure 7.8. Integral nonlinearity of ADCs.

the smallest input that the ADC can sense is $\text{LSB} = \frac{\text{FSR}}{2^N}$, the dynamic range of the ADC is obtained from

$$\text{DR} = \frac{\text{FSR}}{\text{LSB}} = 2^N. \tag{7.4}$$

In unit of decibels,

$$\mathrm{DR} = 20\log_{10}\left(2^N\right) = 6.02N \quad \text{(dB)}. \tag{7.5}$$

When spurs are present, since the amplitude of the spurs is typically larger than quantization error, spurious-free dynamic range (SRDR) should be used. In this case, the dynamic range is defined as the ratio of the full-scale range value of the input signal to the largest amplitude of the spurs.

7.1.7 Signal-to-Noise Ratio

The signal-to-noise ratio (SNR) of an ADC is defined as the ratio of the power of the full-scale input signal to that of the quantization error. Equivalently, SNR is the ratio of the rms value of the full-scale input signal to the rms value of the quantization error. Note the rms value of the quantization error is typically much larger as compared with that of device noise such as thermal and flicker noise. If the input voltage of the ADC is a sinusoid with its amplitude $\frac{\mathrm{FSR}}{2}$, since its rms value is given by $\frac{\mathrm{FSR}}{2\sqrt{2}}$, the SNR is obtained from

$$\mathrm{SNR} = 20\log_{10}\left(\frac{\frac{\mathrm{FSR}}{2\sqrt{2}}}{\frac{\mathrm{FSR}}{2^N\sqrt{12}}}\right) = 6.02N + 1.76 \quad \text{(dB)}. \tag{7.6}$$

The preceding SNR does not include the effect of the distortion of ADCs, which manifests itself as spurs in the spectrum of the output voltage of the ADCs. It is therefore the maximum SNR [209, 210].

7.1.8 Signal-to-Noise-and-Distortion Ratio

Since the amplitude of spurs, which are stemmed from the nonlinearity of ADCs, is typically higher than the noise floor of the ADCs, the signal-to-noise-and-distortion (SINAD) ratio of an ADC defined as the ratio of the power of the full-scale input to the sum of the power of the noise and that of the harmonics of the ADC

$$\mathrm{SINAD} = 20\log\left(\frac{P_s}{P_n + P_d}\right), \tag{7.7}$$

where P_s, P_n, and P_d are the power of the signal, noise, and harmonics of the ADC, respectively, should be used to quantify the performance of ADCs. Note that SINAD is typically smaller than SNR.

7.1.9 Effective Number of Bits

When the effect of distortion is not considered, the effective number of bits (ENOB) is obtained from (7.6)

$$\text{ENOB} = \frac{\text{SNR} - 1.76}{6.02}. \tag{7.8}$$

It is the maximum number of the bits that an ADC can provide for a given signal-to-noise ratio.

When the effect of the nonlinearities of the ADC is accounted for, SINAD provides a better picture of the difference between the power of the signal and that of the disturbances and noise of the ADC. In this case, ENOB is obtained from

$$\text{ENOB} = \frac{\text{SINAD} - 1.76}{6.02}. \tag{7.9}$$

7.2 Integrating ADCs

In this section, we focus on a special class of low-power ADCs called integrating ADCs. Integrating ADCs offer the key advantages of a simple configuration subsequently low power consumption, a good resolution, a high conversion speed, and the ability to suppress transient disturbances present at the input, making them particularly suitable for passive wireless microsystems. Depending upon the number of integration operation, integrating ADCs can be categorized into single-slope integrating ADCs, dual-slope integrating ADCs, and multi-slope integrating ADCs. Multi-slope integrating ADCs are rarely used due to their complex configurations subsequently high power consumption. Only single-slope integrating ADCs and dual-slope integrating ADCs will therefore be studied in this section.

7.2.1 Single-Slope Integrating ADCs

The block diagram of a single-slope integrating ADC is shown in Fig.7.9. When the ADC starts, it first resets the ramp generator, typically an integrator where a ramping voltage is generated. Because initially $v_{in} > v_{ramp}$, the output of the comparator is set to high, allowing the clock to pass through the AND2 gate and the counter increments. The ramping voltage rises with time linearly and continues to climb until it reaches v_{in}. At that time, the output of the comparator changes its logic state and the output of the AND2 gate becomes low, disabling the counter. The content of the counter provides the binary representation of the input voltage. Since the ramping voltage only has a single slope, this ADC is termed the single-slope integrating ADC.

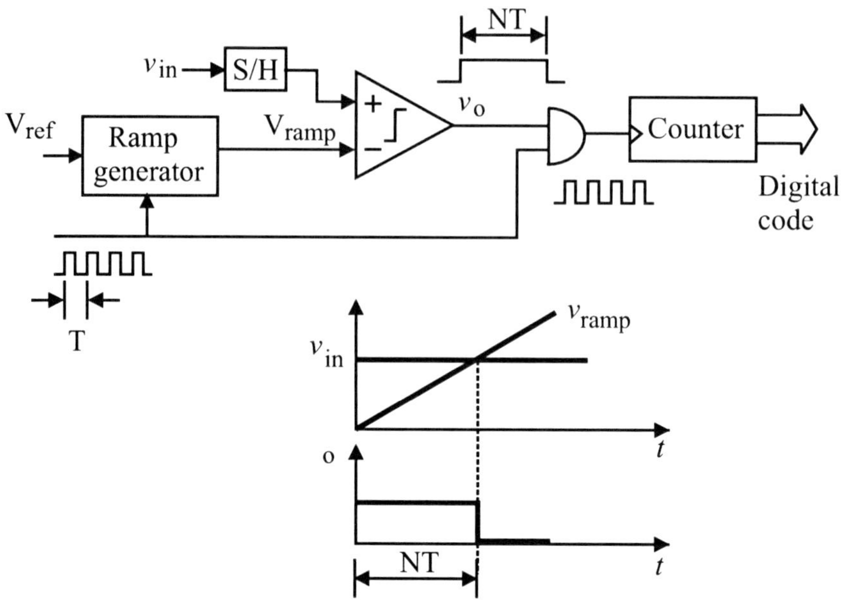

Figure 7.9. Single-slope integrating ADCs.

As an example, Fig.7.10 shows the simplified schematic of the single-slope integrating ADC proposed by Kocer and Flynn for wireless sensors [11, 12]. The ramping voltage is generated by charging capacitor C with a constant current I. Initially the capacitor is fully discharged and $v_c = 0$. When the conversion starts, the capacitor is charged by I and v_c rises with time linearly as per

$$v_c = \frac{I}{C}t. \tag{7.10}$$

At the same time, the counter starts to increment from zero. When v_c reaches v_{in}, the output of the comparator changes its logic state and the counter stops counting. The content of the counter provides the digital representation of the input.

The main advantage of single-slope integrating ADCs is their simple configuration subsequently low power consumption. The performance of single-slope ADCs critically depends upon the slope of the ramping voltage, the linearity of the ramping voltage, and the stability of the clock frequency. For a given clock rate, the smaller the slope of the ramping voltage, the larger the content of the counter at the end of the conversion and the higher the resolution. This, however, is at the cost of a longer conversion time. On the other hand, the higher

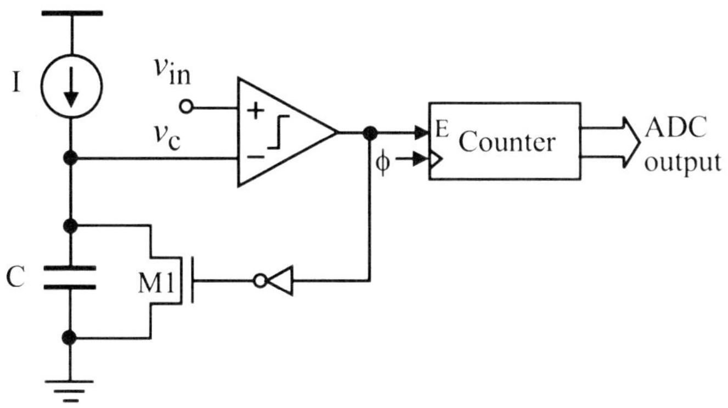

Figure 7.10. Single-slope integrating ADC proposed by Kocer and Flynn [11, 12].

the clock rate, the better the resolution. This is at the expense of increased power consumption.

Another disadvantage of single-slope integrating ADCs is that integration is performed on the reference voltage rather than the input voltage. The value of the input voltage must therefore be held by a S/H block for an accurate conversion. As a result, single-slope integrating ADCs are subject to the effect of the transient disturbances present at the input.

7.2.2 Dual-Slope Integrating ADCs

To minimize the dependence of the performance of single-slope integrating ADCs on the ramping voltage and the clock frequency, dual-slope integrating ADCs with their block diagram shown in Fig.7.11 can be used [210]. The operation of dual-slope integrating ADCs is depicted as follows : At the start of the conversion, v_{in} is applied to the input of the integrator and the counter starts to increment from zero. After a pre-determined amount of time t_c, the content of the counter is stored and the counter is reset. At the same time, the input is disconnected from the integrator and a reference voltage V_{ref} of the opposite polarity is applied to the input of the integrator. The output voltage of the integrator starts to decrease and the counter resumes counting from zero. When the output voltage of the integrator reduces to zero, the output of the comparator changes its logic state and the counter stops counting. Because the charge stored in the capacitor of the integrator from $t = 0$ to $t = t_c$ is proportional to $v_{in}t_c$ while the amount of the charge that the capacitor lost from $t = 0$ to $t = t_d$ at which $v^+ = 0$ is proportional to $V_{ref}t_d$. From the principle of charge conservation, we have

$$v_{in} = \frac{t_d}{t_c}V_{ref}. \qquad (7.11)$$

The intrinsic advantages of dual-slope integration ADCs over their single-slope counterparts become apparent from the preceding presentation. The most important advantage of dual-slope ADCs is that the conversion accuracy of dual-slope integrating ADCs is less affected by the parameters of the integrator and the clock frequency as they are used in both the charging and discharging processes. To illustrate this, let Δt_d and Δt_c be the error of t_d and t_c, respectively with $\Delta t_d \ll t_d$ and $\Delta t_c \ll t_c$. Eq.(7.11) becomes

$$\begin{aligned} v_{in} &= \left(\frac{t_d + \Delta t_d}{t_c + \Delta t_c}\right) V_{ref} \\ &= \left(\frac{t_d}{t_c}\right) V_{ref} \left(\frac{1 + \frac{\Delta t_d}{t_d}}{1 + \frac{\Delta t_c}{t_c}}\right). \end{aligned} \tag{7.12}$$

Making use of the first-order approximation

$$\frac{1}{1 + \frac{\Delta t_c}{t_c}} \approx 1 - \frac{\Delta t_c}{t_c}, \tag{7.13}$$

(7.12) becomes

$$v_{in} \approx \left(\frac{t_d}{t_c}\right) V_{ref} \left(1 + \frac{\Delta t_d}{t_d} - \frac{\Delta t_c}{t_c}\right). \tag{7.14}$$

Since the term in the secondary parenthesis of (7.14) is approximately unity, the effect of the error of the counter becomes negligible.

Another distinct characteristic of dual-slope integrating ADCs is that integration operation is performed on v_{in}. The integration of v_{in} over the interval $[0, t_c]$ results in the effective suppression of the transient disturbances whose duration is much smaller than t_c at the input. This unique characteristic of dual-slope integrating ADCs is of a practical importance as it warrants these integrating ADCs to be used in applications where transient disturbances are present.

The choice of the value of t_c is based on the following considerations : If t_c is too small, the ADC will have a poor resolution and a poor capability to suppress transient disturbances present at the input. If t_c is too large, the conversion time will be overly long.

Although dual-slope integrating ADCs require a long conversion time, their excellent linearity and resolution make them a preferred choice when speed constraints can be relaxed. The nonlinearity errors of dual-slope ADCs can be kept below 0.01% and the resolution can be pushed to above 20 bits [210].

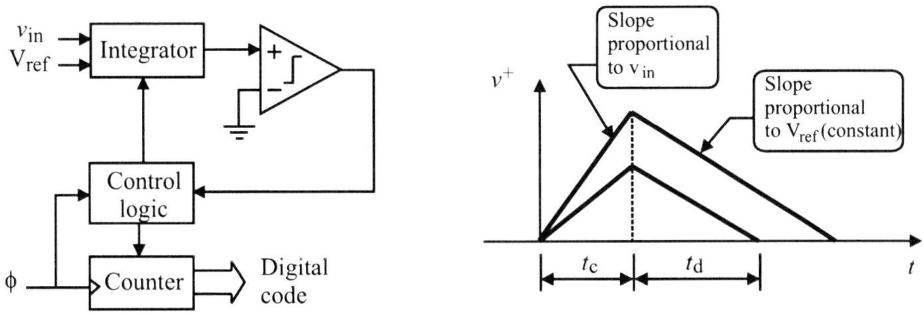

Figure 7.11. Dual-slope integrating ADCs.

7.3 Oscillator-Based Temperature ADCs

When the control voltage of a voltage-controlled oscillator or the control current of a current-controlled oscillator varies with temperature, the frequency of the oscillator will also be a function of temperature. A one-to-one mapping between temperature and the frequency of the oscillator will exist. This temperature-to-frequency relation forms the foundation of oscillator-based temperature ADCs. By measuring the frequency of the oscillator or equivalently by counting the number of the oscillation cycles of the oscillator within a given time interval, temperature can be obtained.

A key advantage of oscillator-based temperature ADCs is their simple configurations subsequently low-power consumption. In this section, we look into the operation of two special types of oscillator-based temperature ADCs : relaxation oscillator-based temperature ADCs and ring oscillator-based temperature ADCs.

7.3.1 Relaxation Oscillator-Based Temperature ADCs

Relaxation oscillators offer the intrinsic advantages of a low frequency sensitivity to PVT effects as the frequency of these oscillator is mainly determined by the charging and discharging processes of passive capacitors by constant current sources. Fig.7.12 shows the simplified schematic of the relaxation oscillator-based temperature ADC proposed by Zhou and Wu [211]. The schematic of the PTAT oscillator is detailed in Fig.7.13. A PTAT oscillator and a timing oscillator are used. The frequency of the PTAT oscillator varies with temperature in a linear fashion while that of the timing oscillator is independent of temperature. The frequency of the timing oscillator is set to be much lower than that of the PTAT oscillator such that a unique relation between the number of the oscillation cycles of the PTAT oscillator within each oscillation period of the timing oscillator and temperature can be established.

The timing oscillator is a relaxation oscillator with its frequency controlled by V_{ref}, I, and C. Its operation is the same as that of the single-slope integrating ADC proposed by Kocer and Flynn for wireless sensors depicted earlier [11, 12]. To minimize power consumption, a clocked comparator that does not consume static power can be used.

The PTAT oscillator is also a relaxation oscillator with its operation depicted as follows : Assume V_{c1}=0 and V_{c2}=1 initially. In this case, M3 is off and M4 is on. C_1 is now charged and V_{c1} rises with time with the rising rate set by the PTAT current and the value of C_1. C_2 at the same time is discharged through the path provided by M4 with discharge time constant $R_{ds4}C_2$ where R_{ds4} is the on-resistance of M4. When V_{c1} climbs to the threshold voltage of the inverter, it will trigger the RS register and change the logic state of its output. As a result, M3 will turn on and M4 will turn off. C_1 will be discharged through M3 and C_2 will be charged by the PTAT current. This process repeats and oscillation is established.

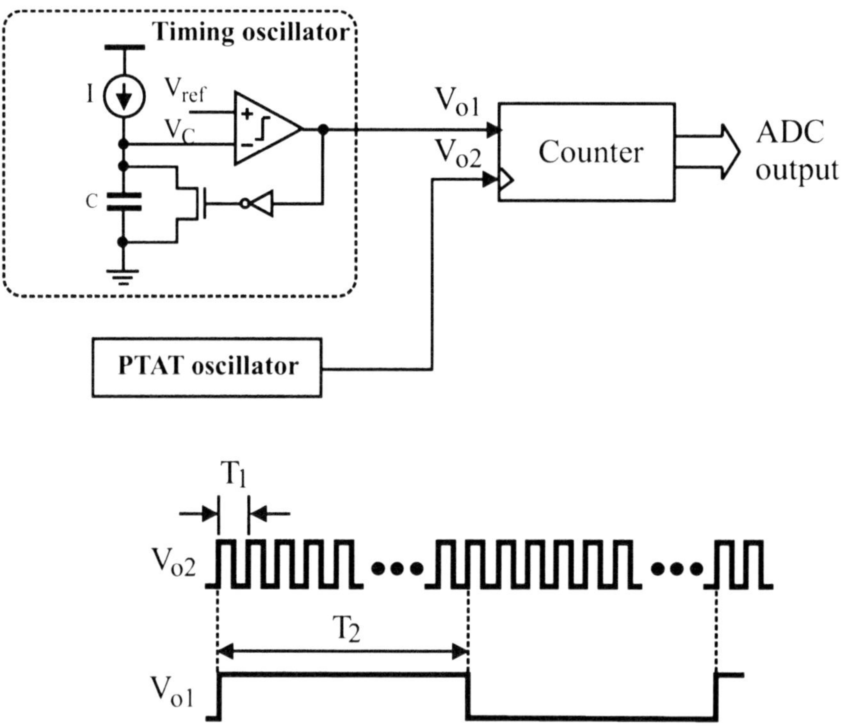

Figure 7.12. Relaxation oscillator-based temperature ADC proposed by Zhou and Wu [211].

Having studied the operation of the PTAT oscillator, let us now turn our attention to the operation of the relaxation oscillator-based temperature ADC. Assume initially V_c of the timing oscillator is zero and V_{o1}=1. The counter

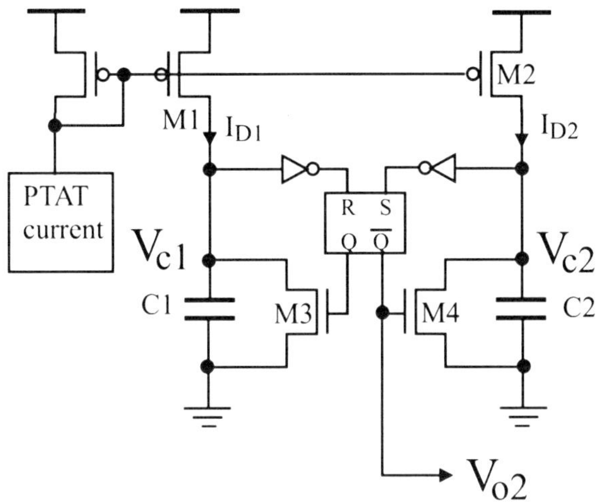

Figure 7.13. PTAT oscillator of the relaxation oscillator-based temperature ADC proposed by Zhou and Wu [211].

is reset and V_c rises from $V_c = 0$ with time. The counter counts the number of the cycles of the output voltage of the PTAT oscillator. Since the control current of the oscillator is PTAT, the higher the temperature, the higher the frequency subsequently the larger the content of the counter. When V_c exceeds V_{ref}, $V_{o1} = 0$, and the counter stops. the content of the counter corresponds to temperature.

There are a number of issues that must be considered to ensure the proper operation of the ADC. The frequency of the timing oscillator must be stable and independent of temperature variation and supply voltage fluctuation. This requires both V_{ref} and I be stable. They should therefore be provided by temperature compensated references. The accuracy of the temperature ADC is set by the smallest frequency variation that can be detected. For a given frequency of the timing oscillator, the higher the frequency of the PTAT oscillator, the better the resolution. This, however, is at the cost of increased power consumption. The conversion time of the ADC is set by the oscillation period of the timing oscillator. For a fixed frequency of the PTAT oscillator, the higher the frequency of the timing oscillator, the shorter the conversion time and the poorer the resolution of the ADC. Clearly depending upon applications, trade-offs need to be made.

7.3.2 Ring Oscillator-Based Temperature ADCs

Ring oscillators have also been used as temperature ADCs to take the advantages of their simple configurations. The simplified schematic of the ring-oscillator based temperature ADC proposed by Kim *et al.* is shown in Fig.7.14

[212]. The operation principle is similar to that of Zhou-Wu temperature ADC of Fig.7.12. The frequency of the ring oscillator with a PTAT biasing current (PTAT oscillator) is directly proportional to temperature while that of the ring oscillator with a constant biasing current (timing oscillator) is independent of temperature. The biasing circuit of the timing oscillator is a generic current source [73]. The output of the PTAT oscillator is fed to a pulse generator whose output is a pulse with pulse width proportional to temperature. The frequency of the PTAT oscillator is lower than that of the timing oscillator. The timing oscillator starts to oscillate at the rising edge of the pulse from the pulse generator and seizes oscillation at the falling edge of the pulse, as illustrated in Fig.7.14. The counter records the number of the oscillation cycles of the timing oscillator within the time duration of the pulse. Since the pulse width is proportional to temperature, the content of the counter provides a measure of temperature.

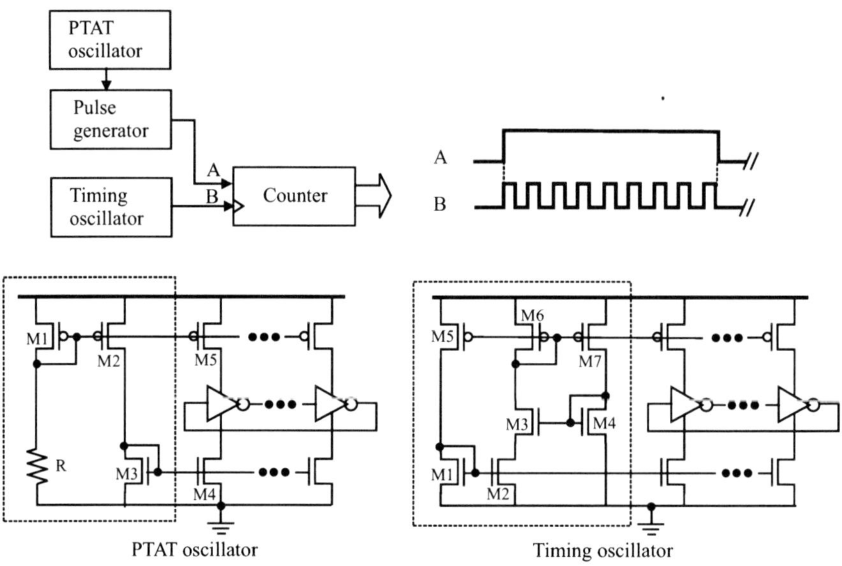

Figure 7.14. Ring oscillator-based temperature ADC proposed by Kim *et al.* [212].

The resolution of the temperature ADC depends upon the width of the pulse from the pulse generator and the width of the pulse of the timing oscillator. This is similar to the preceding relaxation oscillator based temperature ADC. It is interesting to note that in the preceding Zhou-Wu temperature ADC, the frequency of its timing oscillator is fixed and is much smaller as compared with that of its PTAT oscillator. In Kim ring oscillator-based temperature ADC, the period of the timing oscillator is fixed and is much smaller as compared with that of its PTAT oscillator, as shown in Fig.7.15.

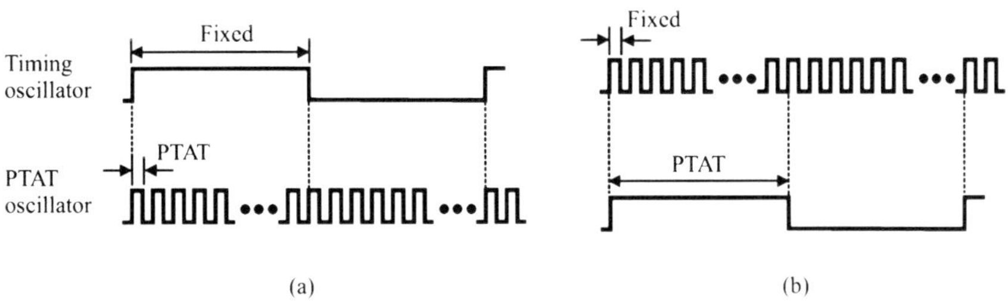

Figure 7.15. Waveforms of oscillator-based ADCs. (a) Relaxation oscillator-based temperature ADC proposed by Zhou and Wu [211]. (b) Ring-oscillator based temperature ADC proposed by Kim *et al.* [212].

The control current of the PTAT oscillator is set by its biasing circuit. The temperature-dependent biasing current is generated by utilizing the temperature-dependent resistance of poly resistor R. As pointed out in Chapter 5, the resistance of poly resistors typically increases with temperature. Since

$$RI_{D1} + |V_{tp}| + \sqrt{\frac{2I_{D1}}{\mu_p C_{ox}(W/L)_1}} = V_{DD}, \tag{7.15}$$

I_{D1} will vary with temperature in a nonlinear fashion as R, V_{tp}, and μ_p are all functions of temperature. The results presented in [212] show that the relation between the frequency of the PTAT oscillator and temperature is fairly linear over the temperature range $0\sim100°$C.

The oscillator-based temperature ADC proposed by Park *et al.* utilizes the temperature-dependence of the frequency of a generic ring oscillator with devices operating in both weak inversion and strong inversion regions [213]. Fig.7.16 shows the simplified schematic of the temperature ADC. The timing signal V_{in2} is provided externally. Since no external control current exists, a large number of delay stages are required such that the oscillation period of the oscillator is large enough in order to be measured accurately. In addition, digitally-controlled binary capacitor-arrays are used as the load of each delay stage to further increase the period of the oscillator and at the same time to add the tunability of the period. This is a drawback of ring oscillator based temperature ADCs as the oscillation period of relaxation oscillators can be conveniently lowered by either reducing the charging and discharging currents or increasing the capacitance of the timing capacitors.

The full-swing operation of the inverting stages requires that the transistors of the inverting stages operate through three regions, namely cut-off, weak inversion, and strong inversion regions. The temperature coefficient of the channel current of MOSFETs in saturation, triode and sub-threshold regions were investigated in detail by Socher *et. al* [214]. The temperature dependence of the channel current of MOSFETs originates from the temperature dependence of the mobility of minority charge carriers and that of the threshold voltage of MOSFETs. Recall that the channel current of a MOSFET in strong inversion is given by

$$I_{D,sat} = \frac{1}{2}\mu C_{ox}\frac{W}{L}(V_{GS} - V_T)^2 \quad \text{(Saturation)}, \tag{7.16}$$

and

$$I_{D,tri} = \mu C_{ox}\frac{W}{L}\left[(V_{GS} - V_T)\,V_{DS} - \frac{V_{DS}^2}{2}\right] \quad \text{(Triode)}. \tag{7.17}$$

When the device is in weak inversion, its channel current is given by (5.43) and is re-written here for convenience

$$I_{D,sub} \approx I_{D0} e^{\frac{V_{GS}-V_T}{nV_t}}, \tag{7.18}$$

where

$$I_{D0} = 2n\mu C_{ox} S V_t^2, \tag{7.19}$$

and $V_{DS} \gg V_t$ is assumed. The normalized temperature coefficient of the channel current, denoted by TCC, in strong inversion (saturation and triode) and weak inversion is obtained from

$$\begin{aligned} \text{TCC}_{sat} &= \frac{1}{I_{D,sat}}\frac{\partial I_{D,sat}}{\partial T} \\ &= \frac{1}{\mu}\frac{\partial \mu}{\partial T} - \frac{2}{V_{GS} - V_T}\frac{\partial V_T}{\partial T}, \end{aligned} \tag{7.20}$$

$$\begin{aligned} \text{TCC}_{tri} &= \frac{1}{I_{D,tri}}\frac{\partial I_{D,tri}}{\partial T} \\ &\approx \frac{1}{\mu}\frac{\partial \mu}{\partial T} - \frac{V_{DS}}{(V_{GS} - V_T)\,V_{DS} - \frac{V_{DS}^2}{2}}\frac{\partial V_T}{\partial T}, \end{aligned} \tag{7.21}$$

$$\begin{aligned} \text{TCC}_{sub} &= \frac{1}{I_{D,sub}}\frac{\partial I_{D,sub}}{\partial T} \\ &= \frac{1}{\mu}\frac{\partial \mu}{\partial T} + \frac{2}{T} + \frac{\partial}{\partial T}\left(\frac{V_{GS} - V_T}{nV_t}\right) \\ &= \frac{1}{\mu}\frac{\partial \mu}{\partial T} + \frac{2}{T} - \frac{1}{nV_t}\left(\frac{\partial V_T}{\partial T} - \frac{V_T}{T}\right). \end{aligned} \tag{7.22}$$

The channel current of MOSFETs in weak inversion exhibits a much larger thermal sensitivity as compared with that in strong inversion [214]. The temperature dependence of the mobility of minority charge carriers and threshold voltage gives rise to the dependence of the channel current subsequently the dependence of the frequency of the ring oscillator on temperature. The measurement results from [213] demonstrate that the frequency of the ring oscillator increases with the rise of temperature in a nonlinear fashion.

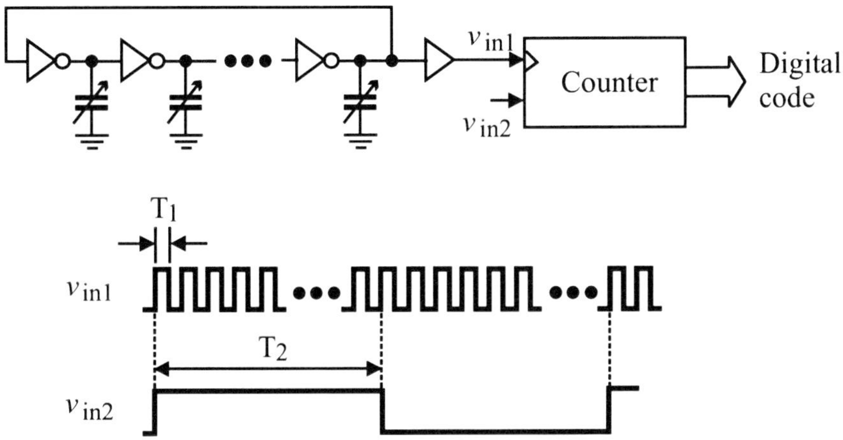

Figure 7.16. Ring oscillator-based temperature ADC proposed by Park *et al.* [213].

7.4 Time-to-Digital Converter Based Temperature ADCs

The power consumption of the preceding oscillator-based temperature ADCs can be lowered by removing the oscillators. The time-to-digital-converter (TDC) based temperature ADC proposed by Chen *et al.* measures temperature by utilizing a generator that generates pulses with pulse width proportional to temperature [215, 216]. The pulse width is determined by using a cyclic time-to-digital converter rather than an oscillator that oscillates at a much higher frequency. Fig.7.17 is the simplified schematic of Chen temperature ADC.

Delay line 1 consists of an even number of inverter stages. The average propagation delay of each delay stage of the delay line is a function of temperature due to the temperature-dependent characteristics of both the mobility of minority charge carriers and the threshold voltage of MOSFETs. As a result, the overall time delay of delay line 1 varies with temperature.

The time delay of delay line 2 is independent of temperature. This is achieved by using a special delay stage with its schematic shown in the figure. The delay cell is a current-starving inverter with the charging and discharging currents controlled by M5-M7. By neglecting the effect of channel length modulation and making use of the temperature-dependent characteristics of the mobility of minority of change carriers and the threshold voltage of MOSFETs given by (5.9) and (5.42), we have

$$I_{D6} = \frac{1}{2}\mu_p(T_o)\left(\frac{T}{T_o}\right)^m C_{ox}\left(\frac{W}{L}\right)_6 \left[V_{SG6} - V_T(T_o) - \alpha_{V_T}(T - T_o)\right]^2. \tag{7.23}$$

Since we want I_{D6} to be independent of temperature,

$$\frac{\partial I_{D6}}{\partial T} = 0 \tag{7.24}$$

is imposed. As a result,

$$V_{SG6} = V_T(T_o) + \alpha_{V_T}(T - T_o) + \frac{2\alpha_{V_T}T}{m}. \tag{7.25}$$

The channel current in this case becomes

$$I_{D6} = \frac{1}{2}\mu_p(T_o)\left(\frac{T}{T_o}\right)^m C_{ox}\left(\frac{W}{L}\right)_6 \left(\frac{2\alpha_{V_T}T}{m}\right)^2. \tag{7.26}$$

Recall in studying the zero-temperature-coefficient bias point of MOSFETs in Chapter 5, we observed over there that the value of m is in the vicinity of -2. If we let $m = -2$, (7.26) is simplified to

$$I_{D6} = \frac{1}{2}\mu_p(T_o)C_{ox}\left(\frac{W}{L}\right)_6 (\alpha_{V_T}T_o)^2. \tag{7.27}$$

It becomes apparent from (7.27) that the charging and discharging currents of the delay stage of delay line 2 can be made independent of temperature by operating M6 at its ZTC bias point. The same approach was also used in [217] where a frequency-to-digital converter based temperature ADC was proposed.

A notable advantage of using dual delay lines rather than a single delay line to generate pulses with pulse width proportional to temperature is that the effect of supply voltage fluctuation and process variation on the pulse width is minimized. This is because the pulse width is the difference between the time delays of the two delay lines. The dual delay line configuration therefore effectively suppresses the effect of common-mode disturbances.

The pulse width is measured using the cyclic TDC shown in Fig.7.18 where six inverting stages of the same dimension are used as an example to illustrate the operation of the cyclic TDC. Assume initially $v_{in} = 0$. When RESET is

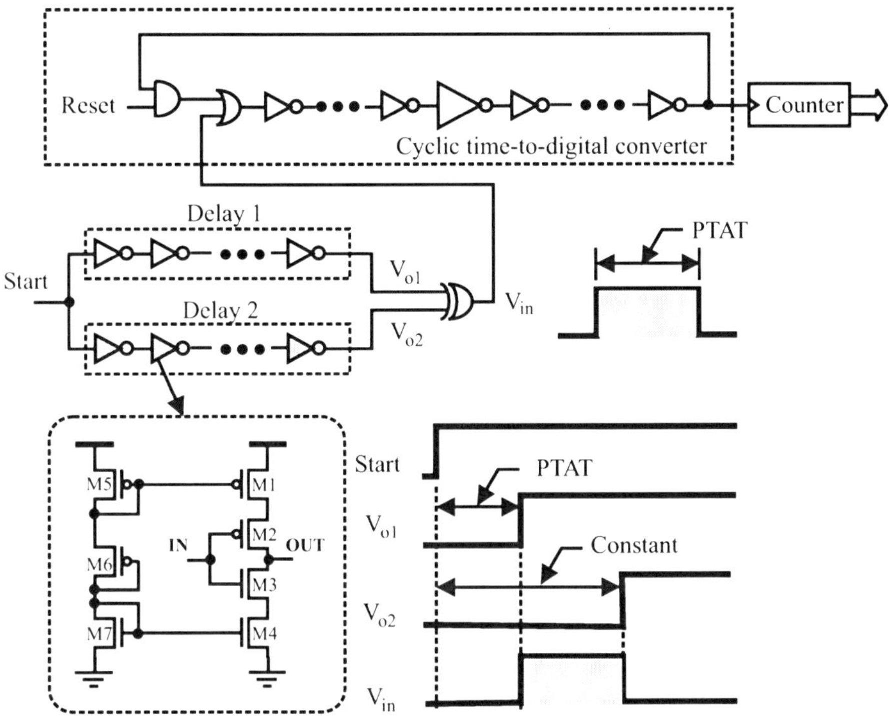

Figure 7.17. Time-to-digital converter-based temperature ADC proposed by Chen *et al.* [215, 216].

asserted (RESET=0), $v_{o1,o3,o5} = 1$ and $v_{o2,o4,o6} = 0$. The logic states of these logic gates will be held unchanged even when RESET becomes 1. Now, when v_{in} becomes 1, $v_{o1,o3,o5} = 0$ and $v_{o2,o4,o6} = 1$. The content of the counter will increment by 1. As long as $v_{in} = 1$, v_{o4} will force $v_{o1} = 1$. The logic state of the output of the following inverters will change subsequently and the counter will continue to increment. When $v_{in} = 0$, the cyclic operation of the loop will stop. Clearly, the time duration between each increment of the counter is the total time delay of the loop. In this case, if we assume that all stages including the OR2 gate and the AND2 gate have the same average propagation delay t_d, then the time duration between each increment of the counter is $7t_d$. This is the resolution of the time-to-digital converter. Clearly, to improve the resolution, reducing t_d is required.

An effective way to reduce t_d is to change the dimension of one delay stage. Now let the dimension of inverter 3 be k time that of other identical inverters, i.e. $W_3 = kW_{1,2,4,5,6} = kW$ and $L_j = L$, $j = 1, 2, ..., 5$. The high-to-low

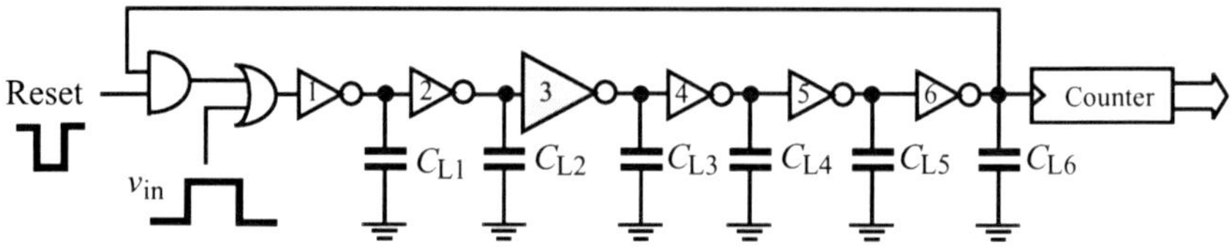

Figure 7.18. Cyclic time-to-digital converter proposed by Chen *et al.* $(W/L)_3 = k(W/L)_{1,2,4,5,6}$ [218].

and low-to-high propagation delays of CMOS static inverters with a step input, denoted by τ_{PHL} and τ_{PLH}, respectively, are given by [219]

$$\tau_{PHL} = \frac{C_L}{k_n(V_{DD} - V_T)}\left[\frac{2V_T}{V_{DD} - V_T} + \ln\left(\frac{3V_{DD} - 4V_T}{V_{DD}}\right)\right], \quad (7.28)$$

and

$$\tau_{PHL} = \frac{C_L}{k_p(V_{DD} - V_T)}\left[\frac{2V_T}{V_{DD} - V_T} + \ln\left(\frac{3V_{DD} - 4V_T}{V_{DD}}\right)\right], \quad (7.29)$$

where C_L is the load of the regular inverter stages, i.e. inverters 1, 2, 4, and 5, $k_n = \mu_n C_{ox}\left(\frac{W}{L}\right)_n$, $k_p = \mu_p C_{ox}\left(\frac{W}{L}\right)_p$, and the threshold voltage of nMOS transistors and that of pMOS transistors are assumed to be the same to simplify analysis and are denoted by V_T. The width shrinkage of the output voltage of inverter 2 due to the increased load capacitance is obtained from

$$\begin{aligned}\Delta W_2 &= \tau_{PLH,2} - \tau_{PHL,2} \\ &= \left(\frac{1}{k_{p2}} - \frac{1}{k_{n2}}\right) C_{L2} B,\end{aligned} \quad (7.30)$$

where C_{L2} is the load capacitance of inverter 2 and

$$B = \frac{2V_T}{(V_{DD} - V_T)^2} + \frac{1}{V_{DD} - V_T}\ln\left(\frac{3V_{DD} - 4V_T}{V_{DD}}\right). \quad (7.31)$$

In a similar manner, one can show that the width shrinkage of the output voltage of inverter 3 is given by

$$\Delta W_3 = -\left(\frac{1}{k_{p3}} - \frac{1}{k_{n3}}\right) C_{L3} B, \quad (7.32)$$

where C_{L3} is the load capacitance of inverter 3. The overall shrinkage of the pulse width from stage 2 to stage 4 is given by

$$\Delta W = \Delta W_2 + \Delta W_3 = \left(k - \frac{1}{k}\right)\left(\frac{1}{k_p} - \frac{1}{k_n}\right) C_L B. \tag{7.33}$$

7.5 Frequency-to-Digital Converter Based Temperature ADCs

The preceding time-to-digital converter based temperature ADCs have the drawback that a large number of inverter delay stages are needed in order to generate a sufficiently large pulse delay subsequently a high temperature resolution. To overcome this drawback, a frequency-to-digital converter (FDC) based temperature ADC was proposed by Kim *et al.* with its block diagram shown in Fig.7.19 [217]. Two current-starving ring oscillators are used. The frequency of one oscillator varies with temperature linearly while the frequency of the other is independent of frequency. The former is called temperature-sensitive oscillator (TSO) while the latter is termed temperature-insensitive oscillator (TIO). The difference between the frequencies of the two oscillators is measured using a frequency-to-digital converter. The biasing for the temperature-insensitive oscillator is similar to that of the temperature ADC in Fig.7.17 where devices are biased at their ZTC bias points to generate a temperature-independent control current. The temperature-sensitive oscillator is a generic current-starving ring VCO whose oscillation frequency varies with temperature. To save power, TIO and TSO operate alternatively. Frequency difference is measured as follows : Assume initially TSO is selected by the controller. The output of the TSO is routed to the FDC by the multiplexer. The counter counts the number of the rising edge of the output voltage of the TSO for a given time interval. The content of the counter is sampled by the sampler and stored. Next, the TIO is selected and the counter counts the number of the rising edge of the output voltage of the TIO. The content of the counter is then sampled and stored. By comparing the stored content of the counter in these two operations, the difference between the frequencies of the two oscillators can be obtained. Since the frequency difference is directly related to the temperature, a measure of temperature is available.

7.6 Charge Redistribution Successive Approximation ADCs

A successive approximation ADC converts an analog waveform to a digital code via a binary search through all possible quantization levels. Fig.7.20 shows the basic configuration of successive approximation ADCs. At the start of conversion, the input is sampled, held, and applied to the non-inverting

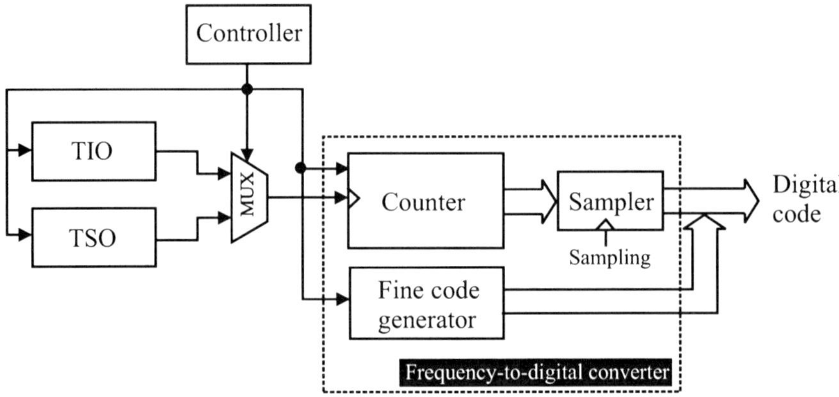

Figure 7.19. Frequency-to-digital converter based temperature ADC proposed by Kim *et al.* [217]. Legends : TIO - temperature insensitive oscillator. TSO - temperature sensitive oscillator.

Table 7.1. Performance comparison of oscillator-based temperature ADCs.

Ref.	Technology	Temp. range	Sensitivity	Power
Zhou & Wu (07) [211]	$0.18\mu m$	$27\sim47°C$	$\pm1°C$	$0.9\mu W$
Park *et al.* (09) [213]	$0.13\mu m$	$8\sim85°C$	$0.4°C$	95 nW
Kim *et al.* (09) [217]	65nm	$-40\sim110°C$	$0.34°C$	$400\ \mu W$

Sampling rate of ADC by Kim *et al.* is 300 KHz.

terminal of the comparator. The successive approximation register (SAR) is initialized such that its MSB=1 while all other bits of the SAR are set to zero. This code is fed to the DAC. The corresponding output of the DAC, which is $\frac{V_{ref}}{2}$, is compared with v_{in}. If $v_{in} < \frac{V_{ref}}{2}$, the SAR will set MSB=0. Otherwise, MSB=1 will remain unchanged. The next bit of the SAR is then tested and set in a similar way. This process continues until all bits of SAR are set. The resultant code of the SAR is the digital representation of the analog input. The conversion time of an N-bit successive approximation ADC is N comparison periods, each set by the conversion time of the DAC and comparator.

7.6.1 Charge-Scaling DACs

Although there are a number of ways to conduct digital-to-analog conversion, charge scaling DACs are widely favored in low-power successive approximation ADCs due to their low static power consumption and the fact that the

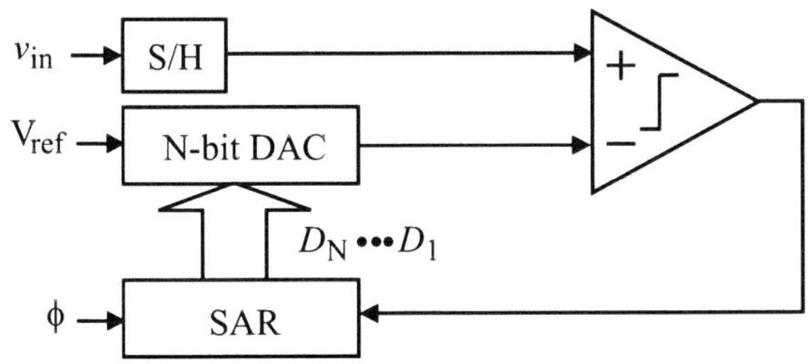

Figure 7.20. Block diagram of successive approximation ADCs.

performance of the DACs is determined by capacitance ratios rather than the absolute value of capacitances [220, 221]. Fig.7.21 is the basic configuration of charge scaling DACs. It consists of a binary-weighted capacitor array, a terminating capacitor, a reference voltage V_{ref}, and a switching network controlled by a SAR and control logic. The operation of the charge scaling DAC is depicted as follows : In the reset phase, both the top and bottom plates of the capacitors are shorted to the ground and the capacitors are discharged completely. In the following evaluation phase, the top plates of the capacitors are disconnected from the ground while the bottom plates of the capacitors are connected to V_{ref}, depending upon the value of their respective binary input from the SAR. Specifically, the bottom plate of the kth-capacitor will be connected to V_{ref} if the corresponding k^{th} digit of the SAR is 1. Otherwise, it will remain connected to the ground. Note that the bottom plate of the terminating capacitor is grounded all the time.

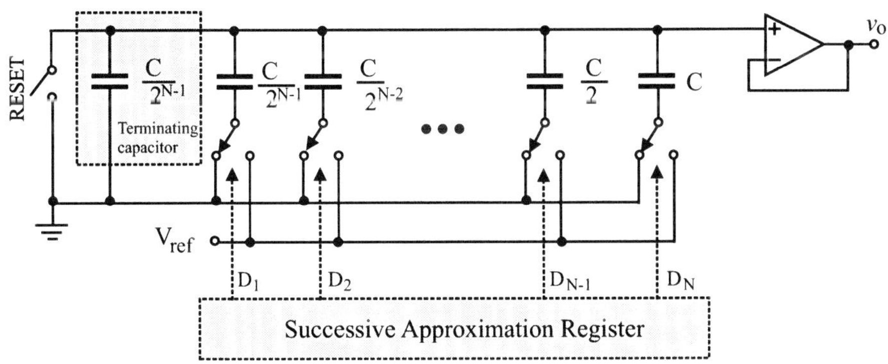

Figure 7.21. Basic configuration of charge-scaling DACs with a single-stage binary-weighted capacitor array.

To illustrate the operation of the charge scaling DAC in a more intuitive way, let us assume that only the kth digit of the SAR is 1, i.e. $D_k = 1$ while all

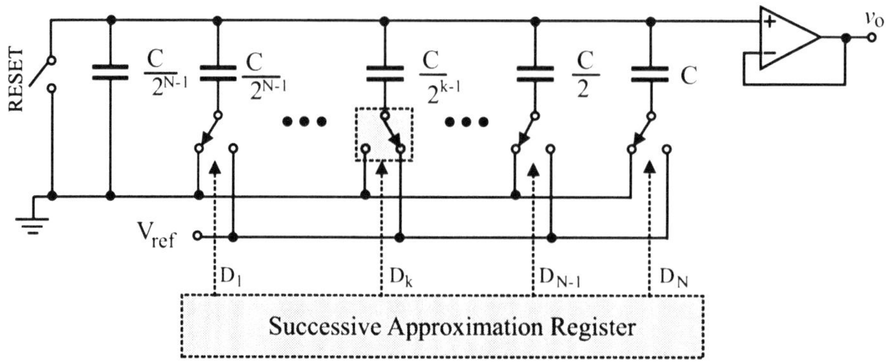

Figure 7.22. Charge-scaling DACs when $D_k = 1$ and the rest of the bits of SAR are 0.

other bits of the SAR are zero, as shown in Fig.7.22. The capacitor of D_k is connected to V_{ref} while all others are connected to the ground. Fig.7.23 shows the corresponding schematic of the DAC in the evaluation phase. The only floating capacitor in this case is $\frac{C}{2^{k-1}}$. All other capacitors, namely C, $\frac{C}{2}$, ..., $\frac{C}{2^{k-2}}$, $\frac{C}{2^k}$, ... ,$\frac{C}{2^{N-1}}$, and $\frac{C}{2^{N-1}}$, are grounded. The output voltage is obtained from the capacitor voltage divider

$$
\begin{aligned}
v_o &= \left(\frac{\frac{C}{2^{k-1}}}{C + \frac{C}{2} + ... + \frac{C}{2^{N-1}} + \frac{C}{2^{N-1}}} \right) V_{ref} \\
&= \left(\frac{\frac{C}{2^{k-1}}}{2C} \right) V_{ref} \\
&= \frac{V_{ref}}{2^k}, \qquad k = 1, 2, ..., N. \qquad (7.34)
\end{aligned}
$$

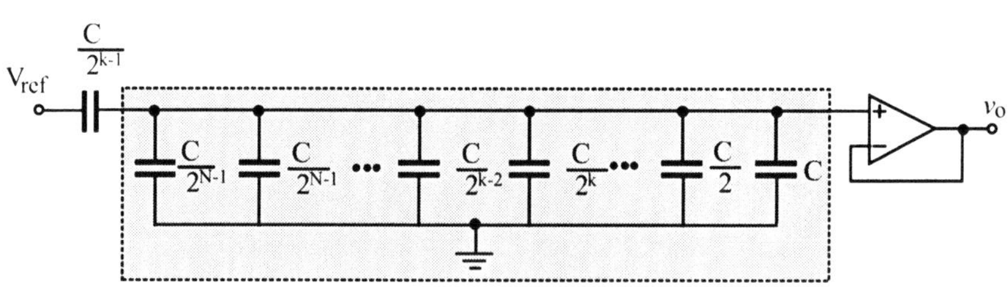

Figure 7.23. Charge scaling DACs with $D_k = 1$ and $D_i = 0$, $i = 1, 2, ..., N$ and $i \neq k$.

To find out the pattern of v_o when more digits of the SAR are set to 1, let us now assume that both the kth and $(k+1)^{th}$ digits of the SAR are 1, i.e. $D_k = D_{k+1} = 1$ while the rest of the digits are zero. In this case, there are two floating capacitors $\frac{C}{2^{k-1}}$ and $\frac{C}{2^k}$ that are connected in parallel. The grounded capacitors are C, $\frac{C}{2}$, ...$\frac{C}{2^{k-2}}$, $\frac{C}{2^{k+1}}$,...,$\frac{C}{2^{N-1}}$, and the terminating capacitor is $\frac{C}{2^{N-1}}$.

$$\begin{aligned} v_o &= \left(\frac{\frac{C}{2^{k-1}} + \frac{C}{2^k}}{C + \frac{C}{2} + ... + \frac{C}{2^{N-1}} + \frac{C}{2^{N-1}}}\right) V_{ref} \\ &= \left(\frac{\frac{C}{2^{k-1}} + \frac{C}{2^k}}{2C}\right) V_{ref} \\ &= \left(\frac{1}{2^k} + \frac{1}{2^{k+1}}\right) V_{ref}, \quad k = 1, 2, ..., N-1. \end{aligned} \tag{7.35}$$

The preceding results can be generalized when an arbitrary number of the digits of the SAR are set to 1. The output in this case is given by

$$v_o = \left(\frac{D_N}{2} + \frac{D_{N-1}}{2^2} + ... + \frac{D_2}{2^{N-1}} + \frac{D_1}{2^N}\right) V_{ref}. \tag{7.36}$$

Although the capacitor array does not consume static power, its dynamic power consumption, arising from the charging and discharging the capacitors, is significant. The dynamic power consumption of charging and discharging a shunt capacitor C between 0 and V_{ref} is given by

$$P_d = CfV_{ref}^2, \tag{7.37}$$

where f is the frequency. Clearly the shunt capacitor of the MSB consumes the most dynamic power when charged and discharged at the same frequency as that for other capacitors. Also observed is that the capacitor array consumes the most power at the transition from $D_N...D_0 = 10...0$ to $D_N...D_0 = 011...1$ where a total of $N-1$ shunt capacitors are charged and the MSB capacitor is discharged.

Also observed is that the higher the conversion speed, the higher the dynamic power consumption. For a given conversion speed, to lower the dynamic power consumption, the capacitance of the capacitors should be reduced. Minimizing the unit capacitance is constrained by both capacitance tolerance and $\frac{kT}{C}$ effect. The former determines the accuracy of DACs, as to be detailed in the following section, while the latter sets the noise floor of the converter. As to be seen shortly that there are a number of capacitor array configurations distinct from the preceding binary-weighted capacitor array that can greatly reduce the total capacitance subsequently the dynamic power consumption of capacitor arrays.

Another unique characteristic of the charge-scaling DAC is that the sample-and-hold operation on the input is carried out naturally by the capacitor array. As a result, there is no need for a separate sample-and-hold block.

The speed of the charge-scaling DAC is determined by the settling time of the charging and discharging processes of the capacitors, which is set by the time constant of the RC networks formed by the switching network and the capacitor array.

7.6.2 Accuracy of Charge-Scaling DACs

The accuracy of charge scaling DACs is set by the tolerance of the capacitance of the capacitor array. To illustrate this, we follow the approach by Allen and Holberg [209]. Assume that only the kth digit of the SAR is 1, i.e. $D_k = 1$, while all other digits of the SAR are zero. Since the only floating capacitor is the kth capacitor whose capacitance is $\frac{C}{2^{k-1}}$, the output voltage is given by (7.34)

$$v_o = \left(\frac{V_{ref}}{2^k}\right) = \frac{2^N}{2^k}\frac{V_{ref}}{2^N} = \left(\frac{2^N}{2^k}\right)\text{LSB}. \tag{7.38}$$

Now let us assume that the capacitor C has a capacitance tolerance $\frac{\Delta C}{C}$, i.e., $C_{max} = C + \Delta C$ and $C_{min} = C - \Delta C$. The output voltage in this case becomes

$$\begin{aligned} v_o &= \left(\frac{\frac{C \pm \Delta C}{2^{k-1}}}{2C}\right) V_{ref} \\ &= \left(1 \pm \frac{\Delta C}{C}\right)\frac{2^N}{2^k}\text{LSB}. \end{aligned} \tag{7.39}$$

Note that we have used $2C$ for the total capacitance of the terminated binary-weighted capacitor array when capacitance variations are present. Such a treatment is justified as the sum of the capacitances of the capacitors averages out the effect of the random variation of the total capacitance. The integral nonlinearity is obtained from

$$\text{INL} = \left|\frac{\Delta C}{C}\right|\frac{2^N}{2^k}\text{LSB}. \tag{7.40}$$

Since DNL at $v_{in} = 0$ and $v_{in} = V_{ref}$ is typically zero due to the fact that DACs are usually calibrated at $v_{in} = 0$ and $v_{in} = V_{ref}$, the worst-case DNL often occurs at the middle point of the input range, i.e., at the transition from

$D_N...D_1 = 01...1$ to $D_N...D_1 = 10...0$ [221, 209]. The worst-case DNL is obtained from

$$\text{DNL} = \left| \frac{v_o(10...0) - v(01...1)}{\text{LSB}} - 1 \right|, \tag{7.41}$$

where $v_o(10...0)$ and $v_o(01...1)$ are the output voltage of the DAC when $D_N...D_1 = 10...0$ and $D_N...D_1 = 01...1$, respectively. To calculate $v_o(10...0)$, since the only floating capacitor in this case is $C + \Delta C$, we have

$$v_o(10...0) = \left(\frac{C + \Delta C}{2C} \right) V_{ref}. \tag{7.42}$$

For $v_o(01...1)$, there are a total of $N - 1$ floating capacitors $\frac{C-\Delta C}{2}$, $\frac{C-\Delta C}{2^2}$,..., $\frac{C-\Delta C}{2^{N-1}}$ that are connected in parallel. Note that we have used $C - \Delta C$ to account for the worst-case variation of the capacitance. The only grounded capacitor in the capacitor array is C. The terminating capacitor is, of course, grounded. Denoting C_f the total capacitance of the floating capacitors, i.e.

$$\begin{aligned} C_f &= \left(\frac{C - \Delta C}{2} \right) + \left(\frac{C - \Delta C}{2^2} \right) + \left(\frac{C - \Delta C}{2^3} \right) + ... + \left(\frac{C - \Delta C}{2^{N-1}} \right) \\ &= (C - \Delta C) \left(1 - \frac{1}{2^{N-1}} \right), \end{aligned} \tag{7.43}$$

we have

$$v_o(01...1) = \left(\frac{C_f}{2C} \right) V_{ref} = \left(\frac{C - \Delta C}{2C} \right) \left(1 - \frac{1}{2^{N-1}} \right) V_{ref}. \tag{7.44}$$

Substituting these results into (7.41) yields

$$\text{DNL} = \left(2^N - 1 \right) \frac{\Delta C}{C}. \tag{7.45}$$

Eqs.(7.40) and (7.45) reveal that both INL and DNL are lower-bounded by the capacitance tolerance $\frac{\Delta C}{C}$. Reducing $\Delta C/C$ will lower DNL. As a result, more bits can be incorporated. A large C is therefore desirable from a resolution point of view. For a given $\Delta C/C$, DNL will rise exponentially with the number of the bits.

It was shown that a reference voltage V_{ref} is required and V_{ref} defines the full-scale input signal. Clearly $V_{ref} = V_{DD}$ is preferred in order to maximize the dynamic range of ADCs. Although V_{ref} can be generated using a reference circuit, the additional power consumption of the reference circuit will increase the burden of the power consumption of the ADCs. In [220–222], $V_{ref} = V_{DD}$ is used directly to eliminate the need for a reference circuit and to maximize the dynamic range. One drawback of using $V_{ref} = V_{DD}$ is the reduced rejection of the effect of the fluctuation of the supply voltage of ADCs. This is especially prominent for passive wireless microsystems as the power of these systems is harvested from sources such as radio-frequency waves or solar energy and varies substantially with both the sources and the operation of these microsystems.

7.6.3 Charge Redistribution ADCs

Having studied charge-scaling DACs, let us now turn our attention to charge redistribution successive approximation ADCs where charge-scaling DACs are the key components. Charge redistribution successive approximation ADCs was first proposed by McCreamy and Gray [223]. Fig.7.24 shows the basic configuration of an N-bit charge redistribution successive approximation ADC. It consists of a successive approximation register, a single-stage binary-weighted capacitor array, a terminating capacitor whose capacitance is the same as that of the capacitor corresponding to the least significant bit, a comparator, a voltage doubler, and a switching network.

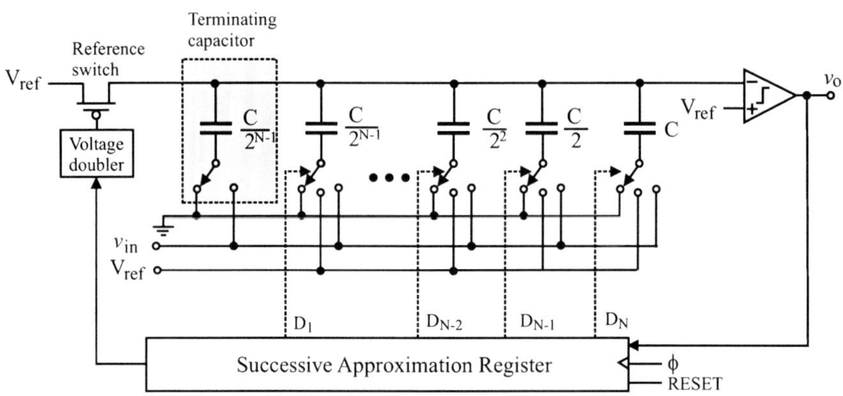

Figure 7.24. Configuration of charge redistribution successive approximation ADCs with a single-stage binary-weighted capacitor array.

The operation of the ADC consists of a sampling phase, a charge holding phase, and a charge redistribution phase. During the sampling phase, the reference switch is closed and the bottom plates of the capacitor array are connected to the input voltage v_{in} via the switching network. The capacitor array is charged to $V_{ref} - v_{in}$, as shown in Fig.7.25. Note that the voltage drop across the reference switch must be sufficiently small such that the voltage of the top plates of the capacitor array is V_{ref}. This requires that the reference switch be in deep triode where $V_{DS}\approx 0$. Since the voltage of the top plates of the capacitor array can be as high as $1.5V_{ref}$, as to be seen shortly, the gate voltage of the reference switch must be sufficiently high to ensure that the reference switch be in deep triode when switched on. A high gate voltage can be obtained from a voltage doubler.

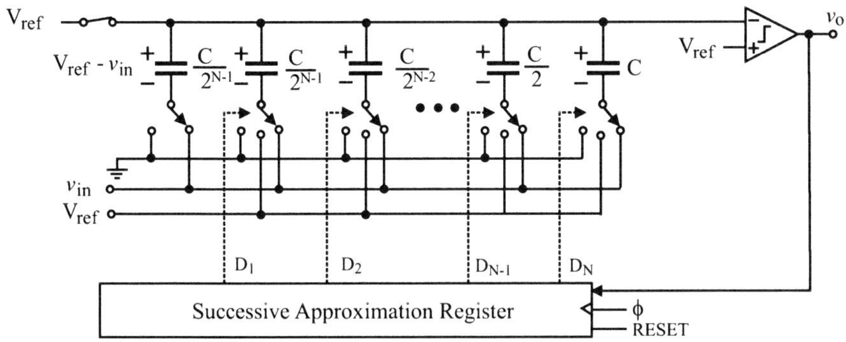

Figure 7.25. Charge redistribution successive approximation ADCs in sampling phase. The reference switch is closed and the bottom plates of the capacitors are connected to v_{in}.

In the following charge holding phase, the reference switch is open and the bottom plates of the capacitor array including the terminating capacitor are connected to the ground, as shown in Fig.7.26. The voltage across the capacitors is held at $V_{ref} - v_{in}$ in this phase. Note that this step is important as it resets the parasitic capacitors between the bottom plate of the capacitor array and the substrate. Although the charge injection from the reference switch during its turn-off will affect the voltage of the top plates of the capacitor array, given the fact that all the capacitors in the capacitor array are connected in parallel, the effect of the charge injection from the reference switch on the voltage of the top plates of the capacitors is rather small.

The last phase of the data conversion is the charge redistribution phase, also known as the bit conversion phase. In this phase, the MSB of the SAR is first set to 1 and the rest of the bits are set to 0. The switch corresponding to the MSB is connected to V_{ref} while all others remain grounded, as shown in Fig.7.27.

Since the capacitor array is charged with a initial voltage $V_{ref} - v_{in}$ in the previous phase, their Thévenin equivalent circuits can be used to represent the capacitors, as shown in Fig.7.28. The voltage of the inverting terminal of the comparator is obtained using the principle of superposition

$$v^- = \left(\frac{C}{C+\frac{C}{2}+\frac{C}{2^2}+...+\frac{C}{2^{N-1}}+\frac{C}{2^{N-1}}}\right)(2V_{ref} - v_{in})$$

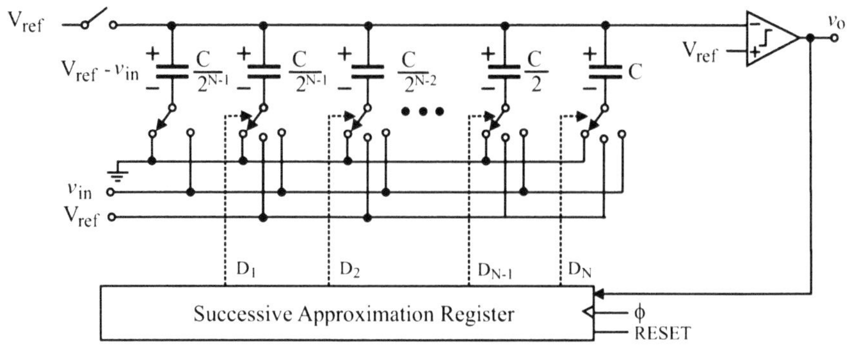

Figure 7.26. Charge redistribution successive approximation ADCs in charge holding phase. The reference switch is open and the bottom plates of the capacitors are connected to the ground.

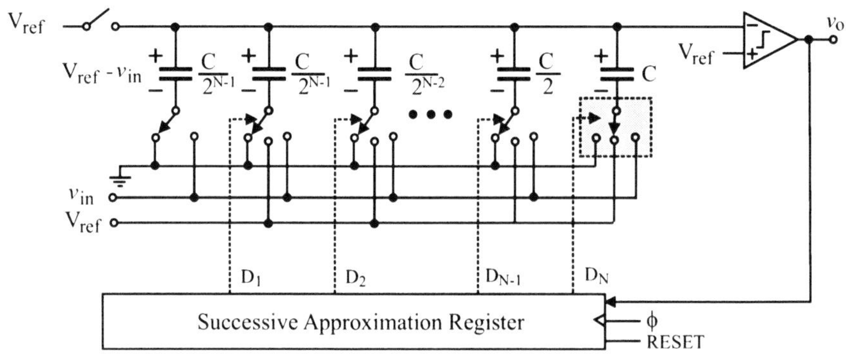

Figure 7.27. Charge redistribution successive approximation ADCs in charge redistribution phase with the MSB=1 and the rest are set to 0. The reference switch is open and the switch of the MSB is connected to V_{ref} while all others are grounded.

$$
\begin{aligned}
& + \left(\frac{\frac{C}{2} + \frac{C}{2^2} + \ldots + \frac{C}{2^{N-1}} + \frac{C}{2^{N-1}}}{C + \frac{C}{2} + \frac{C}{2^2} + \ldots + \frac{C}{2^{N-1}} + \frac{C}{2^{N-1}}} \right) (V_{ref} - v_{in}) \\
& = \frac{3}{2} V_{ref} - v_{in}. \qquad (7.46)
\end{aligned}
$$

It is seen from (7.46) that the voltage of the non-inverting terminal can be as high as $1.5V_{ref}$. This is the reason why a high gate voltage is required to ensure that the reference switch will be in deep triode when switched on. The input of the comparator becomes

$$
\Delta v = v^+ - v^- = v_{in} - \frac{V_{ref}}{2}. \qquad (7.47)
$$

Eq.(7.47) shows that if $v_{in} > \frac{V_{ref}}{2}$, the output of the comparator is high and the position of the switch corresponding to the MSB will remain to be connected to V_{ref}. Otherwise, the switch will be connected to the ground. Once the MSB is set, the next bit is tested and set in a similar manner. This process continues until all bits of the SAR are set.

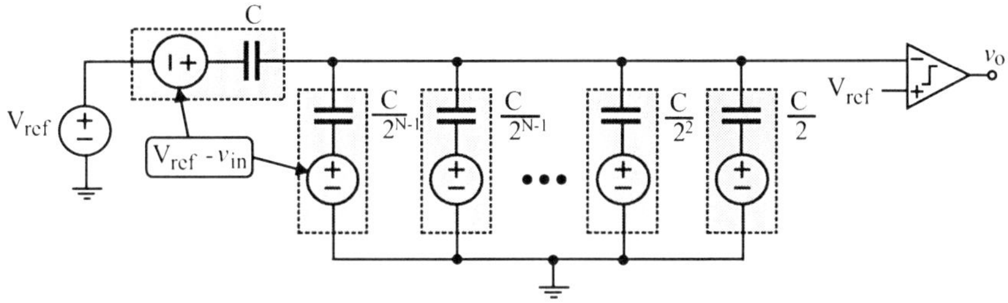

Figure 7.28. Charge redistribution successive approximation ADCs in charge redistribution phase with the MSB of the SAR set to 1 while all other bits of the SAR set to 0. Charged capacitors are replaced with their Thévenin equivalent circuits.

7.6.4 Single-Stage Binary-Weighted Capacitor Arrays

The preceding charge-scaling DACs use a single-stage binary-weighted capacitor array. These capacitors can be implemented using either metal-insulator-metal (MIM) capacitors [224] or metal-metal capacitors [225], depending upon the unit capacitance $\frac{C}{2^{N-1}}$, the number of the bits of ADCs, and the budget on silicon area.

The choice of the unit capacitance is guided by the following considerations : the overall silicon consumption of the capacitor array, capacitance tolerance $\frac{\Delta C}{C}$, power consumption, conversion speed, and $\frac{kT}{C}$. Unit capacitance should be as small as possible such that the overall silicon consumption of the capacitor array is low and yet as large as possible such that the effect of the capacitance tolerance $\frac{\Delta C}{C}$, the effect of parasitic interconnect capacitances, and the effect of $\frac{kT}{C}$ are minimized. Lowering the capacitance of the capacitors is also desirable from a power consumption and conversion speed point of view.

When the unit capacitance is small, the capacitance of the interconnects connecting the capacitors becomes comparable to the unit capacitance. In this case, a careful placement of the small capacitors of the capacitor array is required to ensure that the effect of the capacitance of the interconnects is small. For example, in [225], the small capacitors of the capacitor array are placed close to the edge of the capacitor array such that the length subsequently the capacitance of the interconnects connecting these capacitors is minimized. Also, centroid layout techniques should be used to minimize the effect of oxide gradient variation on the capacitance of the capacitors. Dummy capacitors should also be placed around the capacitor array to minimize the effect of the neighboring devices on the capacitance of the capacitors. Fig.7.29 sketches the layout of the capacitor array of the 8-bit successive approximation charge redistribution ADC used in [225].

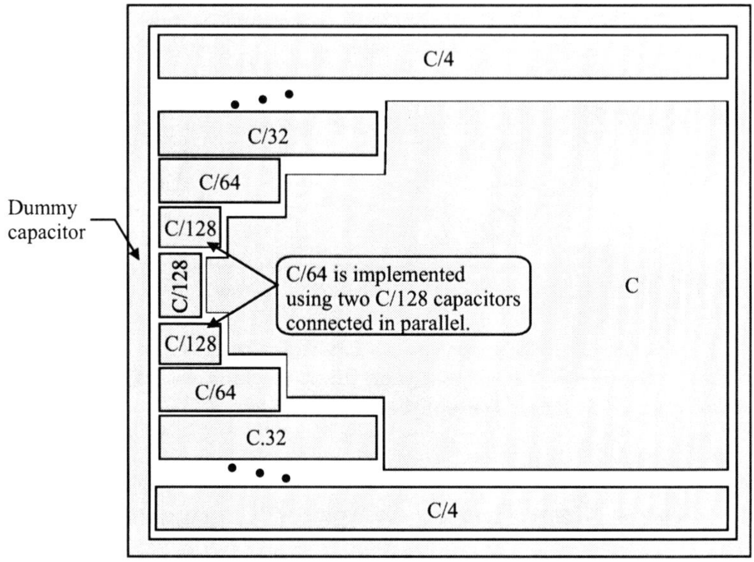

Figure 7.29. Layout of the capacitor array of 8-bit ADC used in [225].

To ensure that the capacitance ratio is independent of the absolute value of capacitances, large capacitors should be implemented using multiple unit capacitors that are connected in parallel such that the effect of the capacitor under-cutting on the capacitance ratio is minimized. To illustrate this, let the tolerance of the unit capacitor C_1 be $\frac{\Delta C}{C}$, i.e. $C_{1,max} = C + \Delta C$ and $C_{1,min} = C - \Delta C$ with $\Delta C \ll C$. Let us first consider a capacitor C_2 of capacitance $2C$. The capacitor is implemented using two unit capacitors connected in parallel such that $C_{2,max} = 2C + 2\Delta C$ and $C_{2,min} = 2C - 2\Delta C$. The ratio of the capacitance of C_2 to that of the unit capacitor is obtained from

$$\begin{aligned} \frac{C_{2,max}}{C_{1,min}} &= \frac{2C + 2\Delta C}{C - \Delta C} \\ &= 2\left(\frac{1 + \frac{\Delta C}{C}}{1 - \frac{\Delta C}{C}}\right) \\ &\approx 2\left(1 + \frac{\Delta C}{C}\right)^2. \end{aligned} \tag{7.48}$$

Note that we have utilized the first-order approximation

$$\frac{1}{1-\frac{\Delta C}{C}} \approx 1+\frac{\Delta C}{C} \tag{7.49}$$

in derivation of (7.48). Similarly one can show that

$$\begin{aligned} \frac{C_{2,min}}{C_{1,max}} &= \frac{2C-2\Delta C}{C+\Delta C} \\ &= 2\left(\frac{1-\frac{\Delta C}{C}}{1+\frac{\Delta C}{C}}\right) \\ &\approx 2\left(1-\frac{\Delta C}{C}\right)^2. \end{aligned} \tag{7.50}$$

Now let us consider a capacitor C_k of capacitance kC. The capacitor is implemented using k unit capacitors connected in parallel. As a result, $C_{k,max} = kC + k\Delta C$ and $C_{k,min} = kC - k\Delta C$. Following the similar procedures as those for C_2, one can show that the ratios of the capacitance of C_k to that of the unit capacitor are given by

$$\frac{C_{k,max}}{C_{k,min}} \approx k\left(1+\frac{\Delta C}{C}\right)^2. \tag{7.51}$$

$$\frac{C_{k,min}}{C_{1,max}} \approx k\left(1-\frac{\Delta C}{C}\right)^2. \tag{7.52}$$

The preceding analysis demonstrates that once large capacitors are implemented using multiple unit capacitors that are connected in parallel, the effect of the capacitance tolerance on capacitance ratios is the same regardless of the absolute value of the capacitances.

7.6.5 Two-Stage Binary-Weighted Capacitor Array

It is seen from Fig.7.29 that the silicon consumption of single-stage binary-weighted capacitor arrays rises exponentially with the number of the bits of ADCs. To alleviate the burden of the large silicon consumption of single-stage binary-weighted capacitor arrays when the number of the bits of ADCs is large, a two-stage binary-weighted capacitor array technique was proposed by Ohri and Callahan [226, 227]. Shown in Fig.7.30 is a two-stage binary-weighted capacitor array for an 8-bit ADC. It consists of a 4-bit single-stage binary-weighted capacitor array called the MSB array, a 4-bit single-stage binary-weighted capacitor array called the LSB array, a scaling capacitor also known as attenuation capacitor C_s, a terminating capacitor whose capacitance is equal to the capacitance of the capacitor of the least significant bit of the LSB array, a comparator, and a switching network. The value of the scaling capacitor C_s must be such that the series combination of the LSB array, the terminating capacitor, and C_s will terminate the MSB array, i.e., the overall capacitance of the series combination of the LSB array, the terminating capacitor, and C_s will equal to the capacitance of the capacitor of the least significant bit of the MSB array, mathematically

$$\frac{1}{C_s} + \frac{1}{C_{LSB} + \frac{C}{2^3}} = \frac{1}{\frac{C}{2^3}}, \tag{7.53}$$

where C_{LSB} is the total capacitance of the LSB array obtained from

$$C_{LSB} = \frac{C}{2^3} + \frac{C}{2^2} + \frac{C}{2} + C = \frac{15}{8}C.$$

Solving (7.53) yields

$$C_s = \frac{2}{15}C.$$

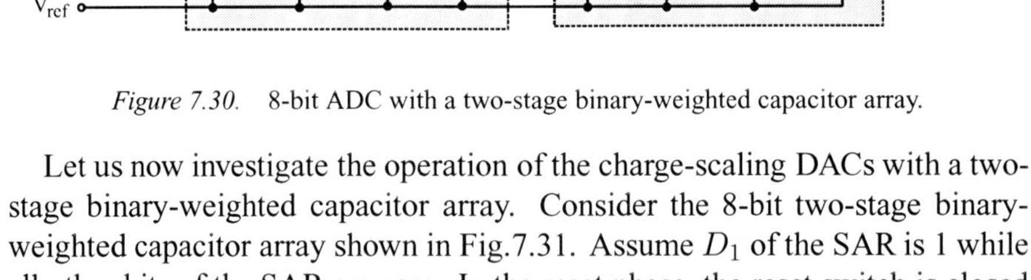

Figure 7.30. 8-bit ADC with a two-stage binary-weighted capacitor array.

Let us now investigate the operation of the charge-scaling DACs with a two-stage binary-weighted capacitor array. Consider the 8-bit two-stage binary-weighted capacitor array shown in Fig.7.31. Assume D_1 of the SAR is 1 while all other bits of the SAR are zero. In the reset phase, the reset switch is closed and the bottom plates of the capacitors are connected to the ground. All the capacitors are fully discharged in this phase.

In the following evaluation phase, the bottom plate of the capacitor corresponding to D_1 is connected to V_{ref} while all others remain connected to the ground. The corresponding circuit is shown in Fig.7.32. To facilitate analysis, the sub-circuit consisting of the reference voltage, the LSB array, and the

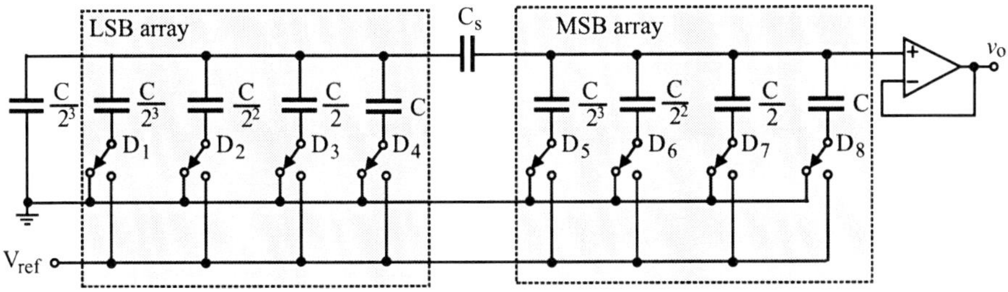

Figure 7.31. 8-bit DAC with a two-stage binary-weighted capacitor array.

terminating capacitor is replaced with its Thévenin equivalent circuit shown in the figure with Thévenin capacitance $C_{T,1}$ and Thévenin voltage $V_{T,1}$ given by

$$C_{T,1} = C_{LSB} + \frac{C}{2^3} = 2C, \tag{7.54}$$

$$V_{T,1} = \frac{\frac{C}{2^3}}{C_{T,1}} V_{ref} = \frac{V_{ref}}{2^4}. \tag{7.55}$$

Note the second subscript in (7.54) and (7.55) specifies that $C_{T,1}$ and $V_{T,1}$ are Thévenin capacitance and Thévenin voltage when $D_1 = 1$ and all other bits of the SAR are zero. It is readily to verify that the capacitance of the sub-network consisting of $C_{T,1}$ in series with C_s, denoted by C_{eq}, is obtained from

$$\frac{1}{C_{eq}} = \frac{1}{C_{T,1}} + \frac{1}{C_s}. \tag{7.56}$$

The result is given by

$$C_{eq} = \frac{C}{8}.$$

C_{eq} thus satisfies the termination criterion of the MSB array. The total capacitance of the MSB array, denoted by C_{MSB}, is given by

$$C_{MSB} = C + \frac{C}{2} + \frac{C}{4} + \frac{C}{8} = \frac{15}{8}C.$$

The output voltage of the DAC, denoted by v_o, is obtained from

$$v_o = \frac{C_{eq}}{C_{eq} + C_{MSB}} V_{T,1} = \frac{V_{T,1}}{2^4} = \frac{1}{2^4}\left(\frac{V_{ref}}{2^4}\right). \tag{7.57}$$

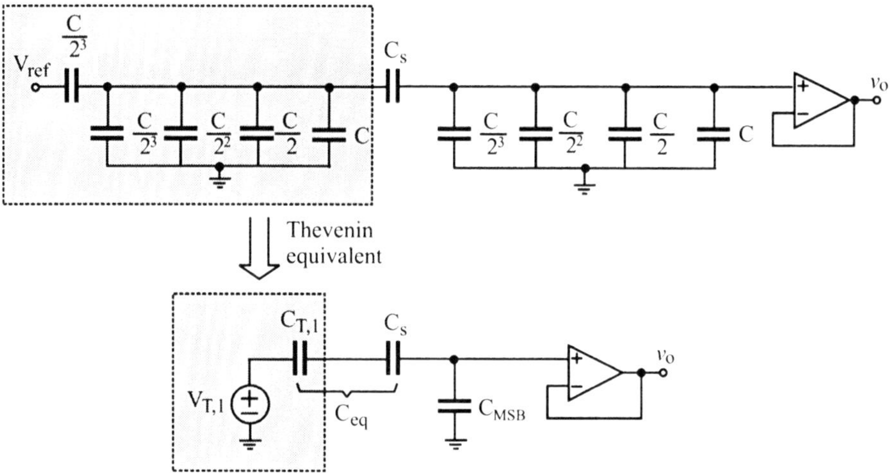

Figure 7.32. 8-bit DAC with a two-stage binary-weighted capacitor array when $D_1 = 1$ while all other bits of the SAR are zero.

Let us consider another case where $D_8 = 1$ while all other bits of the SAR are zero. In the evaluation phase, the bottom plate of the capacitor corresponding to D_8 is connected to V_{ref} while all others remain connected to the ground, as shown in Fig. 7.33. Following the similar steps as those for D_1, one can show that

$$v_o = \frac{V_{ref}}{2}. \tag{7.58}$$

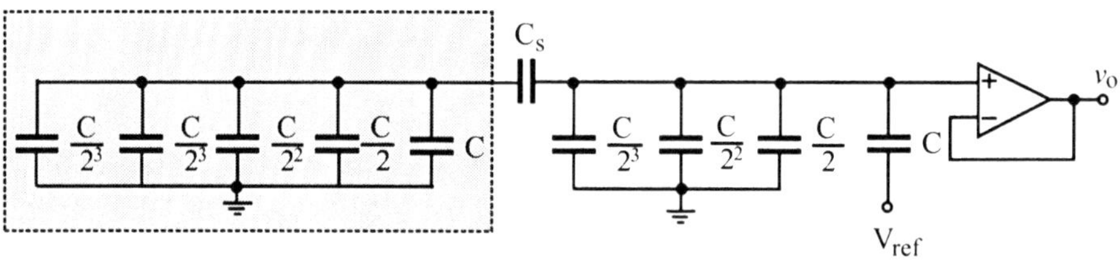

Figure 7.33. 8-bit DAC with a two-stage binary-weighted capacitor array when $D_8 = 1$ while all other bits of the SAR are zero.

Having derived the output voltage of the DAC for $D_1 = 1$ and $D_8 = 1$, let us now consider a general case where $D_k = 1$ while all other bits of the SAR are zero. The bottom plate of the capacitor corresponding to D_k is connected to V_{ref} while the bottom plates of all others remain connected to the ground. Two cases need to be investigated. The first one is where D_k is located in the LSB array. The sub-circuit consisting of the LSB array, the terminating capacitor, and the reference voltage V_{ref} is replaced with its Thévenin equivalent circuit with

$$C_{T,k} = 2C, \tag{7.59}$$

and

$$V_{T,k} = \frac{\frac{C}{2^{k-1}}}{2C} V_{ref} = \frac{V_{ref}}{2^k}, \quad k = 1, 2, 3, 4. \tag{7.60}$$

The output voltage in this case is obtained from

$$\begin{aligned} v_o &= \left(\frac{C_{eq}}{C_{eq} + C_{MSB}} \right) V_{T,k} \\ &= \left(\frac{\frac{C}{8}}{\frac{C}{8} + \frac{15C}{8}} \right) V_{T,k} \\ &= \frac{1}{2^4} \left(\frac{V_{ref}}{2^k} \right), \quad k = 1, 2, 3, 4. \end{aligned} \tag{7.61}$$

The second case is where $D_k = 1$ is located in the MSB array while all other bits of the SAR are zero. It can be shown that the total capacitance of the network consisting of the LSB array, the terminating capacitor, and the scaling capacitor is $\frac{C}{2^3}$, which terminates the MSB array. The output voltage is given by

$$v_o = \left(\frac{\frac{C}{2^{k-1}}}{2C} \right) V_{ref} = \frac{V_{ref}}{2^k}, \quad k = 1, 2, 3, 4. \tag{7.62}$$

Combining (7.61) and (7.62) yields the general expression of v_o when an arbitrary number of the bits of the SAR are 1

$$\begin{aligned} v_o &= \frac{1}{2^4} \left(\frac{D_4}{2} + \frac{D_3}{2^2} + \frac{D_2}{2^3} + \frac{D_1}{2^4} \right) V_{ref} + \left(\frac{D_8}{2} + \frac{D_7}{2^2} + \frac{D_6}{2^3} + \frac{D_5}{2^4} \right) V_{ref}. \\ &= \left(\frac{D_1}{2^8} + \frac{D_2}{2^7} + \frac{D_3}{2^6} + \frac{D_4}{2^5} + \frac{D_5}{2^4} + \frac{D_6}{2^3} + \frac{D_7}{2^2} + \frac{D_8}{2^1} \right) V_{ref}. \end{aligned} \tag{7.63}$$

Eq.(7.63) demonstrates that the DAC with a two-stage binary-weighted capacitor array behaves in a similar way as the corresponding DAC with a single-stage binary-weighted capacitor array.

Two-stage binary-weighted capacitor arrays suffer from two main drawbacks : (i) Because the scaling capacitor is floating, the parasitic capacitance associated with the bottom plate of the scaling capacitor affects the accuracy of the DAC. The parasitic capacitance of the top plate of the scaling capacitor is much smaller as compared with that of the bottom plate and is not of a critical concern. (ii) It is difficult to construct the scaling capacitor preciously. This is because the capacitance of the scaling capacitor is not a multiple of the unit capacitance. For example, in the preceding 8-bit single-stage binary-weighted DAC, the unit capacitance is $\frac{C}{8}$ while C_s of the corresponding 2-stage binary-weighted DAC is $\frac{2C}{15}$. The capacitance error of the scaling capacitor will limit the performance of DACs. In what follows we examine these two drawbacks in detail.

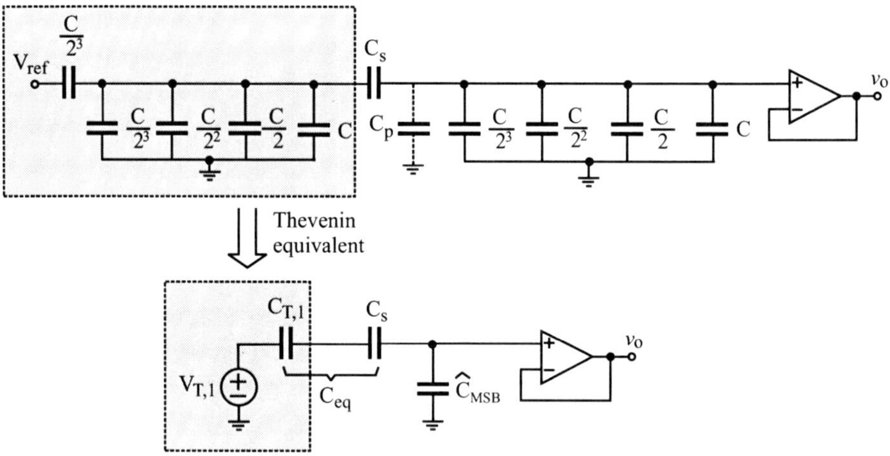

Figure 7.34. Bottom plate of scaling capacitor is on MSB array side.

Let us first investigate the effect of the parasitic capacitance between the bottom plate of the scaling capacitor and the substrate on the performance of the DAC with a two-stage binary-weighted capacitor array. Depending upon the wiring of the scaling capacitor, two cases might exist : (i) The bottom plate of the scaling capacitor is on the MSB array side and (ii) the bottom plate of the scaling capacitor is on the LSB array side.

Consider the first case where the bottom plate of the scaling capacitor is on the MSB array side, as shown in Fig.7.34 where $D_1 = 1$ and all other bits of the SAR are zero. Let C_p denote the parasitic capacitance between the bottom plate of the scaling capacitor and the substrate. Since $V_{T,1} = \frac{V_{ref}}{2^4}$, $C_{T,1} = 2C$, $C_{eq} = \frac{C}{8}$, and $\hat{C}_{MSB} = C_{MSB} + C_p$, we have the output voltage

$$\begin{aligned} v_o &= \frac{C_{eq}}{C_{eq} + \hat{C}_{MSB}} V_{T,1} \\ &= \frac{V_{T,1}}{2^4} \left(\frac{1}{1 + \frac{C_p}{2C}} \right). \end{aligned} \tag{7.64}$$

Note $2C$ is the total capacitance of a terminated MSB array or that of a terminated LSB array. Since $C_s = \frac{2}{15}C$, if $C_p = 0.1C_s$, i.e. the bottom-plate parasitic capacitance is 10% of the inter-plate capacitance, $C_p = \frac{C}{75} \ll C$ holds. Since

$$\frac{C_p}{2C} = \frac{1}{150} \ll 1, \tag{7.65}$$

making use of the first-order approximation

$$\frac{1}{1 + \frac{C_p}{2C}} \approx 1 - \frac{C_p}{2C}, \tag{7.66}$$

we have

$$v_o = \frac{1}{2^4} \left(\frac{V_{ref}}{2^4} \right) \left(1 - \frac{C_p}{2C} \right). \tag{7.67}$$

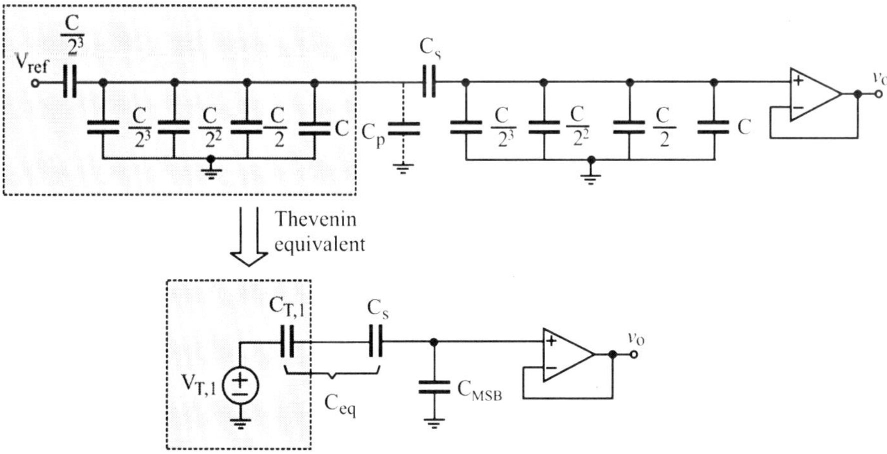

Figure 7.35. Bottom plate of scaling capacitor is on LSB array side.

Now let us consider the second case where the bottom plate of the scaling capacitor is on the LSB array side, as shown in Fig.7.35. Again, $D_1 = 1$ and all other bits of the SAR are zero. Thévenin capacitance is given by

$$C_{T,1} = 2C + C_p. \tag{7.68}$$

Thévenin voltage in this case is obtained from

$$\begin{aligned} V_{T,1} &= \left(\frac{\frac{C}{8}}{\frac{C}{8} + \frac{15C}{8} + C_p} \right) V_{ref} \\ &= \frac{V_{ref}}{2^4} \left(\frac{1}{1 + \frac{C_p}{2C}} \right) \\ &\approx \frac{V_{ref}}{2^4} \left(1 - \frac{C_p}{2C} \right). \end{aligned} \tag{7.69}$$

Further

$$\begin{aligned} \frac{1}{C_{eq}} &= \frac{1}{C_s} + \frac{1}{C_{T,1}} \\ &= \frac{15}{2C} + \frac{1}{2C} \left(\frac{1}{1 + \frac{C_p}{2C}} \right) \\ &\approx \frac{15}{2C} + \frac{1}{2C} \left(1 - \frac{C_p}{2C} \right) \\ &= \frac{8}{C} - \frac{C_p}{4C^2}. \end{aligned} \tag{7.70}$$

As a result,

$$\begin{aligned} C_{eq} &= \frac{C}{8} \left(\frac{1}{1 - \frac{C_p}{32C}} \right) \\ &\approx \frac{C}{8} \left(1 + \frac{C_p}{32C} \right). \end{aligned} \tag{7.71}$$

Finally, the output voltage is obtained from

$$\begin{aligned} v_o &= \frac{C_{eq}}{C_{eq} + C_{MSB}} V_{T,1} \\ &= \frac{1}{2^4} V_{T,1} \left(\frac{1 + \frac{C_p}{32C}}{1 + \frac{C_p}{512C}} \right) \end{aligned}$$

(7.72)

Making use of the following approximations

$$1 + \frac{C_p}{32C} \approx 1, \qquad 1 + \frac{C_p}{512C} \approx 1 \tag{7.73}$$

we obtain from (7.72)

$$v_o \approx \frac{1}{2^4}\left(\frac{V_{ref}}{2^4}\right)\left(1 - \frac{C_p}{2C}\right). \tag{7.74}$$

Eqs.(7.67) and (7.72) reveal that the effect of the parasitic capacitance of the bottom plate of the scaling capacitor on the performance of the DAC is the same regardless of whether the bottom plate of the scaling capacitor is on the MSB side or the LSB side.

Since $C_s = \frac{2}{15}C$, if $C_p = 0.1C_s = \frac{C}{75}$ i.e. the capacitance between the bottom plate of the scaling capacitor and the substrate is 10% of the inter-plate capacitance, the error term becomes $\frac{C_p}{2C} = \frac{1}{150}$ =0.67%. The preceding analysis demonstrates that the effect of the parasitic bottom plate capacitance of the scaling capacitor on the performance of the DAC is rather small.

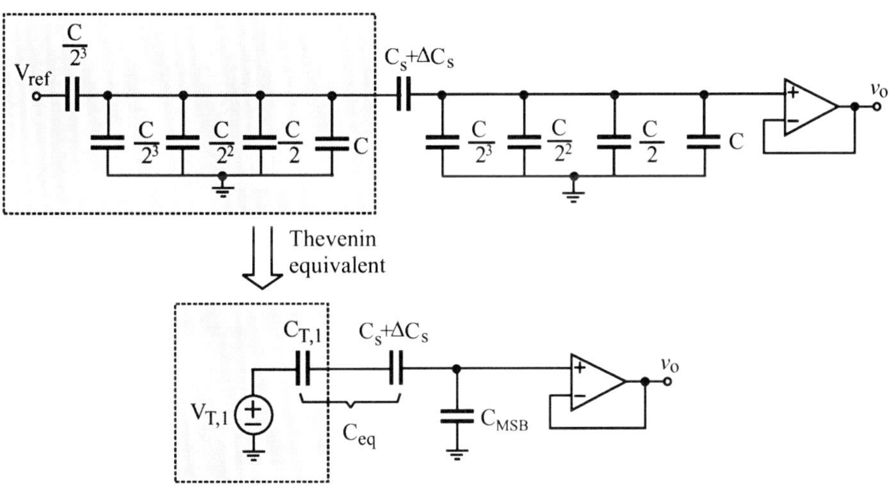

Figure 7.36. Analysis of the effect of the capacitance error of scaling capacitor.

Let us now turn our attention to the effect of the capacitance error of the scaling capacitor on the performance of the DAC. As pointed out earlier that the capacitance error of the scaling capacitor rises from the fact that the capacitance of the scaling capacitor is not a multiple of the unit capacitance. Let the capacitance error of C_s be ΔC_s with $\Delta C_s \ll C_s$, as shown in Fig.7.36. Again, let $D_1 = 1$ while all other bits of the SAR are zero. Because

$$\begin{aligned}\frac{1}{C_{eq}} &= \frac{1}{C_{T,1}} + \frac{1}{C_s + \Delta C_s} \\ &= \frac{1}{2C} + \frac{15}{2C}\left(\frac{1}{1 + \frac{15\Delta C_s}{2C}}\right) \\ &\approx \frac{1}{2C} + \frac{15}{2C}\left(1 - \frac{15\Delta C_s}{2C}\right) \\ &= \frac{8}{C}\left(1 - \frac{225\Delta C_s}{32C}\right),\end{aligned} \tag{7.75}$$

utilizing

$$\frac{1}{1 - \frac{225\Delta C_s}{32C}} \approx 1 + \frac{225\Delta C_s}{32C}, \tag{7.76}$$

we have

$$C_{eq} \approx \frac{C}{8}\left(1 + \frac{225\Delta C_s}{32C}\right). \tag{7.77}$$

The output voltage is obtained from

$$\begin{aligned}v_o &= \frac{C_{eq}}{C_{eq} + C_{MSB}} V_{T,1} \\ &= \frac{V_{T,1}}{2^4}\left(\frac{1 + \frac{225\Delta C_s}{32C}}{1 + \frac{225\Delta C_s}{512C}}\right).\end{aligned} \tag{7.78}$$

Since $C_s = \frac{2}{15}C$, if we let $\Delta C_s = \alpha C_s = \frac{2C}{15}\alpha$, (7.78) becomes

$$\begin{aligned}v_o &= \frac{V_{T,1}}{2^4}\left(\frac{1 + \frac{15}{16}\alpha}{1 + \frac{15}{256}\alpha}\right) \\ &\approx \frac{1}{2^4}\left(\frac{V_{ref}}{2^4}\right)\left(1 + \frac{15}{16}\alpha\right).\end{aligned} \tag{7.79}$$

Note we have utilized

$$1 + \frac{15}{256}\alpha \approx 1 \tag{7.80}$$

in simplifying (7.79). As an example, if the capacitance error of the scaling capacitor is 10%, i.e. $\alpha = 0.1$, the error term of v_o will be $\frac{15}{16}\alpha = 0.094$. The error at the output caused by a 10% capacitance error of C_s will be 9.4%, nearly the same as the capacitance error of the scaling capacitor. Eq.(7.79) reveals that the capacitance error of the scaling capacitor has a strong impact on the performance of the DAC. To minimize performance degradation due to the capacitance error of the scaling capacitor, a digitally-controlled tunable capacitor array was used for the scaling capacitor in [228]. By adjusting the capacitance of the scaling capacitor, the optimal performance of the ADC can be obtained.

7.6.6 C-2C Capacitor Arrays

To further lower the silicon consumption of capacitor arrays, a C-2C technique was proposed by McCreamy [229] and further developed by a number of authors including Singh, Prabhakar, and Cong [230–233]. Fig.7.37 shows the configuration of an C-2C capacitor array for an 8-bit ADC.

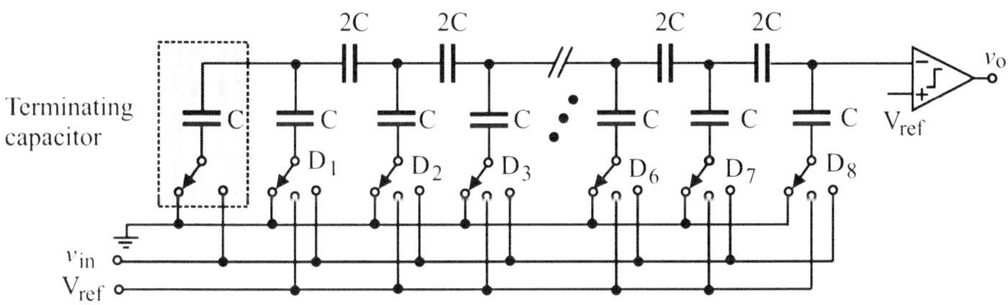

Figure 7.37. C-2C capacitor array for 8-bit ADC.

C-2C capacitor array technique is evolved from the preceding 2-stage binary-weighted capacitor arrays and can be considered as a multi-stage binary-weighted capacitor array with the maximum number of stages. To demonstrate this, consider a 4-bit 2-stage binary-weighted capacitor array shown in Fig.7.38. The value of the scaling capacitor C_s is determined from

$$\frac{1}{C_s} + \frac{1}{2C} = \frac{1}{\frac{C}{2}}, \qquad (7.81)$$

from which we obtain

$$C_s = \frac{2}{3}C.$$

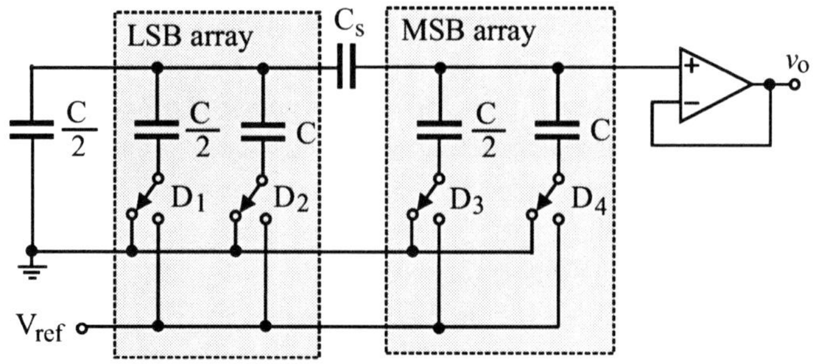

Figure 7.38. 4-bit DAC with a 2-stage binary-weighted capacitor array.

Now let us continue this by examining the 2-bit 2-stage binary-weighted capacitor array shown in Fig.7.39. The value of the scaling capacitor C_s in this case is given by $C_s = 2C$. Clearly, this is a 2-bit C-2C capacitor array.

It is important to note that the total capacitance looking into the capacitor array from the buffer is $2C$ regardless whether the capacitor array is a single-stage binary-weighted capacitor array, a two-stage binary-weighted capacitor array, or a C-2C capacitor array. Also, the total capacitance, which is $2C$, is independent of the number of the stages of the capacitor ladder.

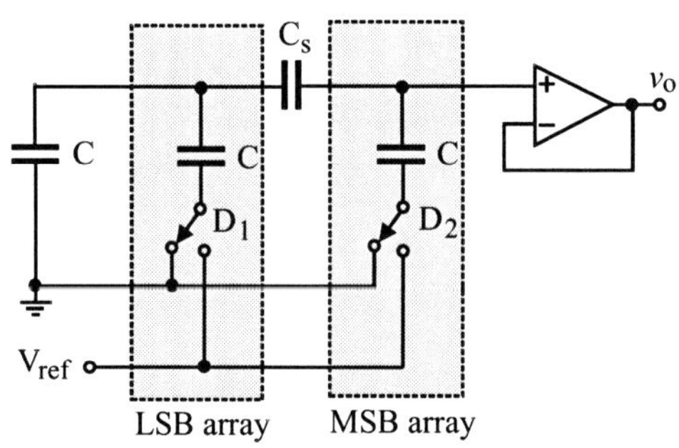

Figure 7.39. 2-bit DAC with a 2-stage binary-weighted capacitor array.

To examine the operation of C-2C capacitor arrays in DACs, consider the 8-bit DAC using a C-2C capacitor array shown in Fig.7.40. Assume $D_6 = 1$ while all other bits of the SAR are zero. The switch of D_6 is connected to V_{ref} while all others are connected to the ground via the switching network, as shown in Fig.7.41

It is readily to verify that the total capacitance of the sub-network on the left of D_6 switch is C. The sub-network consisting of the shunt capacitor of D_6, V_{ref}, and the capacitor network on the left of D_6 switch is represented by its Thévenin equivalent circuit shown in Fig.7.41. The value of Thévenin capacitance and that of Thévenin voltage are given explicitly in the figure.

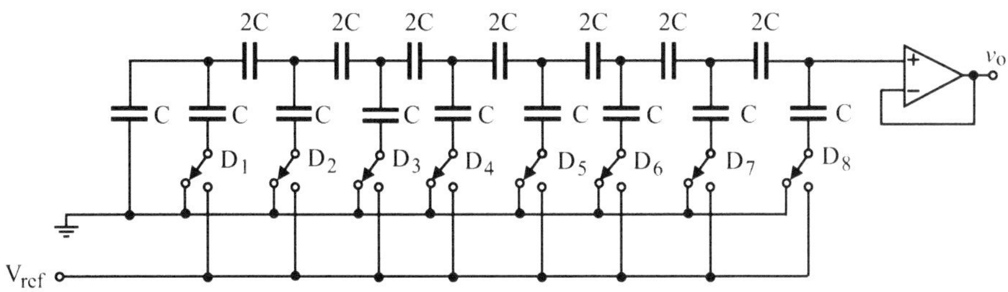

Figure 7.40. 8-bit DAC with a C-2C capacitor array.

Continuing the process of Thévenin equivalent circuit representation shown in the figure, one can show that the output voltage is given by

$$v_o = \frac{V_{ref}}{2^3}. \tag{7.82}$$

Generalizing the preceding result yields the output voltage of the DAC when an arbitrary number of the bits of the SAR are set to 1.

$$v_o = \left(\frac{D_1}{2^8} + \frac{D_2}{2^7} + \frac{D_3}{2^6} + \frac{D_4}{2^5} + \frac{D_5}{2^4} + \frac{D_6}{2^3} + \frac{D_7}{2^2} + \frac{D_8}{2^1}\right) V_{ref}. \tag{7.83}$$

Eq.(7.83) confirms that the C-2C capacitor array has the same transfer function as that of the corresponding DAC with a single-stage binary-weighted capacitor array. Because only capacitors of capacitances C and $2C$ are used, C-2C capacitor arrays are less affected by capacitance error. This differs from the two-stage binary-weighted capacitor arrays where a scaling capacitor whose capacitance is not a multiple of the unit capacitance is needed. The overall capacitance of a C-2C capacitor array is much smaller as compared with the corresponding two-stage binary-weighted capacitor array. As a result, the silicon consumption is greatly reduced. The smaller RC time constant of each ladder of C-2C capacitor arrays also enables a faster charge redistribution process subsequently a higher conversion speed.

The main drawback of C-2C capacitor arrays is that the effect of the parasitic capacitance between the bottom plate of the floating 2C capacitors and the substrate has a severe impact on the performance of DACs. In what follows, we use the 4-bit C-2C DAC shown in Fig.7.42 to investigate this effect in detail. For convenience, let $D_3 = 1$ and all other bits of the SAR be zero. Further neglect the parasitic capacitance between the top plate of the floating capacitors and the substrate and only consider the parasitic capacitance between the bottom plates of the floating capacitors and the substrate. Because

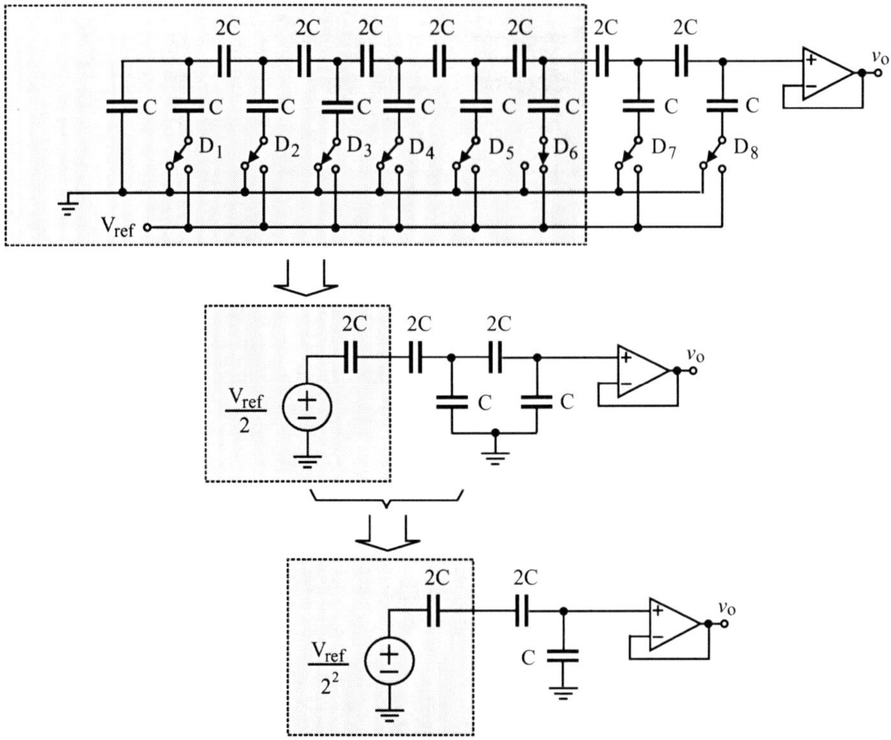

Figure 7.41. Analysis of a 8-bit DAC using C-2C capacitor array.

$$\begin{aligned}\frac{1}{C_1} &= \frac{1}{2C} + \frac{1}{2C + C_p} \\ &= \frac{1}{2C} + \frac{1}{2C}\frac{1}{1 + \frac{C_p}{2C}} \\ &\approx \frac{1}{2C} + \frac{1}{2C}\left(1 - \frac{C_p}{2C}\right) \\ &= \frac{1}{C} - \frac{C_p}{4C^2},\end{aligned} \tag{7.84}$$

C_1 is thus obtained from

$$\begin{aligned}C_1 &= \frac{C}{1 - \frac{C_p}{4C}} \\ &\approx C\left(1 + \frac{C_p}{4C}\right).\end{aligned} \tag{7.85}$$

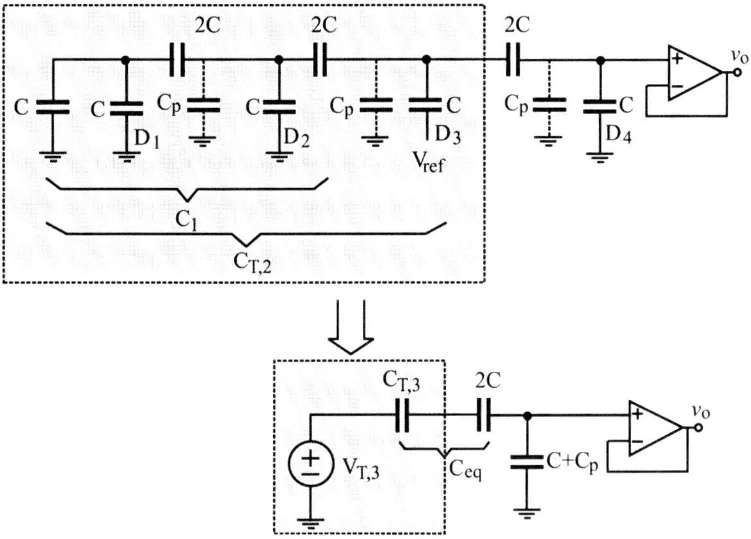

Figure 7.42. Analysis of the error due to the bottom-plate parasitic capacitance of floating capacitors of DACs using C-2C capacitor array.

Note we have utilized $\frac{C_p}{2C} \ll 1$ in derivation of (7.84). Thévenin voltage and Thévenin capacitance of the sub-circuit highlighted in Fig.7.42 are given by

$$\begin{aligned} V_{T,3} &= \frac{C}{C+C_1+C_p}V_{ref} \\ &= \frac{V_{ref}}{2}\frac{1}{1+\frac{5C_p}{8C}} \\ &\approx \frac{V_{ref}}{2}\left(1-\frac{5C_p}{8C}\right). \end{aligned} \tag{7.86}$$

and

$$\begin{aligned} C_{T,3} &= C+C_1+C_p \\ &= 2C+\frac{5C_p}{4}. \end{aligned} \tag{7.87}$$

The equivalent capacitance of $C_{T,3}$ in series with $2C$, denoted by C_{eq}, is obtained from

$$\frac{1}{C_{eq}} = \frac{1}{2C}+\frac{1}{2C+\frac{5C_p}{4}}$$

$$
\begin{aligned}
&= \frac{1}{2C} + \frac{1}{2C}\left(\frac{1}{1+\frac{5C_p}{8C}}\right) \\
&\approx \frac{1}{2C} + \frac{1}{2C}\left(1-\frac{5C_p}{8C}\right) \\
&= \frac{1}{C} - \frac{5C_p}{16C^2},
\end{aligned} \tag{7.88}
$$

and the result is given by

$$
\begin{aligned}
C_{eq} &= \frac{C}{1-\frac{5C_p}{16C}} \\
&\approx C\left(1+\frac{5C_p}{16C}\right).
\end{aligned} \tag{7.89}
$$

The output voltage is obtained from

$$
\begin{aligned}
v_o &= \frac{C_{eq}}{C_{eq}+C+C_p}V_{T,3} \\
&= \frac{V_{ref}}{2}\frac{C\left(1+\frac{5C_p}{16C}\right)}{2C\left(1+\frac{21C_p}{32C}\right)}\left(1-\frac{5C_p}{8C}\right) \\
&\approx \frac{V_{ref}}{2^2}\left(1+\frac{5C_p}{16C}\right)\left(1-\frac{5C_p}{8C}\right)\left(1-\frac{21C_p}{32C}\right) \\
&\approx \frac{V_{ref}}{2^2}\left(1-\frac{31}{32}\frac{C_p}{C}\right).
\end{aligned} \tag{7.90}
$$

As an example, if C_p is 10% of the inter-plate capacitance, i.e. $C_p = 0.1\times 2C = 0.2C$, the error given by $\frac{31C_p}{32C}$ is approximately 19.38%. The preceding analysis reveals that the bottom-plate parasitic capacitance of the floating capacitors will have a severe impact on the performance of the DAC.

Recall that in analysis of two-stage binary-weighted capacitor arrays, the bottom-plate parasitic capacitance of the scaling capacitor of two-stage binary-weighted capacitor arrays has a negligible effect on the performance of the DAC. The reason for that is that C_s of two-stage binary-weighted capacitor arrays is much smaller than C. For example, $C_s = \frac{2}{15}C \approx 0.13C$ for a 8-bit two-stage binary-weighted capacitor array. If the parasitic bottom-plate capacitance is 10% of C_s, i.e. $C_p = 0.1C_s = 0.013C$. Since the capacitance of the terminated LSB array is $2C$ and that of the MSB array is $\frac{15}{8}C$, C_p is therefore only 0.67% of the shunt capacitance. On top of this, there is only one floating capacitor in a two-stage binary-weighted capacitor array. As a result, the bottom-plate parasitic capacitance of C_s will have a negligible impact on the performance of the DAC.

Unlike the scaling capacitor of two-stage binary-weighted capacitor arrays, the capacitance of the floating capacitors of C-2C capacitor arrays is twice that of the capacitance of the shunt capacitors. If the parasitic bottom-plate capacitance is 10% of the inter-plate capacitance, i.e. $C_p = 0.2C$, it is 20% of the shunt capacitance. Also, for each floating capacitor, there is a corresponding bottom-plate parasitic capacitance. As a result, the capacitance of the bottom-plate parasitic capacitors of the floating capacitors of C-2C capacitor arrays is comparable to the shunt capacitance, resulting in a strong effect on the performance of the DAC.

To minimize the effect of the parasitic bottom-plate capacitance of the floating capacitors of C-2C capacitor arrays, Cong proposed a pseudo C-2C ladder-based data converter technique, as shown in Fig.7.43 [233]. In this approach, the C-2C ladder is replaced with a C-2αC ladder. Note that the capacitance of the terminating capacitor and that of the MSB capacitor are αC. The capacitance of the floating capacitors is $2\alpha C$. Also, the reference voltage connected to the MSB capacitor is V_{ref} while that connected to all other capacitors is αV_{ref}. By adjusting the capacitance ratio α, the effect of the parasitic bottom-plate capacitances is reduced. The simulation results of a 12-bit DAC show that when the parasitic bottom-plate capacitance is 14% of the inter-plate capacitance, DNL is reduced from over 1.5 with a conventional C-2C capacitor array to approximately 0.8 with a pseudo C-2C capacitor array. INL is also reduced from over 1.5 with a conventional C-2C capacitor array to below 0.5 with a pseudo C-2C capacitor array.

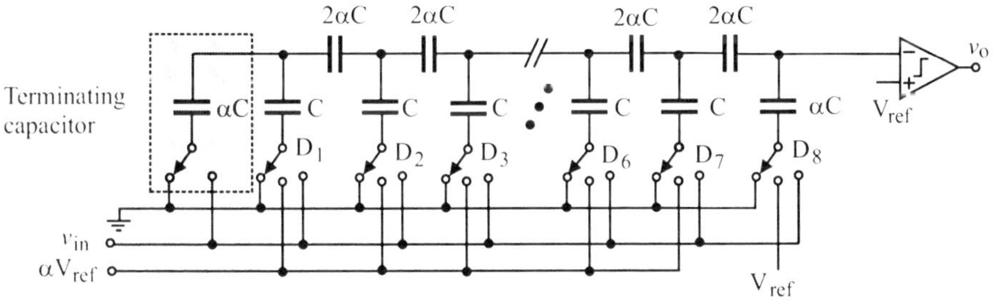

Figure 7.43. DAC with a pseudo C-2C capacitor array proposed by Cong [233].

7.6.7 Switching Network

There are three types of switches, namely nMOS switches, pMOS switches, and transmission-gate (TG) switches that are available in standard CMOS technologies.

Consider first a nMOS switch. Assume that the voltage at the gate and drain of the nMOS transistor is V_{DD}. The voltage at the source of the nMOS transistor must not exceed $V_{DD} - V_{tn}$ in order to guarantee that the transistor stays on. In other words, although the voltage at the drain is V_{DD}, the maximum voltage that can be passed on to the drain of the transistor is only $V_{DD} - V_{tn}$. A voltage loss of V_{tn} therefore occurs at nMOS switches.

pMOS switches can be analyzed in a similar way. Let the gate voltage of a pMOS switch be zero. The lowest voltage at the source of the pMOS switch in order to ensure that the transistor is on is $|V_{tp}|$. Thus, the lowest voltage that a pMOS switch can pass from the source to the drain is $|V_{tp}|$. Also, it is well understood that the channel resistance of either nMOS or pMOS switches is not constant and varies with the voltages of the terminals of the switches [138]. This will result in different charging and discharging times of capacitor arrays.

In comparison, TG-based switches offer a much smaller and nearly constant channel resistance. In addition, the voltage loss across the switches is also nearly zero, making them a preferred choice for the switching network. The only two constraints affecting the use of TG switches are the need for a complementary clock and the large capacitance due to the use of two transistors.

7.6.8 Hybrid Charge-Scaling DACs

It has been shown that DACs with a single-stage binary-weighted capacitor array are not subject to the effect of top / bottom plate parasitic capacitances whiles DACs with a C-2C capacitor array suffers from the drawback of degraded performance caused by the parasitic bottom-plate capacitance of the floating capacitors. The performance of DACs with a two-stage binary-weight capacitor array is largely affected by the capacitance error of the scaling capacitor. Several hybrid capacitor arrays that are stemmed from these three generic configurations of capacitor arrays have emerged to improve the performance. In [222], Kim *et al.* proposed a 10-bit successive approximation ADC with the LSB portion of the capacitor array implemented using a C-2C capacitor array and the MSB portion implemented using a single-stage binary-weighted capacitor array, as shown in Fig.7.44. Since the accuracy of the MSB portion is more critical as compared with that of the LSB portion, the MSB portion is implemented using a binary-weighted configuration while the LSB is implemented using a C-2C configuration. A very similar configuration was used in [234, 235].

7.6.9 Multi-Stage Charge-Scaling DACs

The two-stage binary-weighted capacitor array approach was further developed by Kuramochi *et al.* to have multiple binary-weighted capacitor arrays to reduce silicon consumption, as shown in Fig.7.45 [236]. The principle of multi-stage binary-weighted capacitor arrays is the same as that of two-stage

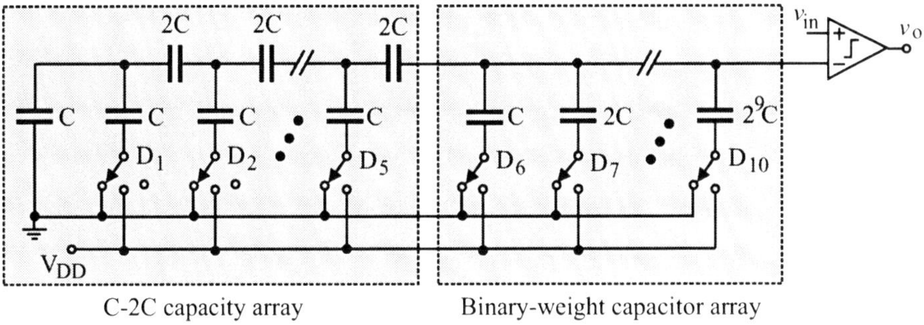

Figure 7.44. DAC with a pseudo C-2C capacitor array proposed by Kim *et al.* [222].

binary-weighted capacitor arrays investigated earlier. With the increase of the number of the floating capacitors, the effect of the parasitic bottom-plate capacitance of the floating capacitors becomes increasingly critical. A compromise between the performance of DACs and the silicon consumption is therefore needed.

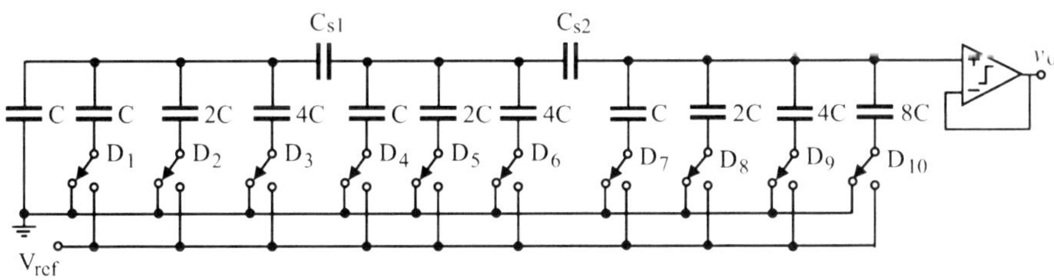

Figure 7.45. DAC with a multi-stage binary-weighted capacitor array [236].

7.7 Performance Comparison

Table 7.2 compares the performance of the recently published low-voltage charge redistribution successive approximation ADCs. It is seen that the number of the bits of these ADCs is generally low, in th range of $8 \sim 10$ bits. The power consumption of these ADCs is sample-rate dependent. The higher the sample rate, the higher the power consumption. When the sample rate is low, the power consumption can be lowered to below 1 μW. The ADC proposed by Abdelhalim *et al.* consumes the lowest amount of power with a sample rate of 20 kS/s.

Table 7.2. Performance comparison of charge redistribution successive approximation ADCs.

Ref.	Technology	Rate [kS/s]	SFDR [dB]	V_{DD} [V]	Power μW (pJ/s)	Remarks
Sauerbrey *et al.* [220](03)	0.18μm	4.1	>60	0.5	0.85(207)	8-bit ST
Scott *et al.* [221](03)	0.25μm	100	60	1	3.1 (31)	8-bit ST
Fu *et al.* [237](06)	0.5μm SOS	4	–	1.1	0.9 (225)	8-bit C2C
Abdelhalim *et al.* [228](07)	0.13μm	20	60.8	0.3	0.213 (10.7)	8-bit DT
Cho *et al.* [234](07)	0.18μm SOI	10	–	1	1.8 (180)	8-bit Hybrid
Hong-Lee [238](07)	0.18μm	200	47.4	0.9	2.47 (12.4)	8-bit ST
Verma-Chandrakasan [239](07)	0.18μm	100	65	1	25 (250)	12-bit ST
Kuramochi [236](07)	0.18μm	10^3	69.8	1.8	198 (198)	10-bit MT
Kim *et al.* [222](08)	0.18μm	137	53.8	1.5	13.4 (97.8)	10-bit Hybrid
Rodriguez-Perez *et al.* [235](08)	0.13μm	100	70.8	1	1.93 (19.3)	10-bit Hybrid
Yang *et al.* [240](09)	65 nm	10^6	31.5	1.2	6.7 mW (6.7)	6-bit C2C

Legends :
ST - Single-stage binary-weighted capacitor arrays.
DT - Double-stage binary-weighted capacitor arrays.
MT - Multi-stage binary-weighted capacitor arrays.
Hybrid - Hybrid (C-2C/Binary-weighted) capacitor arrays.
pJ/s - pico Joule per sample.

7.8 Chapter Summary

A brief review of the widely used figure-of-merits that quantify the performance of ADCs has been provided. It is followed with an investigation of a special class of ADCs, namely integrating ADCs. A key advantage of integrating ADCs, in particular, dual-slope integrating ADCs, is that they can effectively suppress the effect of transient disturbances present at the input of the ADCs. As a result, these ADCs can be used in applications where transient disturbances exist.

Oscillator-based temperature ADCs have also been investigated in this chapter. Both relaxation oscillator-based temperature ADCs and ring oscillator-based temperature ADCs have been studied. The former have the advantage that the performance of the ADCs is less sensitive to PVT effect while the latter offer the advantage of simple configurations, subsequently low power consumption.

Time-to-digital converter based ADCs have also been investigated. These temperature ADCs offer the advantages of low power consumption due to the elimination of high-frequency oscillators.

Charge redistribution successive approximation ADCs are widely favored due to the their low power consumption, a high conversion speed, and a good accuracy. The key component of these ADCs is the charge-scaling DACs implemented using a capacitor array. Three different configurations of capacitor arrays, namely, single-stage binary-weighted capacitor arrays, two-stage binary-weighted capacitor arrays, and C-2C capacitor arrays, have been investigated in detail. DACs with a single-stage binary-weighted capacitor array offer the advantages of a low sensitivity to the parasitic capacitances of the capacitor arrays. These DACs, however, suffer from high power and silicon consumption and a low conversion speed. DACs with a two-stage binary-weighted capacitor array lower the power and silicon consumption by limiting the number of the ladders of each stage of the capacitor arrays. The performance of these DACs is affected by the parasitic capacitance of the bottom plate-substrate capacitance of the scaling capacitor. DACs with a C-2C capacitor array consumes the least amount of silicon area and power, and offers the highest conversion speed among the three types of DACs. The parasitic bottom plate-substrate capacitances of the floating capacitors, however, degrade the performance.

Appendix A
Material and Physical Constants

Table A.1. Value of constants.

Boltsmann's constant	$k = 1.38\times10^{-23}$ V.C/°K
Bandgap voltage of silicon	$E_g = 1.12$ eV at 300°K
Charge of an electron	$q = 1.6\times10^{-16}$ C
Concentation of intrinsic carriers in silicon	$n_i \approx 1.5\times10^{10}/\text{cm}^3$ at 300°K
Dielectric constant of vacuum	$\epsilon_o = 8.85\times10^{-14}$ F/cm
Dielectric constant of oxide	$\epsilon_{ox} = 3.5\times10^{-13}$ F/cm
Dielectric constant of silicon	$\epsilon_{si} = 1.05\times10^{-12}$ F/cm
Room temperatue	300°K
Tari	Time duration from 6.35μs to 25μs.
Thermal voltage	$V_t = 25.9$ mV at 300°K

F. Yuan, *CMOS Circuits for Passive Wireless Microsystems*,
DOI 10.1007/978-1-4419-7680-2,

References

[1] R. Simon, D. Hall, and F. Miranda. "RF telemetry system for an implantable Bio-MEMS sensor". In *Proc. IEEE MTT-S Int'l Microwave Symp. Digest*, volume 3, pages 1433 – 1436, 2004.

[2] C. Sauer, M. Stanacevic, G. Cauwenberghs, and N. Thakor. "Power harvesting and telemetry in CMOS for implanted devices". *IEEE Trans. on Circuits and Systems I.*, 52(12):2605 – 2613, December.

[3] W. Liu, K. Vichienchom, M. Clements, S. DeMarco, C. Hughs, E. McGucken, M. Humayun, E. Juan, J. Weiland, and R. Greenberg. "A neuron-stimulus chip with telemetry unit for retinal prosthetic device". *IEEE J. Solid-State Circuits*, 35(10):1487–1497, October 2000.

[4] G. Wang, W. Liu, M. Sivaprakasam, and G. Kendir. "Design and analysis of an adaptive transcutaneous power telemetry for biomedical implants". *IEEE Trans. on Circuits and Systems I.*, 52(10):2109–2117, October 2005.

[5] S. Lee and S. Lee. "An implantable wireless bidirectional communication microstimulator for neuromuscular stimulation". *IEEE Trans. on Circuits and Systems I.*, 52(12):2526–2538, December 2005.

[6] M. Ghovanloo and S. Atluri. "A wide-band power-efficient inductive wireless link for implantable microelectronic devices using multiple carriers". *IEEE Trans. on Circuits and Systems I.*, 54(10):2211–2221, October 2007.

[7] Arshak, E. Jafer, G. Lyons, D. Morris, and O. Korostynska. "A review of low-power wireless sensor microsystems for biomedical capsule diagnosis". *Microelectronics International*, 2(3):8–19, 2004.

[8] B. Chi, J. Yao, S. Han, X. Xie, G. Li, and Z. Wang. "Low-power transceiver analog front-end circuits for bidirectional high data rate wireless telemetry in medical endoscopy applications". *IEEE Trans. on Biomedical Engineering*, 54(7):1291–1299, July 2007.

[9] B. Chi, J. Yao, S. Han, X. Xie, G. Li, and Z. Wang. "A 2.4 GHz low power wireless transceiver analog front-end for endoscopy capsule system". *Analog Integrated Circuits and Signal Processing*, 51:59–71, 2007.

[10] A.DeHennis and K. Wise. "A fully integrated multi-site pressure sensor for wireless arterial flow characterization". *J. Microelectromechanical Systems*, 15(3):678–685, June 2006.

[11] F. Kocer and M. Flynn. "A new transponder architecture with on-chip ADC for long-range telemetry applications". *IEEE J. Solid-State Circuits*, 41(5):1142–1148, May 2006.

[12] F. Kocer and M. Flynn. "An RF-powered, wireless CMOS temperature sensor". *IEEE Sensors J.*, 6(3):557–564, June 2006.

[13] W. Jeon, J. Melngailis, and R. Newcomb. "Disposable CMOS passive RFID transponder for patient monitoring ". In *Proc. IEEE Int'l Symp. Circuits and Systems*, pages 5071–5074, May 2006.

[14] K. Opasjumruskit, T. Thanthipwan, O. Sathusen, P. Sirinamarattana, P. Gadmanee, E. Pootarapan, N. Wongkomet, A. Thanachayanont, and M. Thamsirianunt. "Self-powered wireless temperature sensors exploit RFID technology". *IEEE Pervasive Computing*, 5(1):54–61, February 2006.

[15] R. Glidden, C. Bockorick, C. Cooper, C. Diorio, D. Dressler, V. Gutnik, C. Hagen, D. Hara, T. Hass, T. Humes, J. Hyde, R. Oliver, O. Onen, A. Pesavento, K. Sundstrom, and M. Thomas. "Design of ultra-low-cost UHF RFID tags for supply chain applications". *IEEE Communications Magazine*, 42(8):140–151, 2004.

[16] V. Chawla and D. Ha. "An overview of passive RFID". *IEEE Communications Magazine*, 45(9):11–17, 2007.

[17] M. Lehtonen, F. Michahelles, and E. Fleisch. "Trust and security in RFID-based product authentication systems". *IEEE Systems Journal*, 1(2):129–144, 2007.

[18] K. Finkenzeller. *RFID handbook.* Wiley & Sons, 2003.

[19] M. Loy, R. Karingattil, and L. Williams. *ISM-band and short range device regulatory compliance overview - Application report SERA048.* Texas Instruments, 2005.

[20] J. Cockcroft and E. Walton. "Production of high velocity positive ions". *Proc. Royal Society of London, Series A*, 136(831):619–630, July 1932.

[21] J. Dickson. "On-chip high-voltage generation in NMOS integrated circuits using an improved voltage multiplier technique". *IEEE J. Solid-State Circuits*, SC-11(3):374–378, June 1976.

[22] J. Wu and K. Chang. "MOS charge pumps for low-voltage operation". *IEEE J. Solid-State Circuits*, 33(4):592–597, April 1998.

[23] H. San, H. Kobayashi, T. Myono, T. Iijima, and N. Kuroiwa. "Highly-efficient low-voltage-operation charge pump circuits using bootstrapped gate transfer switches". *T. IEE Japan*, 120-C(10):1339–1345, 2000.

[24] A. Facen and A. Boni. "CMOS power retriever for UHF RFID tags". *IEE Electronics Letters*, 43(25):1424 – 1425, December 2007.

[25] A. Shameli, A. Safarian, A. Rofougaran, M. Rofougaran, and F. De Flaviis. "Power harvester design for passive UHF RFID tag using a voltage boosting technique". *IEEE Trans. on Microwave Theory and Techniques*, 55(6):1089–1097, June 2007.

[26] S. Mandal and R. Sarpeshkar. "Low-power CMOS rectifier design for RFID applications". *IEEE Trans. on Circuits and Systems I.*, 54(6):1177–1188, June 2007.

[27] T. Umeda, H. Yoshida, S. Sekine, Y. Fujita, T. Suzuki, and S. Otaka. "A 950-MHz rectifier circuit for sensor network tags with 10-m distance ". *IEEE J. Solid-State Circuits*, 41(1):35–41, January 2006.

[28] H. Nakamoto, D. Yamazaki, T. Yamamoto, H. Kurata, S. Yamada, K. Miksida, T. Ninomiya, T. Ohkawa, S. Masui, and K. Gotoh. "A passive RFID RF identification CMOS tag IC using ferroelectric RAM in 0.35-μm technology ". *IEEE J. Solid-State Circuits*, 42(1):101–110, Jan. 2007.

[29] N. Soltani and F. Yuan. "A high-gain power-matching technique for efficient radio-frequency power harvesting of passive wireless transponders". *IEEE Trans. on Circuits and Systems I.*, Accepted for publication in April 2010.

[30] V. Pillai, H. Heinrich, D. Dieska, P. Nikitin, R. Martinez, and K. Rao. "An ultra-low-power long range battery/passive RFID tag for UHF and microwave". *IEEE Trans. on Circuits and Systems I.*, 54(7):1500–1512, July 2007.

[31] K. Rongsawat and A. Thanachayanont. "Ultra low power analog front-end for UHF RFID transponder ". In *Proc. Int'l Symp. Comm. Inform. Tech.*, pages 1195–1198, 2006.

[32] G. De Vita and G. Iannaccone. "A sub-1 V 10 ppm/°C, nanopower voltage reference generator". *IEEE J. Solid-State Circuits*, 42(7):1536–1542, July 2007.

[33] F. Cilek, K. Seeman, G. Holweg, and R. Weigel. "Impact of the local oscillator on baseband processing in RFID transponders". In *Proc. Circuits, Systems, Electronics Conf.*, pages 231–234, August 2007.

[34] U. Karthaus and M. Fischer. "Fully integrated passive UHF RFID transponder IC with 16.7 μW minimum RF input power". *IEEE J. Solid-State Circuits*, 38(10):1602–1608, October 2003.

[35] N. Cho, S. Song, S. Kim, S. Kim, and H. Yoo. "A 5.1 μW UHF RFID tag chip integrated with sensors for wireless environmental monitoring ". In *Proc. European Solid-State Circuit Conf.*, pages 279–282, Grenoble, France, 2005.

[36] N. Cho, S. Song, J. Lee, S. Kim, S. Kim, and H. Yoo. "A 8-μW 0.3-mm^2 RF-powered transponder with temperature sensor for wireless environmental monitoring ". In *Proc. Int'l Symp. Circuits and Systems*, number 5, pages 4763–4766, May 2005.

[37] H. Ryu and J. Moo. "Miniaturization of circular loop antenna using short stub for RFID system". *IEE Electronics Letters*, 42(17):955–956, August 2006.

[38] M. Keskilammi and M. Kivikoski. "Using text as a meander line for RFID transponder antennas". *IEEE Antenna and Wireless Propagation Letters*, 3:372–374, 2004.

[39] S. Chen and P. Hsu. "CPW-fed folded-slot antenna for 5.8 GHz RFID tags". *IEE Electronics Letters*, 40(24):1516–1517, November 2004.

[40] M. Zhang, Y. Jiao, and F. Zhang. "Dual-band CPW-fed folded-slot monopole antenna for RFID applications". *IEE Electronics Letters*, 42(21):1193–1194, 2006.

[41] D. Ma and W. Zhang. "Broadband CPW-fed RFID antenna at 5.8 GHz". *IEE Electronics Letters*, 42(22):1258–1259, October 2006.

[42] G. De Vita and G. Iannaccone. "Design criteria for the RF section of UHF and microwave passive RFID transponders". *IEEE Trans. on Microwave Theory and Techniques*, 53(9):2978–2990, September 2005.

[43] D. Pozar. *Microwave and RF wireless systems*. John Wiley & Sons, New York, 2001.

[44] J. Curty and M. Declercq. *Design and optimization of passive UHF RFID systems*. Springer, New York, 2009.

[45] E. Bergeret, J. Gaubert, and P. Pizzo. "Power generation system for UHF passive RFID". *IEE Electronics Letters*, 42(25):1452–1454, December 2006.

[46] P. Favrat, P. Deval, and M. Declercq. "A high-efficiency CMOS voltage doubler". *IEEE J. Solid-State Circuits*, 33(3):410–416, March 1998.

[47] Y. Moisiadis, I. Bouras, and A. Arapayanni. "A CMOS charge pump for low voltage operation ". In *Proc. Int'l Symp. Circuits and Systems*, volume 5, pages 577–580, Geneva, May 2000.

[48] J. Curty, N. Joehl, C. Dehollain, and M. Declercq. "Remotely powered addressable UHF RFID integrated system ". *IEEE J. Solid-State Circuits*, 40(11):2193–2202, November 2005.

[49] B. Razavi. *Design of CMOS integrated circuits for optical communications*. McGraw-Hill, Boston, 2003.

[50] J. Long. "Monolithic transformers for silicon RF IC design". *IEEE J. Solid-State Circuits*, 35(9):1382–1368, September 2000.

[51] Y. Meyevskiy, A. Watson, P. Francis, K. Hwang, and A. Weisshaar. "A new compact model for monolithic transformers in silicon-based RFICs". *IEEE Microwave and Wireless Components Letters*, 15(6):419–421, June 2005.

[52] F. Yuan. *CMOS current-mode circuits for data communications*. Springer, New York, 2007.

[53] *EPC radio-frequency identity protocols class-1 generation-2 UHF RFID protocol for communications at 860 MHz - 960 MHz, version 1.0.9*. EPCglobal Inc., 2005.

[54] D. Dobkin. *The RF in RFID : Passive UHF RFID in practice*. Newnes, Burlington, MA., USA, 2008.

[55] H. Yu and K. Najafi. "Circuitry for wireless microsystems for neural recording microprobes". In *Proc. Annual Int'l Conf. Embedded Systems*, pages 761–764, Istanbul, October 2001.

[56] Y. Tsai M. Hsia and O. Chen. "An UHF passive RFID transponder using a low-power clock generator without passive components". In *Proc. IEEE Mid-West Symp. Circuits and Systems*, volume 2, pages 11–15, Aug. 2006.

[57] F. Naghmouchi, M. Ghorbel, A. Ben-Hamid, and M. Samet. "CMOS ASK system modulation dedicated to cochlear prosthesis". In *Proc. Int'l Symp. Control, Comm., Signal Processing*, pages 267–270, 2004.

[58] A. Allstot. "A precision variable-supply CMOS comparator". *IEEE J. Solid-State Circuits*, 17(6):1080–1087, December 1982.

[59] O. Schmitt. "A thermionic trigger". *J. Scientific Instruments*, 15:24–26, January 1938.

[60] B. Dokic. "CMOS Schmitt triggers". *IEE Proc - Part G, Circuits, Devices, and Systems*, 131(5):197–202, October 1984.

[61] I. Filanovsky and H. Baltes. "CMOS Schmitt trigger design". *IEEE Trans. on Circuits and Systems I*, 41(1):46–49, January 1994.

[62] M. Steyaert and W. Sansen. "Novel CMOS Schmitt trigger". *IEE Electronics Letters*, 22(4):203–204, February 1986.

[63] Z. Wang. "CMOS adjustable Schmitt trigger". *IEEE Trans. on Instrumentation and Measurement*, 40(3):601–605, June 1991.

[64] D. Kim, J. Kih, and W. Kim. "A new waveform-reshaping circuit : an alternative approach to Schmitt trigger". *IEEE J. Solid-State Circuits*, 28(2):162–164, February 1993.

[65] S. Al-Sarawi. "Low power Schmitt trigger circuit". *IEE Electronics Letters*, 38(18):1009–1010, August 2002.

[66] C. Zhang, A. Srivastava, and P. Ajmera. "Low voltage CMOS Schmitt trigger circuits". *IEE Electronics Letters*, 39(24):1696–1698, November 2003.

[67] V. Pedroni. "Low-voltage high-speed Schmitt trigger and compact window comparator". *IEE Electronics Letters*, 41(22):1213–1214, October 2005.

[68] R. Sapawi, R. Chee, S. Sahari, and N. Julai. "performance of cmos schmitt trigger". In *Proc. Int'l Conf. Computer Comm. Eng.*, pages 1317–1320, May 2008.

[69] A. Pfister. "Novel CMOS Schmitt trigger with controllable hysteresis". *IEE Electronics Letters*, 28(7):639i–641, March 1992.

[70] V. Katyal, R. Geiger, and D. Chen. "Adjustable hysteresis CMOS Schmitt trigger". In *Proc. IEEE Int'l Symp. Circuits and Systems*, pages 1938–1941, May 2008.

[71] B. Choi, J. Yao, S. Han, X. Xie, G. Li, and Z. Wang. "A 2.4 GHz low power wireless transceiver analog front-end for endoscopy capsule system". *Analog Integrated Circuits and Signal Processing*, 51:59–71, 2007.

[72] F. Yuan. "A high-speed differential CMOS Schmitt trigger with regenerative current feedback and adjustable hysteresis ". *Analog Integrated Circuits and Signal Processing*, 63(1):121–127, 2010.

[73] B. Razavi. *Design of analog CMOS integrated circuits*. McGraw-Hill, Boston, 2001.

[74] M. Lee, W. Dally, and P. Chiang. "Low-power area-efficient high-speed I/O circuit techniques". *IEEE J. Solid-State Circuits*, 35(11):1591–1599, Nov. 2000.

[75] M. Lee, W. Dally, R. Farjad-Rad, H. Ng, R. Senthinathan, J. Edmondson, and J. Poulton. "CMOS high-speed I/Os - present and future". In *Proc. 21th Int'l Conf. Computer Design*, pages 454–461, October 2003.

[76] A. Sedra and K. Smith. *Microelectronics circuits*. Oxford University Press, London, 1998.

[77] J. Bouvier, T. Thorigne, S. Hassan, M. Revillet, and P. Senn. "A smart card CMOS circuit with magnetic power and communications interface". In *Proc. IEEE Int'l Solid-State Circuits Conf.*, pages 296–297, 1997.

[78] H. Yu and K. Najafi. "Low-power interface circuits for bio-implantable microsystems". In *Proc. IEEE Int'l Solid-State Circuits Conf.*, volume 1, pages 194 – 487, 2003.

[79] G. Hmida, M. Dhieb, H. Ghariani, and M. Samet. "Transcutaneous power and high data rate transmission for biomedical implants ". In *Proc. Int'l Conf. Design & Test Integrated Systems in Nanoscale Tech.*, pages 374–378, September 2006.

[80] H. Chow and I. Wang. "High performance automatic gain control circuit using a S/H peak detector for ASK receiver ". In *Proc. Int'l Conf. Electronics, Circuits and Systems*, pages 429–432, September 2002.

[81] H. Yu and R. Bashirullah. "A low power ASK clock and data recovery circuit for wireless implantable electronics". In *Proc. IEEE Custom Integrated Circuits Conf.*, pages 249–252, 2006.

[82] K. Kimura. "A CMOS logarithmic IF amplifier with unbalanced source-coupled pairs". *IEEE J. Solid-State Circuits*, 28(1):78–83, January 1993.

[83] C. Wang, Y. Hsueh, U. Chio, and Y. Hsiao. "A C-less ASK demodulator for implantable neural interfacing chips ". In *Proc. IEEE Int'l Symp. Circuits and Systems*, volume 4, pages 57–60, May 2004.

[84] T. Lee, C. Lee, Y. Ciou, C. Huang, and C. Wang. "All-MOS ASK demodulator for low-frequency applications". *IEEE Trans. on Circuits and Systems II.*, 55(5):474–478, May 2008.

[85] N. Cho, S. Song, S. Kim, S. Kim, and H. Yoo. "A 5.1-μW UHF RFID tag chip integrated with sensors for wireless environmental monitoring". In *Proc. European Solid-State Circuits Conf.*, pages 279–282, Grenoble, France, 2005.

[86] A. Djemouai and M. Sawan. "Integrated ASK demodulator dedicated to implantable electronic devices". In *Proc. IEEE Mid-West Symp. Circuits and Systems*, pages 80–83, August 2004.

[87] A. Djemouai and M. Sawan. "New CMOS current-mode amplitude shift keying demodulator dedicated for implantable electronic devices". In *Proc. IEEE Int'l Symp. Circuits and Systems*, volume 1, pages 441–444, May 2004.

[88] G. Gudnason. "A low-power ASK demodulator for inductively coupled implantable electronics". In *Proc. European Solid-State Circuits Conf.*, pages 385–388, September 2000.

[89] O. Oliaei and P. Loumeau. "Current-mode class AB design using floating voltage-source". *IEE Electronics Letters*, 32(17):1526–1527, August 1996.

[90] O. Oliaei and P. Loumeau. "A low-input resistance class AB CMOS current conveyer". In *Proc. IEEE Mid-West Symp. Circuits and Systems*, pages 11–14, 1997.

[91] R. Harjani, O. Birkenes, and J. Kim. "An IF stage design for an ASK-based wireless telemetry system". In *Proc. IEEE Int'l Symp. Circuits and Systems*, volume 1, pages 52–55, Geneva, May 2000.

[92] J. Alegre, S. Celma, M. Sanz, and J. Pozo. "Low-ripple fast-settling envelope detector". *IEE Electronics Letters*, 42(18):1011–1012, August 2006.

[93] J. Alegre, S. Celma, B. Calvo, and J. Pozo. "Design of a novel envelope detector for fast-settling circuits ". *IEEE Trans. on Instrumentation and Measurement*, 57(1):4–9, January 2008.

[94] K. Bult and H. Wallinga. "A class of analog CMOS circuits based on the square-law characteristic of an MOS transistor in saturation". *IEEE J. Solid-State Circuits*, 22(3):357–365, June 1987.

[95] C. Chang and S. Liu. "Current-mode full-wave rectifier and vector summation circuit". *IEE Electronics Letters*, 36(19):1599–1600, September 2000.

[96] S. Khucharoensin and V. Kasemsuwan. "A high performance CMOS current-mode precision full-wave rectifier (PFWR) ". In *Proc. IEEE Int'l Symp. Circuits and Systems*, volume 1, pages 41–45, May 2003.

[97] V. Riewruja and R. Cuntapong. "A low-voltage wide-band CMOS precision full-wave rectifier". *Int'l J. Electronics*, 89(6):467–476, 2002.

[98] M. Ismail and T. Fiez (Ed). *Analog VLSI signal and information processing*. McGraw-Hill, New York, 1994.

[99] K. Koli and K. Halonen. "Low voltage MOS-transistor-only precision current peak detector with signal independent discharge time constant ". In *Proc. IEEE Int'l Symp. Circuits and Systems*, pages 1992–1995, Hong Kong, May 1997.

[100] R. Sarpeshkar, R. Lyon, and C. Mead. "A low-power wide-dynamic-range analog VLSI cochlear". *Analog Integrated Circuits and Signal Processing*, 16:245–274, 1998.

[101] Y. Zhou, G. Huang, S. Nam, and B. Kim. "A novel wide-band envelope detector ". In *Proc. IEEE Radio Frequency Integrated Circuits Symp.*, pages 219–222, 2008.

[102] M. Baker, S. Zhak, and R. Sarpeshkar. "A micro-power envelope detector for audio applications". In *Proc. IEEE Int'l Symp. Circuits and Systems*, volume 5, pages 1–4, May 2003.

[103] M. Baker, S. Zhak, and R. Sarpeshkar. "A low-power wide dynamic range envelope detector". *IEEE J. Solid-State Circuits*, 38(10):1750–1753, October 2003.

[104] Y. Lin, T. Wang, S. Lu, and G. Huang. "A 0.5V 3.1 mW fully monolithic OOK receiver for wireless local area sensor network ". In *Proc. Asia Solid-State Circuits Conf.*, pages 373–376, November 2005.

[105] J. Cinco-Galicia and F. Sandoval-Ibarra. "A low-power 2.7 μW 915-MHz demodulator for RFID applications". In *Proc. Int'l Conf. Electrical & Electronics Engineering*, pages 1–4, September 2006.

[106] N. Tran, B. Lee, and J. Lee. "'Development of long-range UHF-band RFID tag chip using Schottky diodes in standard CMOS technology". In *Proc. IEEE Radio Frequency Integrated Circuits Symp.*, pages 281–284, 2007.

[107] T. Lee, C. Lee, Y. Ciou, C. Huang, and C. Wang. "C-less and R-less low-frequency ASK demodulator for wireless implantable devices". In *Proc. IEEE Int'l Symp. Integrated Circuits*, pages 604–607, 2007.

[108] Y. Zou, Y. Zheng, Y. Jin, and G. Shi. "Signal modulation schemes comparison in the telemetry unit for retinal prosthesis system". In *Proc. IEEE Int'l Conf. Nano/Micro Engineered and Molecular Systems*, pages 632–636, January 2009.

[109] M. Ghovanloo and K. Najafi. "A wideband frequency-shift keying wireless link for inductively powered biomedical implants". *IEEE Trans. Circuits and Systems I.*, 51(12):2374–2383, December 2004.

[110] M. Ghovanloo and K. Najafi. "A modular 32-site wireless neural stimulation microsystem". *IEEE J. Solid-State Circuits*, 39(12):2457–2466, December 2004.

[111] R. Weng, S. Li, and J. Wang. "Low power frequency-shift keying demodulators for biomedical implants". In *Proc. IEEE Conf. Electron Devices and Solid-State Circuits*, pages 1079–1082, 2007.

[112] Y. Hwang and H. Lin. "A new CMOS analog front end for RFID tags". *IEEE Trans. on Industrial Electronics*, 56(7):2299–2307, July 2009.

[113] Y. Hu and M. Sawan. "A fully-integrated low-power BPSK based wireless inductive link for implantable medical devices". In *Proc. IEEE Int'l Mid-West Symp. Circuits and Systems*, volume 3, pages 25–28, 2004.

[114] C. Isert, M. McCoy, D. Jackson, and J. Naber. "Digital and mixed-signal integrated circuits for an RFID telemetry system". In *Proc. IEEE Mid-West Symp. Circuits and Systems*, pages 1197–1200, 2007.

[115] Y. Chuah, P. Chan, and L. Siek. "A wireless telemetry system for strain measurement". In *Proc. IEEE Canadian Conf. Electrical and Computer Engineering*, pages 1018–1021, 2000.

[116] K. Yamamoto, S. Fujii, and K. Matsuoka. "A single chip FSK modem". *IEEE J. Solid-State Circuits*, SC-19(6):855–861, December 1984.

[117] H. Gustat and F. Herzel. "Integrated FSK demodulator with very high sensitivity". *IEEE J. Solid-State Circuits*, 38(2):375–360, February 2003.

[118] J. Wilson, R. Youell, T. Richards, G. Luff, and R. Pilaski. "A single-chip VHF and UHF receiver for radio paging". *IEEE J. Solid-State Circuits*, 26(12):1944–1950, December 1991.

[119] H. Darabi, S. Khorram, H. Chien, M. Pan, S. Wu, S. Moloudi, J. Leete, J. Rael, M. Syed, R. Lee, B. Ibrahim, M. Rofougaran, and A. Rofougaran. "A 2.4 GHz CMOS transceiver for Bluetooth". *IEEE J. Solid-State Circuits*, 36(12):2016–2024, December 2001.

[120] L. Jung, P. Byrnes-Preston, R. Hessler, T. Lehmann, G. Suaning, and N. Lovell. "A dual band wireless power and FSK data telemetry for biomedical implants". In *Proc. IEEE Annual Int'l Conf. EMBS*, pages 6596–6599, 2007.

[121] Y. Hu and M. Sawan. "A fully integrated low-power BPSK demodulator for implantable medical devices". *IEEE Trans. on Circuits and Systems I.*, 52(12):2552–2562, December 2005.

[122] J. Costas. "Synchronous communications". *Proceedings of The IRE*, 90(8):1713–1718, December 1956.

[123] Z. Luo and S. Sonkusale. "A novel BPSK demodulator for biological implants". *IEEE Trans. on Circuits and Systems I.*, 55(6):1478–1484, July 2008.

[124] S. Sonkusale and Z. Lou. "A wireless data and power telemetry system using novel BPSK demodulator for non-destructive evaluation of structures". In *Proc. IEEE Sensors Conf.*, pages 300–303, 2007.

[125] S. Sonkusale and Z. Lou. "A complete data and power telemetry system utilizing BPSK and LSK signaling for biomedical implants". In *Proc. Annual Int'l IEEE EMBS Conf.*, pages 3216–3219, Vancouver, August 2008.

[126] F. Asgarian and A. Sodagar. "A high-data-rate BPSK demodulator and clock recovery circuit for implantable biomedical devices". In *Proc. Int'l IEEE EMBS conf. neural engineering*, pages 407–410, Antalya, Turkey, May 2009.

[127] F. Asgarian and A. Sodagar. "A low-power noncoherent BPSK demodulator and clock recovery circuit for high-data rate biomedical applications". In *Proc. Annual Int'l Conf. of IEEE EMBS*, pages 4840–4843, Minneapolis, September 2009.

[128] C. Gong, M. Shiue, K. Yao, T. Chen, Y. Chang, and C. Su. "A truly low-cost high-efficiency ASK demodulator based on self-sampling scheme for bio-implantable applications ". *IEEE Trans. on Circuits and Systems I.*, 55(6):1464–1477, July 2008.

[129] S. Deng, Y. Hu, and M. Sawan. "A high data rate QPSK demodulator for inductively powered electronics implants" In *Proc. IEEE Int'l Symp. Circuits and Systems*, pages 2577–2580, May 2006.

[130] Z. Lu and M. Sawan. "An 8 Mbps data rate transmission by inductive link dedicated to implantable devices". In *Proc. IEEE Int'l Symp. Circuits and Systems*, pages 3057–3060, May 2008.

[131] P. Gray, P. Hurst, S. Lewis, and R. Meyer. *Analysis and design of analog integrated circuits*. John Wiley and Sons, New York, 4th edition, 2001.

[132] Y. Tsividis. *Operation and modeling of the MOS transistors*. McGraw-Hill, New York, second edition, 1999.

[133] J. Brugler. "Silicon transistor biasing for linear collector current temperature dependence". *IEEE J. Solid State Circuits*, 2(2):57–58, June 1967.

[134] Y. Tsividis. "Accurate analysis of temperature effect in $i_c - v_{BE}$ characteristics with applications to bandgap reference resources". *IEEE J. Solid-State Circuits*, SC-13(6):1076–1084, December 1980.

[135] Y. Varshni. "Temperature dependence of the energy gap in semiconductors". *Physica*, 34:149–154, 1967.

[136] W. Bludau and A. Onton. "Temperature dependence of the band gap silicon". *J. Applied Physics*, 45(4):1846–1848, April 1974.

[137] S. Lin and A. Salama. "A $v_{be}(t)$ model with application to bandgap reference design". *IEEE J. Solid-State Circuits*, SC-20(6):1283–1285, December 1985.

[138] J. Rabaey, A. Chandrakasan, and B. Nikolic. *Digital Integrated Circuits - A design perspective*. Pearson Education, Upper Saddle River, New Jersey, 2nd edition, 2003.

[139] I. Filanovsky and A. Allam. "Mutual compensation of mobility and threshold voltage temperature effects with applications in CMOS circuits". *IEEE Trans. on Circuits and Systems I.*, 48(7):876–884, July 2001.

[140] I. Filanovsky, A. Allam, and S. Lim. "Temperature dependence of output voltage generated by interaction of threshold voltage and mobility of an NMOS transistor". *Analog Integrated Circuits and Signal Processing*, 27:229–238, 2001.

[141] B. Perumana, S. Chakraborty, C. Lee, and J. Laskar. "A fully monolithic 260-μW, 1-GHz sub-threshold low noise amplifier". *IEEE Microwave and Wireless Components Letters*, 15(6):428–430, June 2005.

[142] H. Lee and S. Mohammadi. "A 3 GHz sub-threshold CMOS low noise amplifier". In *Proc. IEEE Radio-Frequency Integrated Circuits Symp.*, pages 4–7, June 2006.

[143] A. Do, C. Boon, M. Do, K. Yeo, and A. Cabuk. "A sub-threshold low-noise amplifier optimized for ultra-low-power applications in ISM band". *IEEE Trans. on Microwave Theory and Techniques*, 56(2):286–292, February 2008.

[144] K. Ueno, T. Hirose, T. Asai, and Y. Amemiya. "A 300 nW, 15 ppm/°C, 20 ppm/V CMOS voltage reference circuit consisting of sub-threshold MOSFETs". *IEEE J. Solid-State Circuits*, 44(7):2047–2054, July 2009.

[145] B. Perumana, R. Mukhopadhyay, S. Chakraborty, C. Lee, and J. Laskar. "A low-power fully monolithic sub-threshold CMOS receiver with integrated LO generation for 2.4 GHz wireless PAN applications". *IEEE J. Solid-State Circuits*, 43(10):2229–2238, October 2008.

[146] A. Wang, B. Calhoun, and A. Chandrakasan. *Sub-threshold design for ultra low-power systems*. Springer, New York, 2006.

[147] D. Foty. *MOSFET modeling with SPICE - principle and practice*. Prentice-Hall, Upper Saddle River, New Jersey, 1997.

[148] C. Enz, F. Krummenacher, and E. Vittoz. "An analytical MOS transistor model valid in all regions of operation and dedicated to low-voltage and low-current applications". *Analog Integrated Circuits and Signal Processing*, 8:83–114, July 1995.

[149] G. Yu and X. Zou. "A novel current reference based on sub-threshold MOSFETs with high PSRR". *Microelectronics J.*, 39:1874–1879, 2008.

[150] A. Hastings. *The art of analog layout*. Prentice-Hall, Upper Saddle River, New Jersey, second edition, 2006.

[151] R. Dehghani and S. Atarodi. "A new low voltage precision CMOS current reference with no external components". *IEEE Trans. on Circuits and Systems II.*, 50(12):928–932, December 2003.

[152] K. Leung, P. Mok, and C. Leung. "A 2-V 23-μA, 5.3-ppm/°C curvature-compensated CMOS bandgap voltage reference". *IEEE J. Solid-State Circuits*, 38(3):561–564, March 2003.

[153] R. Widlar. "Some circuit design techniques for linear integrated circuits". *IEEE Trans. on Circuit Theory*, CT-12(4):586–590, December 1965.

[154] R. Widlar. "New developments in IC voltage regulators". *IEEE J. Solid-State Circuits*, SC-6(1):2–7, February 1971.

[155] A. Brokaw. "A simple three-terminal IC bandgap reference". *IEEE J. Solid-State Circuits*, SC-9(6):388–393, December 1974.

[156] B. Song and P. Gray. "A precision curvature-compensated CMOS bandgap reference". *IEEE J. Solid-State Circuits*, SC-18(6):634–643, December 1983.

[157] M. Pertija, G. Meijer, and J. Huijsing. "Precision temperature measurement using CMOS substrate PNP transistors". *IEEE Sensors J.*, 4(3):294–300, June 2004.

[158] M. Ker and J. Chen. "New curvature-compensation technique for CMOS bandgap reference with sub-1-V operation". *IEEE Trans. on Circuits and Systems II*, 53(8):667–671, August 2006.

[159] H. Lin and C. Liang. "A sub-1V bandgap reference circuit using sub-threshold current". In *Proc. IEEE Int'l Symp. Circuits and Systems*, pages 4253–4256, May 2005.

[160] A. Bendali and Y. Audet. "A 1-V CMOS current reference with temperature and process compensation". *IEEE Trans. on Circuits and Systems I.*, 54(7):1424–1429, July 2007.

[161] M. Waltari and K. Halonen. "Reference voltage driver for low-voltage CMOS A/D converters". In *Proc. Int'l Conf. Electronics, Circuits and Systems*, pages 28–31, December 2000.

[162] S. Wadhwa. "A low voltage CMOS proportional-to-absolute temperature current reference". In *Proc. IEEE 22nd Int'l Conf. VLSI Design*, pages 171–174, 2009.

[163] H. Banba, H. Shiga, A. Umezawa, T. Miyaba, T. Tanzawa, S. Atsumi, and K. Sakui. "A CMOS bandgap reference circuit with sub-1-V operation". *IEEE J. Solid-State Circuits*, 34(5):670–674, May 1999.

[164] K. Leung and P. Mok. "A sub-1-V 15-ppm/°C CMOS bandgap voltage reference without requiring low threshold voltage device". *IEEE J. Solid-State Circuits*, 37(4):526–530, April 2002.

[165] J. Chen and B. Shi. "1 V CMOS current reference with 50 ppm/°C temperature coefficient". *IEE Electronics Letters*, 39(2):209–210, January 2003.

[166] Y. Jiang and E. Lee. "Design of low-voltage bandgap reference using transimpedance amplifier". *IEEE Trans. on Circuits and Systems II*, 47(6):552–555, June 2000.

[167] A. Buck, C. McDonald, S. Lewis, and T. Viswanathan. "A CMOS bandgap reference without resistors". *IEEE J. Solid-State Circuits*, 37(1):81–83, January 2002.

[168] Y. Dai, D. Comer, D. Comer, and C. Petrie. "Threshold voltage based CMOS voltage reference". *IEE Proc. - Circuits, Devices, and Systems*, 151(1):58–62, February 2004.

[169] A. Buck, C. McDonald, S. Lewis, and T. Viswanathan. "A CMOS bandgap reference without resistors". In *Proc. IEEE Int'l Conf. Solid-State Circuits*, pages 442–443, February 2000.

[170] Y. Gu, S. Yueh, T. Chang, and K. Huang. "CMOS voltage reference with multiple outputs". *IET Proc. - Circuits, Devices, and Systems*, 2(2):222–226, April 2008.

[171] J.Chen, X. Ni, B. Mo, and Z. Wang. "A high precision curvature compensated bandgap reference". In *Proc. Int'l Conf. Solid-State and Integrated Circuits*, pages 1748 – 1750, October 2006.

[172] Y. Song, W. Lu, Z. Chen, J. Gao, J. Tang, and L. Ji. "A precise compensated bandgap reference without resistors". In *Proc. Int'l Conf. Solid-State and Integrated Circuits*, pages 1583–1586, October 2004.

[173] H. Wang and Q. Ye. "A CMOS voltage reference without resistors for ultra-low power applications". In *Proc. Int'l Conf. ASIC*, pages 526–529, October 2007.

[174] I. Lee, G. Kim, and W. Kim. "Exponential curvature-compensated BiCMOS bandgap references". *IEEE J. Solid-State Circuits*, 29(11):1396–1403, November 1994.

[175] S. Hoon, J. Chen, and F. Maloberti. "An improved bandgap reference with high power supply rejection". In *Proc. IEEE Int'l Symp. Circuits and Systems*, volume 5, pages 833–836, May 2002.

[176] P. Crovetti and F. Fiori. "Compact, very low voltage, temperature-independent reference circuit". *IET Circuits, Devices, and Systems*, 1(1):63–71, February 2007.

[177] C. Falconi, A. D'Amico, G. Scotti, and A. Trifiletti. "Low voltage CMOS current and voltage references without resistors". In *IEEE Proc. Int'l Symp. Circuits and Systems*, pages 1907–1910, May 2007.

[178] G. Rincon-Mora and P. Allen. "A 1.1-V current-mode and piecewise-linear curvature-corrected bandgap reference". *IEEE J. Solid-State Circuits*, 33(10):1551–1554, October 1998.

[179] H. Huang, R. Wang, and S. Hsu. "Piecewise linear curvature-compensated CMOS bandgap reference". In *Proc. Int'l Conf. Electronics, Circuits and Systems*, pages 308–311, 2008.

[180] P.Malcovati, F. Maloberti, C. Fiocchi, and M. Pruzzi. "Curvature-compensated BiCMOS bandgap with 1-V supply voltage". *IEEE J. Solid-State Circuits*, 36(7):1076–1081, July 2001.

[181] K. Tom and A. Alvandpour. "Curvature compensated CMOS bandgap with sub 1V supply". In *Proc. IEEE Int'l Workshop Electronic Design, Test and Applications*, 2005.

[182] X. Guan, A. Wang, A. Ishikawa, S. Tamura, Z. Wang, and C. Zhang. "A 3V 110 μW 3.1 ppm/C °c curvature-compensated CMOS bandgap reference". In *Proc. IEEE Int'l Symp. Circuits and Systems*, pages 2861–2864, May 2006.

[183] S. Hsiao, Y. Huang, D. Liang, H. Chen, and H. Chen. "A 1.5-V 10 ppm/C° 2nd-order curvature-compensated CMOS bandgap reference with trimming". In *Proc. IEEE Int'l Symp. Circuits and Systems*, pages 565–568, May 2006.

[184] C. Stanescu, R. Jacob, C. Caracas, and P. Cosmin. "Curvature-compensated CMOS bandgap circuit with 1 V reference voltage". In *Proc. Int'l Semiconductor Conf.*, pages 365–368, October 2002.

[185] G. Yu and X. Zou. "CMOS bandgap voltage reference circuit for supply voltage down to 0.6 V". *IEE Electronics Letters*, 39(20):1427 – 1428, October 2003.

[186] M. Cheng and Z. Wu. "Low-power low-voltage reference using peaking current mirror circuit". *IEE Electronics Letters*, 41(10):572 – 573, May 2005.

[187] G. Giustolisi, G. Palumbo, M. Criscione, and F. Cutri. "A low-voltage low-power voltage reference based on sub-threshold MOSFETs". *IEEE J. Solid-State Circuits*, 38(1):151–154, January 2003.

[188] P. Huang, H. Lin, and Y. Lin. "A simple sub-threshold CMOS voltage reference circuit with channel length modulation compensation". *IEEE Trans. on Circuits and Systems II.*, 53(9):882–885, September 2006.

[189] Z. Qu, M. Zhang, J. Wu, and Y. Li. "A sub-0.6V, 34.8 nW, 4.6 ppm/°C CMOS voltage reference using sub-threshold and body effect techniques". In *Proc. Asic Pacific Conf. Postgraduate Research in Microelectronics and Electronics*, pages 275–278, January 2009.

[190] C. Chang, T. Lo, C. Chen, K. Wu, and C. Hung. "A low-power CMOS voltage reference circuit based on sub-threshold operation". In *Proc. IEEE Int'l Symp. Circuits and Systems*, pages 3844–3847, May 2007.

[191] L. Ferreira, T. Pimenta, and R. Moreno. "A CMOS threshold voltage reference source for very-low-voltage applications". *Microelectronics J.*, 39:1867–1873, 2008.

[192] C. Chan, K. Pun, K. Leung, J. Guo, L. Lincoln, and C. Choy. "A low-power continuously-calibrated clock recovery circuit for UHF RFID IC class-1 generation-2 transponders". *IEEE J. Solid-State Circuits*, 45(3):587–599, March 2010.

[193] M. Mohaisen, H. Yoon, and K. Chang. "Radio transmission performance of EPCglobal Gen-2 RFID system ". In *Proc. Conf. Advanced Comm. Tech.*, pages 1423–1428, February 2008.

[194] L. Leung and H. Luong. "A 7 μW clock generator in 0.18 μm CMOS for passive UHF RFID EPC G2 tags". In *Proc. European Solid-State Circuits Conf.*, pages 412–415, 2007.

[195] U. Kaiser and W. Steinhagen. "A low-power transponder IC for high-performance identification systems". *IEEE J. Solid-State Circuits*, 30(3):306–310, March 1995.

[196] C. Ma, X. Wu, C. Zhang, and Z. Wang. "A low-power RF front-end of passive UHF RFID transponder". In *Proc. IEEE Asia Pacific Conf. Circuits and Systems*, pages 73–76, 2008.

[197] S. Sidiropoulos and M. Horowitz. "A 700-Mb/s/pin CMOS signaling interface using current integrating receivers". *IEEE J. Solid-State Circuits*, 32(5):681–690, May 1997.

[198] B. Razavi. *Design of CMOS integrated circuits for optical communications*. McGraw-Hill, Boston, 2003.

[199] V. Najafi, M. Jenabi, S. Mohammadi, A. Fotowat-Ahmady, and M. Marvasti. "A dual-mode EPC Gen-2 UHF RFID transponder in 1.08 μm CMOS". In *Proc. IEEE Int'l Conf. Electronics, Circuits and Systems*, pages 1135–1138, 2008.

[200] J. Lee and B. Lee. "A long-range UHF-band passive RFID tag IC based on high-Q design approach". *IEEE Trans. on Industrial Electronics*, 56(7):2009, July 2308-2316.

[201] B. Fan, Y. Dai, X. Zhang, and Y. Lu. "Low power clock recovery circuit for passive HF RFID tag". *Analog Integrated Circuits and Signal Processing*, 59:207–214, 2009.

[202] N. Soltani and F. Yuan. "Remote frequency calibration of passive wireless microsensors using injection-locked PLL". In *Proc. IEEE Int'l Symp. Circuits and Systems*, pages 541–544, May 2009.

[203] N. Soltani and F. Yuan. "Analysis and design of non-harmonic injection-locked phase-locked loops with applications in remote frequency calibration of passive wireless transponders". *IEEE Trans. on Circuits and Systems I.*, 57(12), 2010 (in press).

[204] F. Yuan and N. Soltani. "A low-voltage low VDD sensitivity relaxation oscillator for passive wireless microsystems ". *IET Electronics Letters*, 45(21):1057–1058, 2010.

[205] R. Barnett and J. Liu. "A 0.8V 1.52MHz MSVC relaxation oscillator with inverted mirror feedback reference for UHF RFID". In *Proc. IEEE Custom Integrated Circuits Conf.*, pages 26/2/1–4, 2006.

[206] S. Gierkink and E. van Tuijl. "A coupled sawtooth oscillator combining low jitter with high control linearity". *IEEE J. Solid-State Circuits*, 37(6):702–710, June 2002.

[207] K. Lasanen and J. Kostamovaara. "A 12-V CMOS RC oscillator for capacitive and resistive sensor applications". *IEEE Trans. on Instrumentation and Measurement*, 57(12):2792–2800, December 2008.

[208] J. Feddeler and B. Lucas. *Application note - ADC definitions and specifications*. Freescale Semiconductor, Chandler, Arizona, 2004.

[209] P. Allen and P. Holberg. *CMOS analog circuit design*. Oxford University Press, London, second edition, 2002.

[210] S. Franco. *Design with operational amplifiers and analog integrated circuits*. McGraw-Hill, Boston, third edition, 2002.

[211] S. Zhou and N. Wu. "A novel ultra low power temperature sensor for UHF RFID tag chip". In *Proc. IEEE Asian Solid-State Circuits Conf.*, pages 464–467, Jeju, Korea, November 2007.

[212] C. Kim, J. Lee, Y. Jun, C. Lee, and B. Kong. "CMOS temperature sensor with ring oscillator for mobile DRAM self-refresh control". *Microelectronics J.*, 38:1042–1049, 2007.

[213] S. Park, C. Min, and S. Cho. "A 95 nW ring oscillator-based temperature sensor for RFID tags in 0.13μm CMOS". In *Proc. IEEE Int'l Symp. Circuits and Systems*, pages 1153–1156, May 2009.

[214] E. Socher, S. Beer, and Y. Nemirovsky. "Temperature sensitivity of SOI-CMOS transistors for use in uncooled thermal sensing". *IEEE Trans. on Electron Devices*, 52(12):2784–2789, December 2005.

[215] P. Chen, C. Chen, C. Tsai, and W. Lu. "A time-to-digital-converter-based CMOS smart temperature sensor". *IEEE J. Solid-State Circuits*, 40(8):1642–1648, August 2005.

[216] P. Chen, C. Hwang, and W. Chang. "A precise cyclic CMOS time-to-digital converter with low thermal sensitivity". *IEEE Trans. on Nuclear Science*, 52(4):834–838, August 2005.

[217] K. Kim, H. Lee, S. Jung, and C. Kim. "A 366kS/s 400 μW 0.0013 mm^2 frequency-to-digital converter based CMOS temperature sensor utilizing multiphase clock". In *Proc. IEEE Custom Integrated Circuits Conf.*, pages 203–206, 2009.

[218] P. Chen, S. Liu, and J. Wu. "A CMOS pulse-shrinking delay element for time interval measurement". *IEEE Trans. on Circuits and Systems II.*, 47(9):954–958, Sept. 2000.

[219] S. Kang and Y. Leblebici. *CMOS digital integrated circuits - analysis and design*. McGraw-Hall, Boston, third edition, 2003.

[220] J. Sauerbrey, D. Schmitt-Landsiedel, and R. Thewes. "A 0.5-V 1-μW successive approximation ADC". *IEEE J. Solid-State Circuits*, 38(7):1261–1265, July 2003.

[221] M. Scott, B. Boser, and K. Pister. "An ultra low-energy ADC for smart dust". *IEEE J. Solid-State Circuits*, 38(7):1123–1129, July 2003.

[222] H. Kim, Y. Min, Y. Kim, and S. Kim. "A low power consumption 10-bit rail-to-rail SAR ADC using a C-2C capacitor array". In *Proc. IEEE Int'l Conf. Electron Devices and Solid-State Circuits*, pages 1–4, 2008.

[223] J. McCreamy and P. Gray. "All MOS charge redistribution analog-to-digital conversion technique - part I". *IEEE J. Solid-State Circuits*, SC-10:371–379, December 1975.

[224] D. Zhang. *Design and evaluation of an ultra-low power successive approximation ADC*. MASc. thesis, Linkopings University, Sweden, 2009.

[225] Y. Chang. *An ultra-low power SAR ADC*. MASc. thesis, University of British Columbia, Vancouver, Canada, 2009.

[226] K. Ohri and M. Claalhan Jr. "Integrated PCM dedec". *IEEE Trans. on Communications*, COM-27(2):296–304, February 1979.

[227] Y. Yee, L. Terman, and L. Heller. "A two-stage weighted capacitor network for D/A-A/D conversion". *IEEE J. Solid-State Circuits*, SC-14(4):778–781, August 1979.

[228] K. Abdelhalim, L. MacEachern, and S. Mahmoud. "A nanowatt successive approximation ADC with offset correction for implantable sensor applications". In *Proc. IEEE Int'l Symp. Circuits and Systems*, pages 2351–2354, May 2007.

[229] J. McCreamy. "Matching properties, and voltage and temperature dependence of MOS capacitors". *IEEE J. Solid-State Circuits*, SC-16(6):608–616, December 1981.

[230] S. Singh, A. Prabhakar, and A. Bhattacharyya. "C-2C ladder voltage dividers for application in all-MOS A/D converters". *IEE Electronics Letters*, 18(12):537–538, June 1982.

[231] S. Singh, A. Prabhakar, and A. Bhattacharyya. "Modified C-2C ladder voltage divider for application in PCM A/D converters". *IEE Electronics Letters*, 19(19):788–789, September 1983.

[232] S. Singh, A. Prabhakar, and A. Bhattacharyya. "C-2C ladder-based D/A converters for PCM codes". *IEEE J. Solid-State Circuits*, SC-22(6):1197–1200, December 1987.

[233] L. Cong. "Psseudo C-2C ladder-based data converter technique". *IEEE Trans. on Circuits and Systems II.*, 48(10):927–929, October 2001.

[234] S. Cho, C. Lee, and J. Song. "Design of a 1-volt and μ-power SARADC for sensor network application". In *IEEE Int'l Symp. Circuits and Systems*, pages 3852–3855, 2007.

[235] A. Rodriguez-Perez, M. Delgado-Restituto, J. Ruiz-Amaya, and F. Medeiro. "An ultra-low power consumption 1-V 10-bit successive approximation ADC". In *Proc. Int'l Conf. Electronics, Circuits and Systems*, pages 634–637, 2008.

[236] Y. Kuramochi, A. Matsuzawa, and M. Kawabata. "A 0.05-mm^2 110-μW 10-b self-calibrating successive approximation ADC core in 0.18-μm CMOS". In *Proc. IEEE Asia Solid-State Circuits Conf.*, pages 224–227, Jeju, Korea, November 2007.

[237] Z. Fu, P. Weerakoon, and E. Culurciello. "Nano-watt silicon-on-sapphire ADC using 2C-1C capacitor chain". *IET Electronics Letters*, 42(6):341–343, March 2006.

[238] H. Hong and G. Lee. "A 65-fJ/conversion-step 0.9-V 200-kS/s rail-to-rail 8-bit successive approximation ADC". *IEEE J. Solid-State Circuits*, 42(10):2161–2168, October 2007.

[239] N. Verma and A. Chandrakasan. "An ultra low energy 12-bit rate-resolution scalable SAR ADC for wireless sensor nodes". *IEEE J. Solid-State Circuits*, 42(6):1196–1205, June 2007.

[240] J. Yang, T. Naing, and B. Brodersen. "A 1-GS/s 6-bit 6.7-mW ADC in 65-nm CMOS". In *Proc. IEEE Customs Integrated Circuits Conf.*, pages 287–290, 2009.

Index

About the Author

Fei Yuan received the Ph.D. degree in electrical engineering from University of Waterloo, Canada in October 1999. He joined the Department of Electrical and Computer Engineering, Ryerson University, Toronto, Canada in 1999 and is currently a Professor. Dr. Yuan is the author of the book *CMOS Active Inductors and Transformers : Principle, Implementation, and Applications* (Springer, New York, 2008), *CMOS Current-Mode Circuits for Data Communications* (Springer, New York, 2006), the co-author of the book *Computer Methods for Analysis of Mixed-Mode Switching Circuits* (with Ajoy Opal, Kluwer Academic Publishers, Boston, 2004), and the author / co-author of 47 research papers in refereed scientific journals and over 90 research papers in refereed conference proceedings in the field of circuits and systems.

Dr. Yuan currently serves the Chair of the Department of Electrical and Computer Engineering, Ryerson University. He also serves as the Associate Editor of *Canadian Journal of Electrical and Computer Engineering* and is on the editorial board of several international journals. He was awarded the Ryerson Research Chair from Ryerson University in Jan. 2005, the Research Excellence Award from the Faculty of Engineering and Applied Science of Ryerson University in 2004, the Early Tenure from Ryerson University in 2002, the Doctoral Scholarship from the Natural Science and Engineering Research Council of Canada during 1997-1998, and the Teaching Excellence Award from Changzhou Institute of Technology, Jiangsu, China in 1988. Dr. Yuan is a senior member of IEEE and a registered professional engineer in the province of Ontario, Canada.

Lightning Source UK Ltd.
Milton Keynes UK
178188UK00006B/26/P

9 781441 976796